# Universitext

# Universitext



*Universitext* is a series of textbooks that presents material from a wide variety of mathematical disciplines at master's level and beyond. The books, often well class-tested by their author, may have an informal, personal even experimental approach to their subject matter. Some of the most successful and established books in the series have evolved through several editions, always following the evolution of teaching curricula, to very polished texts.

Thus as research topics trickle down into graduate-level teaching, first textbooks written for new, cutting-edge courses may make their way into *Universitext.*

More information about this series at http://www.springer.com/series/223

Alexander A. Balinsky • W. Desmond Evans •
Roger T. Lewis

# The Analysis and Geometry of Hardy's Inequality

Springer

Alexander A. Balinsky
School of Mathematics
Cardiff University
Cardiff, United Kingdom

W. Desmond Evans
School of Mathematics
Cardiff University
Cardiff, United Kingdom

Roger T. Lewis
Department of Mathematics
University of Alabama at Birmingham
Birmingham
Alabama, USA



ISSN 0172-5939 ISSN 2191-6675 (electronic)
Universitext
ISBN 978-3-319-22869-3 ISBN 978-3-319-22870-9 (eBook)
DOI 10.1007/978-3-319-22870-9

Library of Congress Control Number: 2015951415

Mathematics Subject Classification (2010): 31A05, 31B05, 34A40, 35A23, 35Q40, 35R45

Springer Cham Heidelberg New York Dordrecht London


Printed on acid-free paper

Springer International Publishing AG Switzerland is part of Springer Science+Business Media (www.springer.com)

*To*

*Helen, Mari and Barbara*

# Preface

This book is a study of the many refinements and incarnations of the Hardy inequality

$$\int_{\Omega} |\nabla u(\mathbf{x})|^p d\mathbf{x} \geq C(p,\Omega) \int_{\Omega} \frac{|u(\mathbf{x})|^p}{\delta(\mathbf{x})^p} d\mathbf{x}, \quad u \in C_0^{\infty}(\Omega), \tag{1}$$

where $\Omega$ is a domain (an open connected set) in $\mathbb{R}^n$, $n \geq 1$, with non-empty boundary $\partial\Omega$, $\delta(\mathbf{x})$ is the distance from $\mathbf{x} \in \Omega$ to $\partial\Omega$, $1 \leq p < \infty$, and $C(p,\Omega)$ is a positive constant depending on $p$ and $\Omega$ in general. The original continuous form of the inequality was for $\Omega = (0,\infty)$, $\delta(x) = |x|$, and appeared in [72], having been motivated by work of Hardy on a discrete analogue and a double series inequality of Hilbert. It attracted the attention of other mathematicians, notably Landau, and was highlighted by Hardy et al. in [75]; see [91] for a detailed account of the history. In its many guises, the inequality has played an important role in mathematical analysis and mathematical physics, which is way beyond what could have been expected at its humble beginning. Extensions and refinements to a multitude of function spaces have been studied extensively, which, apart from their intrinsic interest, have had significant implications for the function spaces and the relationships between them, and important applications to differential equations. The case $p = 2$, $\Omega = \mathbb{R}^n \setminus \{0\}, \delta(\mathbf{x}) = |\mathbf{x}|$ of (1) is a mathematical representation of Heisenberg's uncertainty principle in quantum mechanics, which asserts that the momentum and position of a particle can't be simultaneously determined. Furthermore, the spectral analysis of quantum mechanical systems involving Coulomb forces between constituent particles features this $L^2(\mathbb{R}^n)$ version of Hardy's inequality in a natural way. In his book *A Mathematician's Apology* [74], Hardy expresses the view that for a theorem to be significant, it must have both *generality* and *depth*. He also asserts that nothing he had ever done was *useful*. The role of his inequality in mathematics undoubtedly confirms his notion of "significance", while its implications in quantum mechanics, with its tentacles affecting every aspect of modern life, would contradict Hardy's feeling about its "uselessness".

In the first two sections of Chap. 1, a general form of the Hardy inequality is proved, initially in $\mathbb{R}_+ := (0, \infty)$ and subintervals $(a, b)$ of $\mathbb{R}$ and then in $\mathbb{R}^n$ for $n \geq 2$, optimal constants being obtained. The rest of Chap. 1 is a cornucopia of techniques and results, which will be of subsequent importance. These include a brief description of Sobolev spaces and the inequalities of Sobolev, Friedrichs and Poincaré; Fourier transforms, rearrangements and their application to making a comparison of the Hardy and Sobolev inequalities; the Cwikel, Lieb, Rosenbljum (CLR) inequality concerning the number of negative eigenvalues of the Dirichlet Laplace operator in $L^2(\Omega)$ and a comparison of the CLR and the appropriate Sobolev inequality; Kato's inequality and relativistic analogues of Hardy's inequality.

Chapter 2 is on properties of a general domain $\Omega$, which are of significance to the function $\delta$. For instance, the **skeleton** $\mathcal{S}(\Omega)$ is the subset of $\Omega$ consisting of points, which are equidistant from more than one point on the boundary, and this coincides with the set of points at which $\delta$ is not differentiable. Another important subset of $\Omega$ is the **ridge**, or **central set**, $\mathcal{R}(\Omega)$, which lies between $\mathcal{S}(\Omega)$ and its closure. It is shown that the closure of the ridge is the **cut locus**, which is a concept used extensively by Li and Nirenberg in [108]. Properties of $\delta$ when $\Omega$ is convex, or $\mathbb{R}^n \setminus \overline{\Omega}$ is convex, are established, and when $\Omega$ is a domain with smooth boundary, an explicit formula for $\Delta\delta(\mathbf{x})$ is determined for all $\mathbf{x} \in \Omega \setminus \overline{\mathcal{R}(\Omega)}$. The principal curvatures of a $C^2$ boundary, and the mean curvature of $\partial\Omega$, feature prominently in the second part of the chapter, and indeed, the rest of the book.

The study of inequalities of type

$$\int_\Omega |\nabla u(\mathbf{x})|^p d\mathbf{x} \geq C(p, \Omega) \int_\Omega \left( \frac{|u(\mathbf{x})|^p}{\delta(\mathbf{x})^p} + a(\mathbf{x}, \delta(\mathbf{x}))|u(\mathbf{x})|^p \right) d\mathbf{x} \tag{2}$$

starts in earnest in Chap. 3, the case $a = 0$ being of particular interest and referred to as Hardy's inequality on $\Omega$. We begin with a list of some important results in the literature to set the scene, which bring in, *inter alia*, the notions of capacity and fatness, the Hausdorff and Aikawa dimensions of the boundary, and the mean distance function $\delta_{M,p}$ introduced by Davies in [41] in the case $p = 2$ for an arbitrary $\Omega$. Included subsequently are a proof of the optimal constant for a convex domain $\Omega$, and of Ancona's lower bound for the constant $C(2, \Omega)$ when $\Omega$ is a simply connected planar domain. Some of the main results in this chapter are based on ones from [20, 107], using tools developed in the previous chapter. Of special note is the result from [107] that if $\Omega$ has a $C^2$ boundary and a non-positive mean curvature (a so-called *weakly mean convex domain*), then

$$\mu_p(\Omega) := \inf_{C_0^\infty(\Omega)} \frac{\int_\Omega |\nabla f|^p d\mathbf{x}}{\int_\Omega |f/\delta|^p d\mathbf{x}} = \left(\frac{p-1}{p}\right)^p, \tag{3}$$

which extends a well-known result for convex domains.

In [118], Maz'ya proved the inequality

$$\int_{\mathbb{R}^n_+}\left(|\nabla u|^2 - \frac{|u|^2}{4x_n^2}\right)d\mathbf{x} \geq K_{n,2}\left(\int_{\mathbb{R}^n_+}|u|^{\frac{2n}{n-2}}d\mathbf{x}\right)^{\frac{n-2}{n}}, \tag{4}$$

for $u \in C_0^\infty(\mathbb{R}^n_+)$, where $\mathbb{R}^n_+$ is the half-space $\mathbb{R}^{n-1} \times \mathbb{R}_+$ and $\mathbf{x} = (\mathbf{x}', x_n)$, $\mathbf{x}' \in \mathbb{R}^{n-1}$, $x_n \in \mathbb{R}_+$. This is the prototype of the Hardy-Sobolev-Maz'ya (HSM) inequalities, which combine elements of both the Hardy and Sobolev inequalities, and Chap. 4 is devoted to them. We present the following result of Frank and Loss from [62] for a general domain $\Omega \subsetneqq \mathbb{R}^n$, in which the mean distance function $\delta_{M,p}$ plays the role of the distance function $\delta$ : there exists a constant $K_{n,p}$, depending only on $n$ and $p$, such that for all $u \in C_0^\infty(\Omega)$ and $p \geq 2$,

$$\int_\Omega\left(|\nabla u|^p - \left(\frac{p-1}{p}\right)^p\frac{|u|^p}{\delta_{M,p}^p}\right)d\mathbf{x} \geq K_{n,p}\left(\int_\Omega|u|^{\frac{np}{n-p}}d\mathbf{x}\right)^{\frac{n-p}{n}}. \tag{5}$$

If $\Omega$ is convex, $\delta_{M,p} \leq \delta$, and (5) becomes an extension of the HSM inequality obtained by Filippas et al. in [60] for a bounded convex domain $\Omega$ with a $C^2$ boundary and $p = 2$, and answers in the affirmative their query if the constant can be chosen to be independent of $\Omega$. Chapter 4 also includes HSM inequalities featuring the mean curvature of the boundary of $\Omega$, and one of Gkikas in [69] for exterior domains.

The first part of Chap. 5 is on Schrödinger operators involving magnetic fields of Aharonov-Bohm type. The Laptev-Weidl inequality in $L^2(\mathbb{R}^2)$ is derived, followed by related Sobolev and CLR inequalities. Hardy-type inequalities for Aharonov-Bohm magnetic fields with multiple singularities are proved, and also a generalised Hardy inequality for magnetic Dirichlet forms. Finally in Chap. 5, there is a discussion of Pauli operators in $\mathbb{R}^3$ with magnetic fields, and inequalities of Hardy, Sobolev and CLR type are proved to exist if the Pauli operator has no zero modes.

Chapter 6 is concerned with the Rellich inequality

$$\int_{\mathbb{R}^n}|\Delta u(\mathbf{x})|^2 d\mathbf{x} \geq \frac{n^2(n-4)^2}{16}\int_{\mathbb{R}^n}\frac{|u(\mathbf{x})|^2}{|\mathbf{x}|^4}d\mathbf{x}. \tag{6}$$

A proof of an $L^p(\mathbb{R}^n)$ version of the inequality is given, based on that of Davies and Hinz in [45], and this is followed by a Rellich-Sobolev inequality in $L^2(\Omega)$ for a domain $\Omega \subset \mathbb{R}^n$ due to Frank (private communication, 2007). Inequalities involving Aharonov-Bohm type magnetic potentials in $L^2(\mathbb{R}^n)$ are established, which are analogous to the Laptev-Weidl inequality of Chap. 5, and a CLR-type inequality for associated bi-harmonic operators is proved.

The book is primarily designed for the mathematician, but we hope that it will also appeal to the scientist who has an interest in quantum mechanics. A good basic knowledge of real and complex analysis is a prerequisite. Also, familiarity with the Lebesgue integral, spectral analysis of differential operators, and elementary differential geometry would be helpful, but only the barest essentials of these areas are assumed, and background information is always provided; where necessary, precise references to the literature are given.

Chapters are divided into sections and sections are sometimes divided into subsections. Theorems, Corollaries, Lemmas, Propositions, Remarks and equations are numbered consecutively. At the end of the book, there are author, subject and notation indices.

Cardiff, UK — Alexander A. Balinsky
Cardiff, UK — W. Desmond Evans
Birmingham, AL, USA — Roger T. Lewis

# Basic Notation

| | |
|---|---|
| $\mathbb{R}$ : | Real numbers |
| $\mathbb{R}^n$ : | $n$-Dimensional Euclidean space |
| $\mathbb{N}$: | Positive integers |
| $\mathbb{N}_0 = \mathbb{N} \cup \{0\}$ | |
| $\mathbb{Z}$ : | Integers |
| $\mathbb{C}$ : | Complex numbers |
| $\Omega$ : | Domain—a connected open subset of $\mathbb{R}^n$ |
| $\partial\Omega$ : | Boundary of $\Omega$ |
| $\overline{\Omega}$ : | Closure of $\Omega$ |
| $\rightharpoonup$: | Weak convergence |
| $X \hookrightarrow Y$ : | $X$ is continuously embedded in $Y$ |
| $L^p(\Omega),\ 1 \leq p < \infty$ : | Lebesgue space of functions $f$ with $\lvert f\rvert^p$ integrable on $\Omega$ |
| $\\|\cdot\\|_p$ or $\\|\cdot\\|_{p,\Omega}$ : | Norm on $L^p(\Omega)$ |
| $l^p,\ 1 \leq p < \infty$ : | Space of sequences $\{x_n\}_{n\in\mathbb{N}}$ such that $\sum_{n=1}^{\infty} \lvert x_n\rvert^p < \infty$ |
| $W^{k,p}(\Omega), H^{k,p}(\Omega)$ : | Sobolev spaces |
| $C_0^\infty(\Omega)$ : | Infinitely differentiable functions with compact supports in $\Omega$ |
| $W_0^{k,p}(\Omega)$ : | Closure of $C_0^\infty(\Omega)$ in $W^{k,p}(\Omega)$ |
| $\omega_n = \pi^{n/2}/\Gamma(1+n/2)$ : | Volume of unit ball in $\mathbb{R}^n$ |

# Contents

# Chapter 1
# Hardy, Sobolev, and CLR Inequalities

## 1.1 Introduction

The Hardy and Sobolev inequalities are of fundamental importance in many branches of mathematical analysis and mathematical physics, and have been intensively studied since their discovery. A rich theory has been developed with the original inequalities on $(0, \infty)$ extended and refined in many ways, and an extensive literature on them now exists. We shall be focusing throughout the book on versions of the inequalities in $L^p$ spaces, with $1 < p < \infty$. In this chapter we shall be mainly concerned with the inequalities in $(0, \infty)$ or $\mathbb{R}^n$, $n \geq 1$. Later in the chapter we shall also discuss the CLR (Cwikel, Lieb, Rosenbljum) inequality, which gives an upper bound to the number of negative eigenvalues of a lower semi-bounded Schrödinger operator in $L^2(\mathbb{R}^n)$. This has a natural place with the Hardy and Sobolev inequalities as the three inequalities are intimately related, as we shall show. Where proofs are omitted, e.g., of the Sobolev inequality, precise references are given, but in all cases we have striven to include enough background analysis to enable a reader to understand and appreciate the result.

In [73], Hardy proved the inequality

$$\int_0^\infty \left( \frac{1}{x} \int_0^x f(t)dt \right)^p dx \leq \left( \frac{p}{p-1} \right)^p \int_0^\infty f(x)^p dx \tag{1.1.1}$$

for non-negative functions $f$, with $1 < p < \infty$. Landau showed in [94] that the constant $\left(\frac{p}{p-1}\right)^p$ is sharp, and that equality is only possible if $f = 0$; Hardy had, in fact, drawn attention to the sharpness of the constant in an earlier paper; see the Appendix in [91] where interesting information on the historical background may be found. A more familiar form of the inequality is obtained by setting $F(x) =$



A.A. Balinsky et al., *The Analysis and Geometry of Hardy's Inequality*,
Universitext, DOI 10.1007/978-3-319-22870-9_1

$\int_0^x f(t)dt$, which gives

$$\int_0^\infty \frac{F(x)^p}{x^p}dx \leq \left(\frac{p}{p-1}\right)^p \int_0^\infty F'(x)^p dx. \tag{1.1.2}$$

The analogue of this inequality in $\mathbb{R}^n$ for $n > 1$ is

$$\int_{\mathbb{R}^n} \frac{|f(\mathbf{x})|^p}{|\mathbf{x}|^p}d\mathbf{x} \leq \left|\frac{p}{p-n}\right|^p \int_{\mathbb{R}^n} |\nabla f(\mathbf{x})|^p d\mathbf{x},$$

where $\nabla f = (\partial f/\partial x_1, \cdots, \partial f/\partial x_n)$ is the gradient of $f$; this holds for all $f \in C_0^\infty(\mathbb{R}^n \setminus \{0\})$ if $n < p < \infty$, and for all $f \in C_0^\infty(\mathbb{R}^n)$ if $1 \leq p < n$. The constant is sharp and equality can only be attained by functions $f = 0$ a.e.

After discussing the Hardy inequalities in the first section, we define the Sobolev spaces $W_0^{1,p}(\Omega)$ and $W^{1,p}(\Omega)$ on a domain $\Omega \subseteq \mathbb{R}^n$, and give a brief coverage of embedding theorems, boundary smoothness criteria and the Friedrichs and Poincaré inequalities. Using the theory of rearrangements, the Hardy and Sobolev inequalities on $\mathbb{R}^n$, $n > p$ are then compared.

The background material for the CLR inequality in $L^2(\mathbb{R}^n)$, $n \geq 3$, is provided, but only references to the independent and challenging proofs of Cwikel, Lieb and Rosenbljum are given. Using the approach of Levin and Solomyak, we show how the Sobolev and CLR inequalities in $L^2(\mathbb{R}^n)$, $n \geq 3$, compare.

Finally in this chapter we discuss Kato's inequality, which is a relativistic form of Hardy's inequality.

## 1.2 Hardy's Inequality in $\mathbb{R}^n$

### *1.2.1 The Case $n = 1$*

The first theorem gives general forms of the Hardy inequality involving weighted $L^p$ spaces on $(0, \infty)$. Special choices of the weights will yield the prototypes of the inequalities to be considered throughout the book. The proof uses the full force of Hölder's inequality; that if $1 < p < \infty$ and $p' = p/(p-1)$, then for non-negative functions $f$, $g$,

$$\int_0^\infty |f(x)g(x)|dx \leq \left(\int_0^\infty |f(x)|^p dx\right)^{\frac{1}{p}} \left(\int_0^\infty |g(x)|^{p'} dx\right)^{\frac{1}{p'}},$$

with equality if and only if there exist constants $A, B$, not both zero, such that $A|f(x)|^p = B|g(x)|^{p'}$.

**Theorem 1.2.1** *Let $1 < p < \infty$ and set $F(x) = \int_0^x f(t)dt$. Then for all $f$ such that $x^{\varepsilon} f(x) \in L^p(0, \infty)$, where $\varepsilon < (1/p') = 1 - 1/p$, we have*

$$\int_0^\infty |F(x)|^p x^{p(\varepsilon-1)} dx \le C_{p,\varepsilon} \int_0^\infty |f(x)|^p x^{p\varepsilon} dx \tag{1.2.1}$$

*for some positive constant $C_{p,\varepsilon}$ which is independent of $f$. If $\varepsilon > (1/p')$, then the inequality takes the form*

$$\int_0^\infty |G(x)|^p x^{p(\varepsilon-1)} dx \le C_{p,\varepsilon} \int_0^\infty |f(x)|^p x^{p\varepsilon} dx, \tag{1.2.2}$$

*where $G(x) = \int_x^\infty f(t)dt$. The best possible constants $C_{p,\varepsilon}$ are the same, and given by*

$$C_{p,\varepsilon} = |\varepsilon - 1/p'|^{-p}, \tag{1.2.3}$$

*and equality can only be attained by $f = 0$.*

*Proof* We may assume, without loss of generality, that $f$ is real-valued and non-negative, since the theorem will follow if we prove it for $|f|$. For $\varepsilon < 1/p'$, we have, by Hölder's inequality,

$$\begin{aligned} F(x)x^{(\varepsilon-1/p')} &\le x^{(\varepsilon-1/p')} \left( \int_0^x f^p(t) t^{\varepsilon p} dt \right)^{1/p} \left( \int_0^x t^{-\varepsilon p'} dt \right)^{1/p'} \\ &= (1 - \varepsilon p')^{-1/p'} \left( \int_0^x f^p(t) t^{\varepsilon p} dt \right)^{1/p}, \end{aligned}$$

which tends to zero as $x \to 0$. On integration by parts, it follows that with $0 < X < \infty$,

$$\begin{aligned} \int_0^X F^p(x) x^{p(\varepsilon-1)} dx &= \frac{F^p(X) X^{p(\varepsilon-1/p')}}{p(\varepsilon - 1/p')} \\ &\quad - \frac{p}{p(\varepsilon - 1/p')} \int_0^X F^{p-1}(x) f(x) x^{p(\varepsilon-1/p')} dx \\ &\le \frac{1}{|(\varepsilon - 1/p')|} \int_0^X F^{p-1}(x) f(x) x^{p(\varepsilon-1/p')} dx \\ &= \frac{1}{|(\varepsilon - 1/p')|} \int_0^X \left( F^p(x) x^{p(\varepsilon-1)} \right)^{1/p'} \left( f^p(x) x^{p\varepsilon} \right)^{1/p} dx \end{aligned}$$

and hence, by Hölder's inequality,

$$\int_0^X F^p(x)x^{p(\varepsilon-1)}dx \le \frac{1}{|(\varepsilon - 1/p')|}\left(\int_0^X F^p(x)x^{p(\varepsilon-1)}dx\right)^{1/p'} \times \left(\int_0^X f^p(x)x^{\varepsilon p}dx\right)^{1/p}$$

and

$$\int_0^X F^p(x)x^{p(\varepsilon-1)}dx \le \frac{1}{|(\varepsilon - 1/p')|^p}\int_0^X f^p(x)x^{\varepsilon p}dx.$$

The inequality (1.2.1), with $C_{p,\varepsilon} = |(\varepsilon - 1/p')|^{-p}$, follows on allowing $X \to \infty$. In the penultimate step, in which Hölder's inequality is applied, the resulting inequality is strict, unless there are constants $A, B$, not both zero, such that $AF^p(x)x^{p(\varepsilon-1)} = Bf^p(x)x^{p\varepsilon}$. But this would mean that $f(x) = F'(x)$ is a power of $x$ and $\int_0^\infty f^p(x)x^{p\varepsilon}dx$ is divergent. Consequently, (1.2.1) is a strict inequality for $f \neq 0$.

To prove that the constant (1.2.3) is sharp in (1.2.1), we choose $f(x) = x^{-1/p+\alpha}\chi_{(0,a)}(x)$, where $\alpha + \varepsilon > 0$, $a > 0$, and $\chi_{(0,a)}$ is the characteristic function of $(0, a)$. Then $x^\varepsilon f(x) \in L^p(0,\infty)$ and

$$\int_0^\infty f^p(x)x^{\varepsilon p}dx = \frac{a^{p(\alpha+\varepsilon)}}{p(\alpha+\varepsilon)},$$

$$F(x) = \begin{cases} \frac{x^{\alpha+1/p'}}{(\alpha+1/p')} & \text{if } x \le a, \\ \frac{a^{\alpha+1/p'}}{(\alpha+1/p')} & \text{if } x > a, \end{cases}$$

and

$$\int_0^\infty F^p(x)x^{p(\varepsilon-1)}dx = \frac{a^{p(\alpha+\varepsilon)}}{p(\alpha+1/p')^p}\left\{\frac{1}{\alpha+\varepsilon} + \frac{1}{1/p'-\varepsilon}\right\}.$$

This gives

$$\frac{\int_0^\infty F^p(x)x^{p(\varepsilon-1)}dx}{\int_0^\infty f^p(x)x^{\varepsilon p}dx} = \frac{1}{(\alpha+1/p')^p}\left\{1 + \frac{(\alpha+\varepsilon)}{(1/p'-\varepsilon)}\right\}$$

which tends to $|\varepsilon - 1/p'|^{-p}$ as $\alpha \to -\varepsilon$. It follows that the constant $|\varepsilon - 1/p'|^{-p}$ in (1.2.1) is sharp. The inequality (1.2.2), with sharp constant (1.2.3) is proved similarly, and so is the fact that equality can only be attained if $f = 0$. □

The choices $\varepsilon = 0, 1$, in Theorem 1.2.1 yield the following familiar forms of the Hardy inequality:

**Corollary 1.2.2** *Let $1 < p < \infty$. Let $F$ be a locally absolutely continuous function on $(0, \infty)$ which is such that $F' \in L^p(0, \infty)$ and $\lim_{x\to 0+} F(x) = 0$. Then*

$$\int_0^\infty \frac{|F(x)|^p}{|x|^p} dx \le \left(\frac{p}{p-1}\right)^p \int_0^\infty |F'(x)|^p dx. \tag{1.2.4}$$

*If $G$ is locally absolutely continuous on $(0, \infty)$ and is such that $xG'(x) \in L^p(0, \infty)$ and $\lim_{x\to\infty} G(x) = 0$, then*

$$\int_0^\infty |G(x)|^p dx \le \left(\frac{p}{p-1}\right)^p \int_0^\infty |xG'(x)|^p dx. \tag{1.2.5}$$

*The constant in (1.2.4) and (1.2.5) is sharp, and equality can only be attained in each inequality by the zero function.*

### 1.2.2 Weighted Hardy-Type Inequalities on Intervals

In (1.2.1), consider the following substitutions:

$$h(x) = x^\varepsilon f(x), \quad (Hh)(x) := x^{\varepsilon-1} F(x) = x^{\varepsilon-1} \int_0^x t^{-\varepsilon} h(t)dt.$$

Then Theorem 1.2.1 expresses the fact that $H$ is a bounded linear operator of $L^p(0, \infty)$ into itself, and the best possible constant is given by its norm:

$$C_{p,\varepsilon} = \|H : L^p(0, \infty) \to L^p(0, \infty)\|^p.$$

We now determine a necessary and sufficient condition for a general Hardy-type operator $T$ of the form

$$Tf(x) := v(x) \int_a^x u(t)f(t)dt \tag{1.2.6}$$

to be bounded as a map from $L^p(a, b)$ into itself, for $-\infty \le a < b \le \infty$ and $1 \le p \le \infty$, i.e., for all functions $f \in L^p(a, b)$, there exists a constant $C > 0$ such that

$$\int_a^b |Tf(x)|^p dx \le C \int_a^b |f(x)|^p dx.$$

The associated inequality is therefore

$$\int_a^b \left| v(x) \int_a^x u(t)f(t)dt \right|^p dx \le C \int_a^b |f(x)|^p dx$$

with best constant $C = \|T\|^p$; on setting $F(x) = \int_a^x u(t)f(t)dt$, the last inequality becomes

$$\int_a^b |v(x)F(x)|^p \, dx \le C \int_a^b \left| \frac{F'(x)}{u(x)} \right|^p dx. \tag{1.2.7}$$

We shall assume that $u, v$ are prescribed real-valued functions such that for all $X \in (a, b)$,

$$u \in L^{p'}(a, X), \tag{1.2.8}$$

$$v \in L^p(X, b), \tag{1.2.9}$$

where $p' = p/(p-1)$; thus $u \notin L^{p'}(a, b)$, $v \notin L^p(a, b)$ are possibilities. The following theorem is a special case of general results which may be found in [49], Chap. 2; see the references therein for a comprehensive treatment.

We denote the standard $L^q(I)$ norm on a sub-interval $I \subset (a, b)$ by $\|f\|_{q,I}$ and write $\|f\|_q$ when $I = (a, b)$; thus $\|f\|_{q,I} = \left( \int_I |f(x)|^q dx \right)^{1/q}$ if $1 \le q < \infty$, and $\operatorname{ess\,sup}_I |f(x)|$ if $q = \infty$.

**Theorem 1.2.3** *Let $1 \le p \le \infty$, and suppose that (1.2.8) and (1.2.9) are satisfied for all $X \in (a, b)$. Then the Hardy-type operator $T$ in (1.2.6) is a bounded linear map of $L^p(a, b)$ into itself if and only if*

$$A := \sup_{a<X<b} \left\{ \|u\|_{p',(a,X)} \|v\|_{p,(X,b)} \right\} < \infty. \tag{1.2.10}$$

*In this case*

$$A \le \|T\| \le 4A. \tag{1.2.11}$$

*Proof* We first prove that for all $X \in (a, b)$,

$$\alpha_X := \inf \left\{ \|f\|_p : f \in L^p(a, b), \int_a^X |f(t)u(t)|dt = 1 \right\} = \|u\|_{p',(a,X)}^{-1}. \tag{1.2.12}$$

Since

$$\int_a^X |f(t)u(t)|dt \le \|f\|_p \|u\|_{p',(a,X)},$$

by Hölder's inequality, we have that $\alpha_X \geq \|u\|_{p',(a,X)}^{-1}$ for $1 \leq p \leq \infty$. If $p > 1$, the choice

$$f(x) = |u^{p'-1}(x)|\chi_{(a,X)}(x)\|u\|_{p',(a,X)}^{-p'},$$

where $\chi_{(a,X)}$ is the characteristic function of $(a, X)$, gives

$$\int_a^X |f(t)u(t)|dt = 1$$

and as $p(p'-1) = p'$, this implies that

$$\alpha_X \leq \|f\|_p = \|u\|_{p',(a,X)}^{-1},$$

which proves (1.2.12) for $1 < p \leq \infty$. If $p = 1$, the assumption (1.2.8) becomes $u \in L^\infty(a, X)$, which means that $u$ is bounded a.e. on $[a, X]$. Given $\varepsilon > 0$, there exists a non-null set $S \subset [a, X]$ such that for all $x \in S$, $|u(x)| > (1+\varepsilon)^{-1}\|u\|_{\infty,(a,X)}$. On choosing $f(x) = \chi_S(x)[\int_S |u(t)|dt]^{-1}$, we have

$$\int_a^X |f(t)u(t)|dt = 1.$$

Hence

$$\alpha_X \leq \|f\|_1 = \frac{|S|}{\int_S |u(t)|dt} \leq \frac{1+\varepsilon}{\|u\|_{\infty,(a,X)}},$$

and (1.2.12) again follows, on taking $\varepsilon \to 0$.

Let $p < \infty$ and define $\mathfrak{I} = \mathbb{Z}$ when $u \notin L^{p'}(a, b)$ and $\mathfrak{I} = \{k \in \mathbb{Z} : -\infty < k \leq M\}$ for some $M \in \mathbb{Z}$ when $u \in L^{p'}(a, b)$. For $f \in L^p(a, b)$ and $i \in \mathfrak{I}$, let

$$X_i := \sup\{x \in (a, b) : \int_a^x |f(t)u(t)|dt = 2^i\}. \tag{1.2.13}$$

Then $\{X_i : i \in \mathfrak{I}\}$ generates a partition of $(a, b)$ and we have

$$\begin{aligned}
\|Tf\|_p^p &\leq \sum_{i\in\mathfrak{I}} \int_{X_i}^{X_{i+1}} \left| v(x) \int_a^x |f(t)u(t)|dt \right|^p dx \\
&\leq \sum_{i\in\mathfrak{I}} 2^{p(i+1)} \|v\|_{p,(X,b)}^p \\
&\leq A^p \sum_{i\in\mathfrak{I}} 2^{p(i+1)} \alpha_X^p
\end{aligned}$$

by (1.2.10) and (1.2.12). Since

$$\int_{X_{i-1}}^{X_i} |f(t)u(t)|dt = 2^i - 2^{i-1}$$

it follows that

$$\alpha_{X_i} \leq 2^{1-i} \|f\chi_{[X_{i-1},X_i]}\|_p$$

and consequently

$$\begin{aligned} \|Tf\|_p^p &\leq (4A)^p \sum_{i\in\mathfrak{I}} \|f\chi_{[X_{i-1},X_i]}\|_p^p \\ &= (4A)^p \|f\|_p^p. \end{aligned}$$

We have therefore proved that if $1 \leq p < \infty$, (1.2.10) is sufficient for the boundedness of $T$ and that $\|T\| \leq 4A$. The proof for $p = \infty$ is similar.

To establish the necessity of (1.2.10), we choose, for a given $\varepsilon > 0$ and $X \in (a, b)$, an $f \in L^p(a, b)$ such that $uf \geq 0$, $\int_a^X f(t)u(t)dt = 1$ and $\|f\|_p \leq \alpha_X(1 + \varepsilon)$. Then $|Tf(x)| \geq |v(x)|$ for all $x \in [X, b)$, and if $T : L^p(a, b) \to L^p(a, b)$ is bounded, we have

$$\alpha_X(1 + \varepsilon)\|T\| \geq \|T\|\|f\|_p \geq \|Tf\|_p \geq \|v\|_{p,(X,b)}.$$

This and (1.2.12) yield

$$(1 + \varepsilon)\|T\| \geq \|u\|_{p',(a,X)}\|v\|_{p,(X,b)},$$

whence (1.2.10) and $\|T\| \geq A$. The theorem is therefore proved. □

*Remark 1.2.4* Theorem 1.2.3 was established by Chisholm and Everitt in [37] for the case p = 2, and by Muckenhoupt in [121] for $1 < p < \infty$. The finiteness of the quantity $A$ in (1.2.10) is generally referred to as the Muckenhoupt condition. In [126] (see Comment 3.6 on page 27), the upper bound

$$\|T\| \leq p^{\frac{1}{p}}(p')^{\frac{1}{p'}} A$$

is derived. This is best possible, for on taking $a = 0$, $b = \infty$, $u(t) = t^{\varepsilon}$ and $v(t) = t^{\varepsilon-1}$, with $\varepsilon < 1/p'$, we obtain

$$\left\{p^{\frac{1}{p}}(p')^{\frac{1}{p'}} A\right\}^p = \left(\frac{1}{p'} - \varepsilon\right)^{-p},$$

which was shown to be optimal for the Hardy inequality in Theorem1.2.1.

### 1.2.3 The Case $n > 1$

Our main concern in the book will be with multi-dimensional Hardy inequalities, and we make a start in this section with analogues in $\mathbb{R}^n, n > 1$, of the results in Sect. 1.1.

**Theorem 1.2.5** *Let $1 \le p < \infty, n > 1$, and $\varepsilon - 1 + n/p \neq 0$. Let $f$ be differentiable a.e. in $\mathbb{R}^n$, and such that $|\mathbf{x}|^{\varepsilon-1+n/p} f(\mathbf{x})$ tends to zero as $|\mathbf{x}| \to 0+$ if $\varepsilon - 1 + n/p < 0$, and as $|\mathbf{x}| \to \infty$ if $\varepsilon - 1 + n/p > 0$. Then*

$$\int_{\mathbb{R}^n} \left(|\mathbf{x}|^{\varepsilon-1} |f(\mathbf{x})|\right)^p d\mathbf{x} \le |\varepsilon - 1 + n/p|^{-p} \int_{\mathbb{R}^n} \left(|\mathbf{x}|^{\varepsilon-1} |(\mathbf{x} \cdot \nabla) f(\mathbf{x})|\right)^p d\mathbf{x} \qquad (1.2.14)$$

*where $\nabla f = (\partial f/\partial x_1, \cdots, \partial f/\partial x_n)$. The constant $|\varepsilon - 1 + n/p|^{-p}$ is sharp.*

*Proof* Let $0 < \delta, N < \infty$, and choose polar co-ordinates $\mathbf{x} = r\omega,\ r = |\mathbf{x}|, \omega \in \mathbb{S}^{n-1}$. On integration by parts

$$\begin{aligned}
\int_\delta^N r^{(\varepsilon-1)p} |f(r\omega)|^p r^{n-1} dr &- \left[\frac{r^{(\varepsilon-1)p+n}}{(\varepsilon-1)p+n} |f(r\omega)|^p\right]_\delta^N \\
&= -\int_\delta^\infty \frac{r^{(\varepsilon-1)p+n}}{(\varepsilon-1)p+n} \frac{\partial}{\partial r} |f(r\omega)|^p dr \\
&\le |(\varepsilon-1)p+n|^{-1} \int_0^\infty r^{(\varepsilon-1)p+n} \left|\frac{\partial}{\partial r} |f(r\omega)|^p\right| dr.
\end{aligned}$$

We next let $\delta \to 0$ and $N \to \infty$, and use the hypothesis to obtain

$$\begin{aligned}
&\int_0^\infty r^{(\varepsilon-1)p} |f(r\omega)|^p r^{n-1} dr \\
&\le |\varepsilon - 1 + n/p|^{-1} \int_0^\infty r^{(\varepsilon-1)p+n} |f(r\omega)|^{p-1} \left|\tfrac{\partial}{\partial r}(|f(r\omega)|)\right| dr \\
&\le |\varepsilon - 1 + n/p|^{-1} \int_0^\infty r^{(\varepsilon-1)p+n} |f(r\omega)|^{p-1} \left|\tfrac{\partial}{\partial r}(f(r\omega))\right| dr \\
&\le |\varepsilon - 1 + n/p|^{-1} \left(\int_0^\infty r^{(\varepsilon-1)p} |f(r\omega)|^p r^{n-1} dr\right)^{1/p'} \\
&\quad \times \left(\int_0^\infty r^{\varepsilon p} \left|\tfrac{\partial f(r\omega)}{\partial r}\right|^p r^{n-1} dr\right)^{1/p},
\end{aligned}$$

where we have used the fact that $|\partial/\partial r |f(r\omega)|| \le |\partial f(r\omega)/\partial r|$ (see Theorem 1.3.8 below), and Hölder's inequality. Therefore, since $r\partial f/\partial r = (\mathbf{x} \cdot \nabla) f$,

$$\begin{aligned}
\int_{\mathbb{R}^n} |\mathbf{x}|^{(\varepsilon-1)p} |f(\mathbf{x})|^p d\mathbf{x} &\le |\varepsilon - 1 + n/p|^{-1} \left(\int_{\mathbb{R}^n} |\mathbf{x}|^{(\varepsilon-1)p} |f(\mathbf{x})|^p d\mathbf{x}\right)^{1/p'} \\
&\quad \times \left(\int_{\mathbb{R}^n} |\mathbf{x}|^{(\varepsilon-1)p} |(\mathbf{x} \cdot \nabla) f(\mathbf{x})|^p d\mathbf{x}\right)^{1/p}
\end{aligned}$$

and (1.2.14) follows.

To prove that the constant is sharp, we consider radial functions $f(\mathbf{x}) = f(r)$, $r = |\mathbf{x}|$, which satisfy the hypothesis of the theorem. Then (1.2.14) becomes

$$\int_{\mathbb{S}^{n-1}} \int_0^\infty |r^{\eta-1} f(r)|^p dr d\omega \le |\eta - 1/p'|^{-p} \int_{\mathbb{S}^{n-1}} \int_0^\infty |r^\eta f'(r)|^p dr d\omega, \tag{1.2.15}$$

where $\eta = \varepsilon + (n-1)/p$. Suppose $\eta - 1/p' = \varepsilon - 1 + n/p < 0$, and let $f(r) = \int_0^r \phi(t)dt$, where for some $a > 0$,

$$\phi(t) = t^{-\frac{1}{p}+\alpha} \chi_{(0,a)}(t), \quad \alpha + \eta > 0,$$

so that

$$f(r) = \begin{cases} \frac{r^{\alpha+\frac{1}{p'}}}{(\alpha+\frac{1}{p'})}, & \text{if } r \le a, \\ \frac{a^{\alpha+\frac{1}{p'}}}{(\alpha+\frac{1}{p'})}, & \text{if } r > a. \end{cases}$$

Then $f$ satisfies the hypothesis and, as in the proof of Theorem 1.2.1,

$$\begin{aligned} \lim_{\alpha \to -\eta} \left\{ \frac{\int_0^\infty f^p(r) r^{p(\eta-1)} dr}{\int_0^\infty f'(r)^p r^{p\eta} dr} \right\} &= \frac{1}{(-\eta + \frac{1}{p'})^p} \\ &= \frac{1}{|\varepsilon - 1 + \frac{n}{p}|^p}. \end{aligned}$$

The constant is therefore sharp. The case $\varepsilon - 1 + n/p > 0$ is treated similarly. □

The choices $\varepsilon = 0, 1$ of Theorem 1.2.5 yield the following corollary: we use the notation $|\nabla f(\mathbf{x})| = \left(\sum_{i=1}^n |\partial f/\partial x_i|^2\right)^{1/2}$.

**Corollary 1.2.6** *The inequality*

$$\int_{\mathbb{R}^n} \frac{|f(\mathbf{x})|^p}{|\mathbf{x}|^p} d\mathbf{x} \le \left| \frac{p}{p-n} \right|^p \int_{\mathbb{R}^n} |\nabla f(\mathbf{x})|^p d\mathbf{x} \tag{1.2.16}$$

*holds for all $f \in C_0^\infty(\mathbb{R}^n \setminus \{0\})$ if $n < p < \infty$ and for all $f \in C_0^\infty(\mathbb{R}^n)$ if $1 \le p < n$.*

*Moreover, for all $f \in C_0^\infty(\mathbb{R}^n)$ and $1 \le p < \infty$,*

$$\int_{\mathbb{R}^n} |f(\mathbf{x})|^p d\mathbf{x} \le \left(\frac{p}{n}\right)^p \int_{\mathbb{R}^n} |(\mathbf{x} \cdot \nabla) f(\mathbf{x})|^p \, d\mathbf{x}, \tag{1.2.17}$$

*and hence*

$$\int_{\mathbb{R}^n} |f(\mathbf{x})|^p d\mathbf{x} \le \left(\frac{p}{n}\right)^p \int_{\mathbb{R}^n} (|\mathbf{x}||\nabla f(\mathbf{x})|)^p \, d\mathbf{x}. \tag{1.2.18}$$

*The constants in (1.2.16), (1.2.17) and (1.2.18) are sharp.*

*Remark 1.2.7* The classical Hardy inequality (1.2.16) is invariant under orthogonal transformations and scaling, but is not invariant under general linear transformations. The inequality (1.2.17) is an affine invariant version of (1.2.16). Moreover, (1.2.18) implies (1.2.16) if $1 < p < n$. This follows from

$$\nabla(|\mathbf{x}|f(\mathbf{x})) = \frac{\mathbf{x}}{|\mathbf{x}|}f(\mathbf{x}) + |\mathbf{x}|\nabla f(\mathbf{x}).$$

For if we suppose that $f$ satisfies (1.2.18), then, with $\|\cdot\|$ denoting the $L^p(\mathbb{R}^n)$ norm,

$$\begin{aligned}\|\nabla(|\mathbf{x}|f)\| &\geq \||\mathbf{x}|\nabla f\| - \|f\| \\ &\geq \left(\frac{n-p}{p}\right)\|f\|,\end{aligned}$$

whence (1.2.16) on replacing $f(\mathbf{x})$ by $f(\mathbf{x})/|\mathbf{x}|$.

### 1.2.4 A Weighted Hardy-Type Inequality on $\Omega \subseteq \mathbb{R}^n$, $n > 1$

The inequality in the next theorem will be needed in Chap. 6. It is proved in [45] for the case $\Delta V < 0$, but the case $p = 2$ was proved earlier by the same technique in [105], Lemma 2.

**Theorem 1.2.8** *Let $\Omega$ be a domain in $\mathbb{R}^n$, $n > 1$, and let $V$ be a real-valued function in $L^1_{loc}(\Omega)$ with partial derivatives of order up to 2 in $L^1_{loc}(\Omega)$, and is such that $\Delta V$ is of one sign a.e.. Then, for all $u \in C_0^\infty(\Omega)$,*

$$\int_\Omega |\Delta V||u|^p d\mathbf{x} \leq p^p \int_\Omega \frac{|\nabla V|^p}{|\Delta V|^{p-1}}|\nabla u|^p d\mathbf{x}. \tag{1.2.19}$$

*Proof* Suppose, for definiteness, that $\Delta V < 0$. Let $v_\varepsilon := (|u|^2 + \varepsilon^2)^{1/2} - \varepsilon$. Then $v_\varepsilon^p \in C_0^\infty(\Omega)$ and

$$\begin{aligned}\int_\Omega |\Delta V| v_\varepsilon^p d\mathbf{x} &= -\int_\Omega \Delta V v_\varepsilon^p d\mathbf{x} = \int_\Omega \nabla V \cdot \nabla v_\varepsilon^p d\mathbf{x} \\ &\leq p\int_\Omega \left(\frac{|\nabla V|}{|\Delta V|^{(p-1)/p}}\right)|\Delta V|^{(p-1)/p} v_\varepsilon^{p-1}|\nabla v_\varepsilon| d\mathbf{x}.\end{aligned}$$

Since $0 \le v_\varepsilon \le |u|$ and

$$\nabla v_\varepsilon = (|u|^2 + \varepsilon^2)^{-1/2}|u|\nabla|u|,$$

we have that

$$v_\varepsilon^{p-1}|\nabla v_\varepsilon| \le |u|^{p-1}|\nabla|u||.$$

Also $|\nabla|u|| \le |\nabla u|$ a.e.; see Theorem 1.3.8 below. Hence

$$\begin{aligned}\int_\Omega |\Delta V| v_\varepsilon^p d\mathbf{x} &\le p\int_\Omega \left(\frac{|\nabla V|}{|\Delta V|^{(p-1)/p}}|\nabla u|\right)|\Delta V|^{(p-1)/p}|u|^{p-1}d\mathbf{x}\\ &\le p\left\{\int_\Omega\left(\frac{|\nabla V|^p}{|\Delta V|^{(p-1)}}|\nabla u|^p\right)d\mathbf{x}\right\}^{1/p}\left\{\int_\Omega |\Delta V||u|^p d\mathbf{x}\right\}^{(p-1)/p}\end{aligned}$$

by Hölder's inequality, whence (1.2.19), by dominated convergence, on allowing $\varepsilon \to 0$. □

**Corollary 1.2.9** *For any $\sigma \in (-\infty, \infty)$,*

$$\int_{\mathbb{R}^n} \frac{|u(\mathbf{x})|^p}{|\mathbf{x}|^{\sigma+2}}d\mathbf{x} \le \left|\frac{p}{\sigma - n + 2}\right|^p \int_{\mathbb{R}^n} \frac{|\nabla u(\mathbf{x})|^p}{|\mathbf{x}|^{(\sigma+2-p)}}d\mathbf{x} \qquad (1.2.20)$$

*for all $C_0^\infty(\mathbb{R}^n \setminus \{0\})$ if $\sigma + 2 > n$, and all $u \in C_0^\infty(\Omega)$ if $\sigma + 2 < n$. In particular, when $\sigma = p - 2$, (1.2.20) coincides with (1.2.16).*

*Proof* To deduce (1.2.20), choose $V(\mathbf{x}) = |\mathbf{x}|^{-\sigma}$ for $\sigma \ne 0$ and $V(\mathbf{x}) = \log|\mathbf{x}|$ if $\sigma = 0$. Then, $\Delta V(\mathbf{x}) = \sigma(\sigma + 2 - n)|\mathbf{x}|^{-(\sigma+2)}, (\sigma \ne 0)$, $\Delta V(\mathbf{x}) = (n-2)|\mathbf{x}|^{-2}, (\sigma = 0)$ and the conditions of the theorem are satisfied. □

The inequality (1.2.20) was proved for the case $p = 2$ by Allegretto in [6] where an earlier proof is attributed to Piepenbrink.

### *1.2.5 The Case $n = p$*

We shall now show that when $n = p$, there is no valid Hardy inequality, i.e., there is no positive constant $C$ such that

$$\int_{\mathbb{R}^n} |\nabla f(\mathbf{x})|^n d\mathbf{x} \ge C\int_{\mathbb{R}^n} \frac{|f(\mathbf{x})|^n}{|\mathbf{x}|^n}d\mathbf{x} \quad \text{for all } f \in C_0^\infty(\mathbb{R}^n); \qquad (1.2.21)$$

in fact we shall prove it is invalid on the set of radial functions.

Let $f(\mathbf{x}) = F(r), r = |\mathbf{x}|$, and suppose that $F \in C_0^\infty((a, b))$, where $a > 0, b < \infty$ are arbitrary. Then (1.2.21) implies that

$$\int_a^b |F'(r)|^n r^{n-1} dr \geq C \int_a^b \frac{|F(r)|^n}{r} dr. \tag{1.2.22}$$

Therefore (1.2.7) is satisfied with $p = n, v(r)^n = r^{-1}$, $u(r)^{-n} = r^{n-1}$ and in (1.2.10)

$$\begin{aligned}
\sup_{a<X<b} \|u\|_{p',(a,X)} \|v\|_{p,(X,b)} &= \sup_{a<X<b} \left( \int_a^X \frac{1}{r} dr \right)^{1/n'} \left( \int_X^b \frac{1}{r} dr \right)^{1/n} \\
&= \sup_{0<C<1} C^{1/n'} (1 - C)^{1/n} \int_a^b \frac{1}{r} dr \\
&\geq \frac{1}{2} \int_a^b \frac{1}{r} dr.
\end{aligned}$$

Since $a > 0, b < \infty$ are arbitrary, (1.2.22) is contradicted by Theorem 1.2.3.

In [2], Theorem 4.6, it is proved that in the case $n = p = 2$, for all $f \in C_0^\infty(\mathbb{R}^2 \setminus \{0\})$ satisfying $\int_{1<|\mathbf{x}|<2} f(\mathbf{x}) d\mathbf{x} = 0$, there exists a constant $C > 0$ such that

$$\int_{\mathbb{R}^2} \frac{|f(\mathbf{x})|^2}{|\mathbf{x}|^2 (1 + \log^2 |\mathbf{x}|)} d\mathbf{x} \leq C \int_{\mathbb{R}^2} |\nabla f(\mathbf{x})|^2 d\mathbf{x}. \tag{1.2.23}$$

It is observed in [138] that the logarithmic factor is needed only for radial functions, and can be removed for functions satisfying $\int_{|\mathbf{x}|=r} f(\mathbf{x}) d\mathbf{x} = 0$ for all $r > 0$.

## 1.3 Sobolev Spaces

In this section we give a brief description of those aspects of Sobolev spaces which are relevant to the content of this book. Where proofs of results quoted are not included, precise references are given. Comprehensive treatments may be found in [1, 48] .

### *1.3.1 The Spaces $W^{k,p}(\Omega)$ and $W_0^{k,p}(\Omega)$*

Let $\Omega$ be a non-empty open subset of $\mathbb{R}^n$, $n \geq 1$, with closure $\overline{\Omega}$ and boundary $\partial\Omega$. For $p \in [1, \infty]$, $k \in \mathbb{N}$, points $\mathbf{x} = (x_1, \cdots, x_n) \in \mathbb{R}^n$ and $n$-tuples $\alpha = (\alpha_1, \cdots, \alpha_n) \in$

$\mathbb{N}_0^n$, we write

$$|\mathbf{x}| = \left(\sum_{j=1}^{n} x_j^2\right)^{1/2}, \ |\alpha| = \sum_{j=1}^{n} \alpha_j, \ D^\alpha = \prod_{j=1}^{n} D_j^{\alpha_j},$$

where $D_j = -i\partial_j$, $\partial_j = \frac{\partial}{\partial x_j}$, and the derivatives are taken in the weak sense. We recall that $f \in L^1_{loc}(\Omega)$ has weak derivative $g =: D_j f$ if $g \in L^1_{loc}(\Omega)$ and

$$\int_\Omega f(\mathbf{x}) D_j \varphi(\mathbf{x}) d\mathbf{x} = -\int_\Omega g(\mathbf{x})\varphi(\mathbf{x}) d\mathbf{x}$$

for all $\varphi \in C_0^\infty(\Omega)$; here $L^1_{loc}(\Omega)$ denotes the set of functions which are integrable on all compact subsets of $\Omega$. We define

$$W^{k,p}(\Omega) := \{u : D^\alpha u \in L^p(\Omega) \text{ for } |\alpha| \le k\}$$

endowed with the norm

$$\|u\|_{k,p,\Omega} = \begin{cases} \left(\sum_{0\le|\alpha|\le k} \|D^\alpha\|^p_{p,\Omega}\right)^{1/p}, & \text{for } 1 \le p < \infty, \\ \left(\sum_{0\le|\alpha|\le k} \|D^\alpha\|_{\infty,\Omega}\right), & \text{for } p = \infty. \end{cases},$$

$\|\cdot\|_{p,\Omega}$ being the standard $L^p(\Omega)$ norm, namely

$$\|u\|_{p,\Omega} := \left(\int_\Omega |u(\mathbf{x})|^p d\mathbf{x}\right)^{1/p}.$$

In particular, when $k = 1$ and $1 \le p < \infty$,

$$\|u\|^p_{1,p,\Omega} = \|u\|^p_{p,\Omega} + \|\nabla u\|^p_{p,\Omega},$$

where

$$\|\nabla u\|_{p,\Omega} = \||\nabla u|\|_{p,\Omega}, \ |\nabla u| := \left(\sum_{i=1}^{n} |\partial_i u|^2\right)^{1/2}.$$

When $\Omega = \mathbb{R}^n$, we shall write $\|\cdot\|_{k,p}$ and $\|\cdot\|_p$ for the norms on $W^{k,p}(\mathbb{R}^n)$ and $L^p(\mathbb{R}^n)$ respectively. For $p \in [1,\infty]$, $W^{k,p}(\Omega)$ is a Banach space, being separable if $p \in [1,\infty)$ and reflexive if $p \in (1,\infty)$. For $p \in [1,\infty)$, the linear subspace $C^\infty(\Omega) \cap W^{k,p}(\Omega)$ is dense in $W^{k,p}(\Omega)$; hence $W^{k,p}(\Omega)$ coincides with the completion $H^{k,p}(\Omega)$ in the $W^{k,p}(\Omega)$ norm, of the set of functions $f$ in $C^\infty(\Omega)$ which are such that $\|f\|_{k,p,\Omega} < \infty$; see [48], Theorem V.3.2.

The closure of $C_0^\infty(\Omega)$ in $W^{k,p}(\Omega)$ is denoted by $W_0^{k,p}(\Omega)$. In general $W_0^{k,p}(\Omega) \neq W^{k,p}(\Omega)$, but $W_0^{k,p}(\mathbb{R}^n) = W^{k,p}(\mathbb{R}^n)$; see [48], Sect. V.3.1. Another space with a prominent role is $D_0^{1,p}(\Omega)$, which is defined as the completion of $C_0^\infty(\Omega)$ with respect to the norm

$$\|u\|_{D_0^{1,p}(\Omega)} := \|\nabla u\|_{p,\Omega} = \left(\int_\Omega |\nabla u(\mathbf{x})|^p d\mathbf{x}\right)^{1/p}. \tag{1.3.1}$$

In general, $D_0^{1,p}(\Omega)$ is not embedded in $L^p(\Omega)$ and contains $W_0^{1,p}(\Omega)$ as a proper subspace, $W_0^{1,p}(\Omega)$ being continuously embedded in $D_0^{1,p}(\Omega)$ since $\|\nabla u\|_{p,\Omega} \leq \|u\|_{1,p,\Omega}$ for all $u \in W_0^{1,p}(\Omega)$; for instance, $D_0^{1,p}(\mathbb{R}^n)$ is not embedded in $L^p(\mathbb{R}^n)$. However $D_0^{1,p}(\Omega)$ coincides with $W_0^{1,p}(\Omega)$ if there exists a positive constant $C$ such that

$$\|u\|_{p,\Omega} \leq C\|\nabla u\|_{p,\Omega}, \quad (u \in C_0^\infty(\Omega)). \tag{1.3.2}$$

For then, the norms $\|\cdot\|_{D_0^{1,p}(\Omega)}$ and $\|\cdot\|_{1,p,\Omega}$ on $C_0^\infty(\Omega)$ are equivalent. The inequality (1.3.2) is the **Friedrichs inequality**. Examples of domains on which it is satisfied are given next.

**Proposition 1.3.1**

*(i) Let $\Omega$ be a bounded domain in $\mathbb{R}^n$ with volume $|\Omega|$. Then for $1 \leq p \leq \infty$,*

$$\|u\|_{p,\Omega} \leq \left(\frac{|\Omega|}{\omega_n}\right)^{1/n} \|\nabla u\|_{p,\Omega}, \quad (u \in C_0^\infty(\Omega)). \tag{1.3.3}$$

*(ii) Let $\Omega$ lie between 2 parallel hyperplanes at a distance $\ell$ apart. Then for $1 \leq p < \infty$,*

$$\|u\|_{p,\Omega} \leq \ell\|\nabla u\|_{p,\Omega}, \quad (u \in C_0^\infty(\Omega)). \tag{1.3.4}$$

*Proof*

(i) On setting $u(\mathbf{x}) = 0$ outside $\Omega$, we may suppose that $u \in C_0^\infty(\mathbb{R}^n)$. For all $\mathbf{x} \in \Omega$ and $\nu \in \mathbf{S}^{n-1}$,

$$u(\mathbf{x}) = -\int_0^\infty \frac{\partial}{\partial r} u(\mathbf{x} + r\nu)dr.$$

Hence, on using the Chain Rule,

$$\begin{aligned} n\omega_n u(\mathbf{x}) &= -\int_0^\infty \int_{\mathbb{S}^{n-1}} \frac{\partial}{\partial r} u(\mathbf{x} + r\nu)d\nu dr \\ &= \int_\Omega |\mathbf{x} - \mathbf{y}|^{-n} \sum_{i=1}^n (x_i - y_i)\partial_i u(\mathbf{y})d\mathbf{y} \end{aligned}$$

and so

$$n\omega_n|u(\mathbf{x})| \leq \int_\Omega |\mathbf{x}-\mathbf{y}|^{-n+1}|\nabla u(\mathbf{y})|d\mathbf{y}$$
$$\leq \left(\int_\Omega |\mathbf{x}-\mathbf{y}|^{-n+1}|\nabla u(\mathbf{y})|^p d\mathbf{y}\right)^{1/p}\left(\int_\Omega |\mathbf{x}-\mathbf{y}|^{-n+1}d\mathbf{y}\right)^{1-1/p}.$$

If $B_R(\mathbf{x})$ is the ball centre $\mathbf{x}$, radius $R$ and volume equal to $|\Omega|$, then since $|\mathbf{x}-\mathbf{y}|^{-n+1}$ increases as $\mathbf{y}$ tends to the centre of the ball, we have

$$\int_\Omega |\mathbf{x}-\mathbf{y}|^{-n+1}d\mathbf{y} \leq \int_{B_R(\mathbf{x})} |\mathbf{x}-\mathbf{y}|^{-n+1}d\mathbf{y}$$
$$= nR\omega_n,$$

and $\omega_n R^n = |\Omega|$. Therefore

$$(n\omega_n)^p \int_\Omega |u(\mathbf{x})|^p d\mathbf{x} \leq (nR\omega_n)^{p-1}\int_\Omega\left(\int_\Omega |\mathbf{x}-\mathbf{y}|^{-n+1}|\nabla u(\mathbf{y})|^p d\mathbf{y}\right)d\mathbf{x}$$
$$\leq (nR\omega_n)^p \int_\Omega |\nabla u(\mathbf{y})|^p d\mathbf{y},$$

whence (1.3.3) since $\omega_n R^n = |\Omega|$.

(ii) We again take $u(\mathbf{x}) = 0$ outside $\Omega$, and we assume, without loss of generality, that $\Omega$ lies between the hyperplanes $x_1 = 0$ and $x_1 = \ell$. Then, for all $\mathbf{x} \in \Omega$,

$$|u(\mathbf{x})| = \left|\int_{-\infty}^{x_1} \partial_1 u(t, x_2, \cdots, x_n)dt\right|$$
$$\leq \left(\int_{-\infty}^{x_1} |\partial_1 u(t, x_2, \cdots, x_n)|^p dt\right)^{1/p} \ell^{1-1/p}.$$

Hence

$$\int_0^t |u(x_1, x_2, \cdots, x_n)|^p dx_1 \leq \ell^p \int_{-\infty}^t |\partial_1 u(t, x_2, \cdots, x_n)|^p dt$$

and therefore

$$\int_\Omega |u(\mathbf{x})|^p d\mathbf{x} \leq \ell^p \int_\Omega |\nabla u(\mathbf{x})|^p d\mathbf{x},$$

as asserted. □

A necessary and sufficient condition on $\Omega$ for the Friedrichs inequality to hold is given in [48], Theorem VIII.2.10.

If $\Omega$ has finite measure, the Friedrichs inequality does not hold for $W^{1,p}(\Omega)$ since $W^{1,p}(\Omega)$ contains non-trivial constant functions which invalidate the inequality. By subtracting integral means, an analogue of the Friedrichs inequality for the space $W^{1,p}(\Omega)$ is obtained in the form of the **Poincaré** inequality which we now consider. An example is given in the following proposition.

**Proposition 1.3.2** *Let $\Omega$ be a bounded convex domain in $\mathbb{R}^n$ and $1 \leq p \leq \infty$. Then for all $u \in W^{1,p}(\Omega)$,*

$$\|u - u_\Omega\|_{p,\Omega} \leq \left(\frac{\omega}{|\Omega|}\right)^{1-1/n} [\mathrm{diam}(\Omega)]^n \|\nabla u\|_{p,\Omega}, \tag{1.3.5}$$

*where* $\mathrm{diam}(\Omega)$ *denotes the diameter of $\Omega$ and $u_\Omega$ is the integral mean*

$$u_\Omega := \frac{1}{|\Omega|}\int_\Omega u(\mathbf{x})d\mathbf{x}.$$

*Proof* We again extend $u$ by zero outside $\Omega$. For any $\mathbf{x}, \mathbf{y} \in \Omega$,

$$u(\mathbf{x}) - u(\mathbf{y}) = -\int_0^{|\mathbf{x}-\mathbf{y}|} \frac{\partial}{\partial r} u(\mathbf{x} + r\omega)dr, \quad \omega = \frac{\mathbf{x}-\mathbf{y}}{|\mathbf{x}-\mathbf{y}|},$$

and so

$$u(\mathbf{x}) - u_\Omega = -\frac{1}{|\Omega|}\int_\Omega \int_0^{|\mathbf{x}-\mathbf{y}|} \frac{\partial}{\partial r} u(\mathbf{x} + r\omega)drd\mathbf{y}.$$

Put $\mathbf{y} - \mathbf{x} = \rho\omega$, with $|\omega| = 1$ and $\rho \leq \rho_0 \leq \mathrm{diam}(\Omega)$. Then

$$\begin{aligned} |u(\mathbf{x}) - u_\Omega| &\leq \frac{1}{|\Omega|}\int_0^{\rho_0}\int_{|\omega|=1}\int_0^{\rho_0} \left|\frac{\partial}{\partial r}u(\mathbf{x}+r\omega)\right| \rho^{n-1} drd\omega d\rho \\ &= \frac{[\mathrm{diam}(\Omega)]^n}{n|\Omega|}\int_0^{\rho_0}\int_{|\omega|=1} \left|\frac{\partial}{\partial r}u(\mathbf{x}+r\omega)\right| d\omega dr \\ &\leq \frac{[\mathrm{diam}(\Omega)]^n}{n|\Omega|}\int_\Omega |\mathbf{x}-\mathbf{y}|^{-n+1}|\nabla u(\mathbf{y})|d\mathbf{y}. \end{aligned}$$

The rest of the proof follows that of Proposition 1.3.1(i). □

We refer to [48], Sect. V.5, for an interesting connection between the **Poincaré** inequality and the *measure of non-compactness* of the embedding $W^{1,p}(\Omega) \hookrightarrow L^p(\Omega)$.

In the next theorem, we collect two results of fundamental importance: they are special cases of (i) the **Sobolev embedding theorem**, and (ii) the **Rellich-Kondrachov theorem**; see [48], Theorems V.3.6 and V.3.7.

**Theorem 1.3.3**

**(i)** *Suppose that* $1 \leq p < n$, *and put* $p^* = np/(n-p)$. *Then* $D_0^{1,p}(\Omega)$ *is continuously embedded in* $L^s(\Omega)$ *for any* $s \in [p, p^*]$ *and there is a constant* $C$, *depending only on* $p$ *and* $n$, *such that for all* $u \in D_0^{1,p}(\Omega)$,

$$\|u\|_{p^*,\Omega} \leq C\|\nabla u\|_{p,\Omega}. \tag{1.3.6}$$

**(ii)** *If* $q \in [1, p^*)$ *and* $\Omega$ *is bounded, then* $D_0^{1,p}(\Omega)$ *is compactly embedded in* $L^q(\Omega)$.

The inequality (1.3.6) is the **Sobolev inequality** and $p^*$ is the **Sobolev conjugate** of $p$. Part (i) of the theorem asserts that any $u \in D_0^{1,p}(\Omega)$ (and thus any element of $W_0^{1,p}(\Omega)$ too) can be identified with a unique element of $L^{p^*}(\Omega)$ and that in terms of this identification map (that is, the **embedding**), (1.3.6) is satisfied. To say that the embedding is compact in Part (ii) means that it takes bounded sequences into relatively compact sequences, that is, ones with convergent subsequences. Analogues of Theorem 1.3.3 may also be found in [48], Sect. V.3.3 when $n < p < \infty$ and $p = n$. For instance, if $\Omega$ is bounded, $W_0^{1,p}(\Omega)$ is continuously embedded in the space $C^{0,\gamma}(\overline{\Omega})$, $\gamma = 1 - n/p$, of continuous functions which are locally Hölder continuous with exponent $\gamma$ on $\overline{\Omega}$, the embedding being compact if $\gamma < 1 - n/p$. When $p = n$ and $\Omega$ is bounded, $W_0^{1,p}(\Omega)$ is continuously embedded into $L^q(\Omega)$ for every $q \in [1, \infty)$; more generally the embedding maps into an Orlicz space, but we refer to [48] for any further discussion of this case. There are corresponding results for $W_0^{k,p}(\Omega)$; see [48], Theorem V.3.7. For instance if $kp < n$, $W_0^{k,p}(\Omega)$ is continuously embedded in $L^s(\Omega)$ for $s \in [p, np/(n - kp)]$, the embedding being compact if $\Omega$ is bounded and $s \in [p, np/(n - kp))$.

*Remark 1.3.4* The inequality (1.2.16) holds for all $f \in D_0^{1,p}(\mathbb{R}^n \setminus \{0\})$ if $n < p < \infty$, and all $f \in D_0^{1,p}(\mathbb{R}^n)$ if $1 \leq p < n$.

*Remark 1.3.5* The inequality (1.3.6), with $\Omega = \mathbb{R}^n$, was established by Sobolev in [137] for $1 < p < n$, the case $p = 1$ being later proved by Gagliardo [67] and Nirenberg [124]. The optimal constant for $p = 1$ was determined independently by Federer and Fleming in [59] and by Maz'ya in [118], and is $(n\omega_n^{1/n})^{-1}$; thus

$$\|u\|_{n/(n-1)} \leq (n\omega_n^{1/n})^{-1}\|\nabla u\|_1 \quad (u \in D_0^{1,1}(\mathbb{R}^n)) \tag{1.3.7}$$

It is closely related to the **isoperimetric inequality** for a measurable subset $E$ of $\mathbb{R}^n$ with finite n-dimensional Lebesgue measure $|E|$ and *perimeter* $P(E)$, namely,

$$|E|^{1/n'} \leq (n\omega_n^{1/n})^{-1}P(E), \quad n' = n/(n-1).$$

The inequality (1.3.7) is strict for non-zero $u$ in $D_0^{1,1}(\mathbb{R}^n)$. However it has a natural extension to the space of bounded variation in which characteristic functions of arbitrary balls are extremals. This extension is, in fact, equivalent to the isoperimetric inequality.

For $p > 1$ it was proved independently by Aubin in [9] and Talenti in [140] that the best possible value of the constant $C$ in (1.3.6) is

$$\pi^{-1/2} n^{-1/p} \left( \frac{p-1}{n-p} \right)^{(p-1)/p} \left\{ \frac{\Gamma(1+n/2)\Gamma(n)}{\Gamma(n/p)\Gamma(1+n-n/p)} \right\}, \tag{1.3.8}$$

and equality is attained for functions $u$ of the form

$$u(\mathbf{x}) = [a + b|\mathbf{x}|^{p/(p-1)}]^{1-n/p},$$

where $a$ and $b$ are positive constants.

*Remark 1.3.6* We note for subsequent use, that if $\Omega$ is a bounded open subset of $\mathbb{R}^n$, and $1 \leq p < n$, then $D_0^{1,p}(R^n)$ and $W_0^{1,p}(\mathbb{R}^n)$ are compactly embedded in $L^{p^*}(\Omega), p^* = np/(n-p)$. To see this, let $B$ be an open ball containing $\overline{\Omega}$ and $\varphi \in C_0^1(B)$ such that $\varphi(\mathbf{x}) = 1$ for $\mathbf{x} \in \Omega$. Then multiplication by $\varphi$ is a bounded map of $D_0^{1,p}(R^n)$ into $W_0^{1,p}(B)$. The compactness of the embedding $W_0^{1,p}(B) \hookrightarrow L^{p^*}(B)$ and the fact that $\varphi : L^{p^*}(B) \to L^{p^*}(\Omega)$ is bounded confirms the assertion.

### 1.3.2 Boundary Smoothness and $W^{k,p}(\Omega)$

In order for the space $W^{k,p}(\Omega)$ to have similar embedding properties to those of $W_0^{k,p}(\Omega)$, the boundary $\partial\Omega$ has to have a certain amount of smoothness. A standard smoothness class is the Hölder space $C^{k,\gamma}(\overline{\Omega})$ which we now define. First, we define $C^k(\overline{\Omega})$ to be the vector space of all bounded functions $u \in C^k(\Omega)$ such that $u$ and all its derivatives $D^\alpha u$ with $|\alpha| \leq k$ can be extended so as to be bounded and continuous on $\overline{\Omega}$. Then $C^{k,\gamma}(\overline{\Omega})$ is the space of functions $u \in C^k(\overline{\Omega})$ which satisfy the condition that, given any $\alpha \in \mathbb{N}_0^n$ with $|\alpha| = k$, there exists a constant $C > 0$ such that

$$|D^\alpha u(\mathbf{x}) - D^\alpha u(\mathbf{y})| \leq C|\mathbf{x} - \mathbf{y}|^\gamma, \quad \text{for all } \mathbf{x},\ \mathbf{y} \in \Omega.$$

**Definition 1.3.7** Let $n \geq 2$, $k \in \mathbb{N}_0$ and $\gamma \in [0, 1]$. The boundary $\partial\Omega$ of an open set $\Omega \subset \mathbb{R}^n$ is said to be of class $C^{k,\gamma}$ if:

(i) $\partial\Omega = \partial\overline{\Omega}$;
(ii) given any point $\mathbf{a} \in \partial\Omega$, there exist an open neighbourhood $U(\mathbf{a})$ of $\mathbf{a}$, local Cartesian coordinates $\mathbf{y} = (y_1, y_2, \cdots, y_n) = (\mathbf{y}', y_n)$ (where $\mathbf{y}' = (y_1, \cdots, y_{n-1})$), with $\mathbf{y} = 0$ at $\mathbf{x} = \mathbf{a}$, a convex, open subset $G$ of $\mathbb{R}^{n-1}$ with $0 \in G$, and a function $h \in C^{k,\gamma}(\overline{G})$ such that $\partial\Omega \cap U(\mathbf{a})$ has a representation

$$y_n = h(\mathbf{y}'), \quad \mathbf{y}' \in G.$$

We shall write $C^k$ in place of $C^{k,0}$ and $C$ for $C^{0,0}$.

We refer to [48], Sect. V.4 for a comparison of various smoothness criteria on $\partial\Omega$. It is proved in [48], Theorem V.4.2, that the boundary of a convex open set is in the Lipschitz class $C^{0,1}$. A consequence is that $\Omega$ has the **extension property**, which means the following (see [48], Theorem V.4.12): with

$$W(\Omega) := \bigcup_{k\in\mathbb{N}_0,\ p\in[1,\infty]} W^{k,p}(\Omega),$$

there is a map $E : W(\Omega) \to W(\mathbb{R}^n)$ such that $Eu = u$ for $u \in W(\Omega)$ and given any $k \in \mathbb{N}_0$ and $p \in [1,\infty]$, $E \in \mathcal{B}(W^{k,p}(\Omega), W^{k,p}(\mathbb{R}^n))$, the set of bounded linear operators mapping $W^{k,p}(\Omega)$ into $W^{k,p}(\mathbb{R}^n)$.

If $\Omega$ has the extension property, it is easy to establish embedding results for $W^{k,p}(\Omega)$ which are similar to those for $W_0^{k,p}(\Omega)$ in Sect. 2.1. For suppose $\Omega$ is a bounded open set with the extension property and that $\overline{\Omega}$ is contained in a ball $B$. Let $\varphi \in C_0^\infty(B)$ be such that $0 \le \varphi \le 1$ and $\varphi = 1$ on $\Omega$. Then, if $u \in W^{k,p}(\Omega)$, $Eu \in W^{k,p}(\mathbb{R}^n)$ and

$$\begin{aligned}\|u\|_{k,p,\Omega} &\le \|\varphi Eu\|_{k,p,B} \le C(\varphi)\|Eu\|_{k,p,B} \le C(\varphi)\|Eu\|_{k,p,\mathbb{R}^n}\\ &\le C(\varphi)C(k,p,\Omega)\|u\|_{k,p,\Omega},\end{aligned}$$

where $C(k,p,\Omega)$ denotes the norm of $E$. Thus inequalities known to hold for the element $\varphi Eu$ of $W_0^{k,p}(B)$ may be translated into similar inequalities relating to $u \in W^{k,p}(\Omega)$. For example, if $kp < n$, $W^{k,p}(\Omega)$ is continuously embedded in $L^s(\Omega)$ for all $s \in [p, np/(n-kp)]$, the embedding being compact if $s \in [p, np/(n-kp))$.

### 1.3.3 Truncation Rules

Next, we record an important result concerning the gradient of the absolute value $|u|$ of a function $u \in W^{1,p}(\Omega)$, and truncation rules on $W^{1,p}(\Omega)$. Proofs may be found in [111], Theorem 6.17 and [48], Sect. VI.2.

**Theorem 1.3.8** *Let $u \in W^{1,p}(\Omega)$, $1 \le p \le \infty$, and define $|u|(\mathbf{x}) := |u(\mathbf{x})|$. Then $|u| \in W^{1,p}(\Omega)$ and $|(\nabla|u|)(\mathbf{x})| \le |(\nabla u)(\mathbf{x})|$ a.e. If $u$ is real-valued, then $|(\nabla|u|)(\mathbf{x})| = |(\nabla u)(\mathbf{x})|$ a.e.*

*Let $u, v$ be real-valued members of $W^{1,p}(\Omega)$ and define* $\max(u,v)(\mathbf{x}) := \max\{u(\mathbf{x}), v(\mathbf{x})\}$ *and* $\min(u,v)(\mathbf{x}) := \min\{u(\mathbf{x}), v(\mathbf{x})\}$. *Then* $\max(u,v)$ *and* $\min(u,v)$ *belong to $W^{1,p}(\Omega)$. If $u, v \in W_0^{1,p}(\Omega)$, then* $\max(u,v)$ *and* $\min(u,v)$ *belong to $W_0^{1,p}(\Omega)$.*

*If $\Omega$ is unbounded, $u \in W^{1,p}(\Omega)$ is real-valued, and $\alpha \in \mathbb{R}$, then $u \vee \alpha :=$* $\min\{u,\alpha\} \in W^{1,p}(\Omega)$ *if and only if $\alpha \ge 0$, and $u \wedge \alpha :=$* $\max\{u,\alpha\} \in W^{1,p}(\Omega)$ *if* *and only if* $\alpha \leq 0$. *The gradients are given by*

$$\nabla(u \vee \alpha)(\mathbf{x}) = \begin{cases} \nabla u(\mathbf{x}) & \text{if } u(\mathbf{x}) < \alpha, \\ 0 & \text{if } u(\mathbf{x}) \geq \alpha, \end{cases}$$

*and*

$$\nabla(u \wedge \alpha)(\mathbf{x}) = \begin{cases} \nabla u(\mathbf{x}) & \text{if } u(\mathbf{x}) > \alpha, \\ 0 & \text{if } u(\mathbf{x}) \leq \alpha. \end{cases}$$

*If* $u \in W_0^{1,p}(\Omega)$ *then* $u \vee \alpha \in W_0^{1,p}(\Omega)$ *if* $\alpha \geq 0$ *and* $u \wedge \alpha \in W_0^{1,p}(\Omega)$ *if* $\alpha \leq 0$.

### 1.3.4 Rearrangements

**Definition 1.3.9** The **distribution function** of a Lebesgue measurable function $f$ on an open subset $\Omega$ of $\mathbb{R}^n$ is the map $\mu_f : [0, \infty) \to [0, \infty)$ defined by

$$\mu_f(\lambda) = |\{\mathbf{x} \in \Omega : |f(\mathbf{x})| > \lambda\}|,$$

where $|\{\cdot\}|$ denotes the Lebesgue measure of the set. The **non-increasing rearrangement** of $f$ is the function $f^* : [0, \infty) \to [0, \infty)$ defined by

$$f^*(t) = \inf\{\lambda \in [0, \infty) : \mu_f(\lambda) \leq t\};$$

the convention that $\inf \varnothing = \infty$ is used.

Since the distribution function $\mu_f$ is decreasing we have that

$$f^*(t) = \sup\{\lambda \in [0, \infty) : \mu_f(\lambda) > t\}, \quad t \geq 0,$$

so that $f^*$ is the distribution function of $\mu_f$. It can be shown that $\mu_f$ is right continuous and this implies that in the definition of $f^*$, the infimum is really a minimum. Moreover, $f^*$ is a non-negative, decreasing and right-continuous function on $[0, \infty)$. If $\mu_f$ is continuous and strictly decreasing, then $f^*$ is the inverse of $\mu_f$. The functions $f$ and $f^*$ are **equimeasurable** in the sense that they have the same distribution function; i.e., $\mu_f(\lambda) = \mu_{f^*}(\lambda)$ for all $\lambda \geq 0$. A consequence of this is that if $0 < p < \infty$,

$$\int_\Omega |f(\mathbf{x})|^p d\mathbf{x} = p \int_0^\infty \lambda^{p-1} \mu_f(\lambda) d\lambda = \int_0^\infty f^*(t)^p dt$$

and

$$\operatorname*{ess\,sup}_{\mathbf{x} \in \Omega} |f(\mathbf{x})| = \inf\{\lambda : \mu_f(\lambda) = 0\} = f^*(0).$$

It is sometimes desirable to work with the symmetric non-increasing rearrangement of a function, which we now define. First we define the symmetric rearrangement $A^\bigstar$ of a set $A$ of finite volume as

$$A^\bigstar := \{\mathbf{x} \in \mathbb{R}^n : \omega_n|\mathbf{x}|^n = |A|\},$$

that is, the open ball centred at the origin, whose volume is that of $A$; if $|A| = \infty$ we set $A^\bigstar = \mathbb{R}^n$.

**Definition 1.3.10** The **symmetric non-increasing rearrangement** of $f$ is the function $f^\bigstar$ defined by

$$f^\bigstar(\mathbf{x}) = f^*(\omega_n|\mathbf{x}|^n), \quad \mathbf{x} \in \Omega^\bigstar.$$

The function $f^\bigstar$ is non-negative, radially symmetric and radially non-increasing. Furthermore, $f^\bigstar$ and $|f|$ are equimeasurable and

$$f^\bigstar(\mathbf{x}) = \int_0^\infty \chi_{\{|f|>t\}^\bigstar} dt$$

The following are the main properties of symmetric non-increasing rearrangements that we shall need; for their proofs and other important properties of $\mu_f, f^*$ and $f^\bigstar$ see [49], 3.2 and [111], 3.3.

(i) $f^\bigstar$ is non-negative, radially symmetric and non-increasing, i.e., $f^\bigstar(\mathbf{x}) = f^\bigstar(\mathbf{y})$ if $|\mathbf{x}| = |\mathbf{y}|$ and $f^\bigstar(\mathbf{x}) \geq f^\bigstar(\mathbf{y})$ if $|\mathbf{x}| \leq |\mathbf{y}|$.

(ii) Let $f,\ g$ be Lebesgue measurable on $\Omega$. Then,

$$\int_\Omega |f(\mathbf{x})g(\mathbf{x})|d\mathbf{x} \leq \int_{\Omega^\bigstar} f^\bigstar(\mathbf{x})g^\bigstar(\mathbf{x})d\mathbf{x}.$$

(iii) For any $f \in L^p(\mathbb{R}^n)$ and $1 \leq p \leq \infty$,

$$\|f\|_{L^p(\mathbb{R}^n)} = \|f^\bigstar\|_{L^p(\mathbb{R}^n)},$$

where $\|f^\bigstar\|_{L^\infty(\mathbb{R}^n)} = \text{ess}\sup_{\mathbf{x}\in\mathbb{R}^n} f^\bigstar(\mathbf{x}) = f(0)$.

(iv) For any real-valued function $f \in C_0^1(\mathbb{R}^n)$ and $1 \leq p < \infty$,

$$\|\nabla f\|_{L^p(\mathbb{R}^n)} \geq \|\nabla f^\bigstar\|_{L^p(\mathbb{R}^n)}.$$

In fact, this inequality holds for all $f \in D_0^{1,p}(\mathbb{R}^n)$; see [49], Theorem 3.2.21 and Remark 3.2.23.

### 1.3.5 Fourier Transform

The Fourier transform $\mathbb{F}f$, or $\hat{f}$, of a function $f \in L^1(\mathbb{R}^n)$ is defined by

$$(\mathbb{F}f)(\mathbf{p}) := \frac{1}{(2\pi)^{n/2}} \int_{\mathbb{R}^n} e^{-i\mathbf{p}\cdot\mathbf{x}} f(\mathbf{x}) d\mathbf{x}, \tag{1.3.9}$$

where $\mathbf{p} \cdot \mathbf{x} = \sum_{j=1}^n p_j x_j$ with $\mathbf{p} = (p_1, p_2, \cdots, p_n)$, $\mathbf{x} = (x_1, x_2, \cdots, x_n)$. Let $\mathcal{S}(\mathbb{R}^n)$ denote the *Schwartz space* of $C^\infty(\mathbb{R}^n)$ functions which go to zero at infinity faster than any power of $\mathbf{x}$; thus $f \in \mathcal{S}(\mathbb{R}^n)$ if and only if $f \in C^\infty(\mathbb{R}^n)$, and for all $\alpha, \beta \in \mathbb{N}_0$,

$$\sup\{|\mathbf{x}^\alpha D^\beta f(\mathbf{x})| : \mathbf{x} \in \mathbb{R}^n\} < \infty.$$

We shall need the following basic properties of the space; see [48], V.1.5 for details.

(i) $\mathbb{F}$ is a linear bijection of $\mathcal{S}(\mathbb{R}^n)$ onto itself and its inverse is given by

$$(\mathbb{F}^{-1}f)(\mathbf{p}) := \frac{1}{(2\pi)^{n/2}} \int_{\mathbb{R}^n} e^{i\mathbf{p}\cdot\mathbf{x}} f(\mathbf{x}) d\mathbf{x} \; = (\mathbb{F}f)(-\mathbf{p}). \tag{1.3.10}$$

(ii) $\mathbb{F}$ is a continuous linear map of $L^1(\mathbb{R}^n)$ into $L^\infty$ and $\|\mathbb{F}u\|_{L^\infty(\mathbb{R}^n)} \le \|u\|_{L^1(\mathbb{R}^n)}$; it is not invertible.

(iii) For $f, g \in L^1(\mathbb{R}^n)$,

$$\mathbb{F}(f * g) = \mathbb{F}(f)\mathbb{F}(g), \tag{1.3.11}$$

where $f * g$ is the **convolution**

$$(f * g)(\mathbf{x}) := \frac{1}{(2\pi)^{n/2}} \int_{\mathbb{R}^n} f(\mathbf{x} - \mathbf{y}) g(\mathbf{y}) d\mathbf{y}.$$

Moreover, for $f, g \in \mathcal{S}(\mathbb{R}^n)$, $\hat{f} * \hat{g} \in \mathcal{S}(\mathbb{R}^n)$, and

$$\mathbb{F}^{-1}(\hat{f} * \hat{g}) = fg. \tag{1.3.12}$$

(iv) For $f \in \mathcal{S}(\mathbb{R}^n)$ and $\alpha \in \mathbb{N}_0$,

$$\{\mathbb{F}D^\alpha\mathbb{F}^{-1}\}f(\mathbf{p}) = \mathbf{p}^\alpha f(\mathbf{p}) \tag{1.3.13}$$

(v) **Plancherel's theorem** The map $f \mapsto \hat{f}$ has a unique extension to a unitary isomorphism of $L^2(\mathbb{R}^n)$ onto itself; thus, for all $f \in L^2(\mathbb{R}^n)$,

$$\int_{\mathbb{R}^n} |f(\mathbf{x})|^2 d\mathbf{x} = \int_{\mathbb{R}^n} |\hat{f}(\mathbf{p})|^2 d\mathbf{p}.$$

If $f$ and $g$ belong to $L^2(\mathbb{R}^n)$, **Parseval's formula** holds:

$$\int_{\mathbb{R}^n} f(\mathbf{x})\overline{g}(\mathbf{x})d\mathbf{x} = \int_{\mathbb{R}^n} \hat{f}(\mathbf{p})\overline{\hat{g}}(\mathbf{p})d\mathbf{p}.$$

It is standard for the Sobolev spaces $W_0^{1,2}(\Omega)$, $W^{1,2}(\Omega)$ to be denoted by $H_0^1(\Omega)$, $H^1(\Omega)$, respectively; this is compatible with the comment in Sect. 1.3.1 that $W^{k,p}(\Omega)$ coincides with the completion $H^{k,p}(\Omega)$ in the $W^{k,p}(\Omega)$ norm, of $C^\infty(\Omega) \cap W^{k,p}(\Omega)$. When $\Omega = \mathbb{R}^n$, it follows from (1.3.13) that

$$\|u\|^2_{H^1(\mathbb{R}^n)} = \int_{\mathbb{R}^n} |\hat{u}(\mathbf{p})|^2(1 + |\mathbf{p}|^2)d\mathbf{p}. \tag{1.3.14}$$

### *1.3.6 The Dirichlet and Neumann Laplacians*

We denote by $\Delta_{D,\Omega}u$, the **Dirichlet Laplacian** of $u \in H_0^1(\Omega)$; it is the Laplacian of $u$ in the weak sense, namely that $v = \Delta_{D,\Omega}u \in L^1_{loc}(\Omega)$ and for all $\varphi \in C_0^\infty(\Omega)$,

$$\int_\Omega \nabla u \cdot \nabla\varphi d\mathbf{x} = -\int_\Omega v\varphi d\mathbf{x}.$$

In $L^2(\Omega)$, $T_{D,\Omega} := -\Delta_{D,\Omega}$ is a non-negative self-adjoint operator with domain

$$\mathcal{D}(T_{D,\Omega}) = \{u \in H_0^1(\Omega) \cap W^{2,2}_{loc}(\Omega) : \Delta u \in L^2(\Omega)\}.$$

If $\partial\Omega$ is of class $C^2$ then

$$\mathcal{D}(T_{D,\Omega}) = H_0^1(\Omega) \cap W^{2,2}(\Omega).$$

The **Neumann Laplacian** $\Delta_{N,\Omega}u$ of $u \in H^1(\Omega)$ is defined by the conditions that $v = \Delta_{N,\Omega}u \in L^1_{loc}(\overline{\Omega})$ and for all $\varphi \in C_0^\infty(\mathbb{R}^n)$,

$$\int_\Omega \nabla u \cdot \nabla\varphi d\mathbf{x} = -\int_\Omega v\varphi d\mathbf{x}.$$

In $L^2(\Omega)$, $T_{N,\Omega} := -\Delta_{N,\Omega}$ is a non-negative self-adjoint operator with domain

$$\mathcal{D}(T_{N,\Omega}) = \{u \in H^1(\Omega) \cap W^{2,2}_{loc}(\Omega) : \Delta u \in L^2(\Omega)\}.$$

## 1.4 Comparison of the Hardy and Sobolev Inequalities

We now show how the Hardy and Sobolev's inequalities in $\mathbb{R}^n$ are intimately related. Hardy's inequality will be seen to imply Sobolev's inequality, while Sobolev's inequality only implies a weak form of Hardy's inequality. In each direction the constants determined are not optimal.

**Proposition 1.4.1** *For all $f \in C_0^\infty(\mathbb{R}^n)$ and $1 \le p < n$, Hardy's inequality (1.2.16) implies Sobolev's inequality (1.3.6),*

*Proof* First observe that since $|\nabla|f(\mathbf{x})|| \le |\nabla f(\mathbf{x})|$ a.e. by Theorem 1.3.8, it is sufficient to establish our claim for non-negative functions. Also, as we know from the properties of symmetric non-increasing rearrangements that $\|f\|_{p^*} = \|f^\star\|_{p^*}$ and $\|\nabla f\|_p \ge \|\nabla f^\star\|_p$, we may suppose that our functions $f$ are non-negative, radial and non-increasing. Thus, on mimicking the argument in [134], p. 8, we have that for any $\mathbf{y} \in \mathbb{R}^n$, $0 < a < 1$ and $q \ge 1$ (to be selected later),

$$\int_{\mathbb{R}^n} f(\mathbf{x})^q d\mathbf{x} \ge \int_{a|\mathbf{y}| \le |\mathbf{x}| \le |\mathbf{y}|} f(\mathbf{x})^q d\mathbf{x} \ge C_n |\mathbf{y}|^n f(\mathbf{y})^q,$$

where $C_n = [(1-a^n)/n]\omega_n$. Hence

$$\left(\int_{\mathbb{R}^n} f(\mathbf{x})^q d\mathbf{x}\right)^{p/n} f(\mathbf{y})^p |\mathbf{y}|^{-p} \ge C_n^{p/n} |\mathbf{y}|^p f(\mathbf{y})^{qp/n} f(\mathbf{y})^p |\mathbf{y}|^{-p}$$
$$= C_n^{p/n} f(\mathbf{y})^{qp/n+p}$$

On choosing $q = p^* = np/(n-p)$, we obtain $qp/n + p = p*$ and

$$C_n^{p/n} \int_{\mathbb{R}^n} f(\mathbf{y})^{p^*} d\mathbf{y} \le \left(\int_{\mathbb{R}^n} f(\mathbf{y})^{p^*} d\mathbf{y}\right)^{p/n} \int_{\mathbb{R}^n} \frac{f(\mathbf{y})^p}{|\mathbf{y}|^p} d\mathbf{y}.$$

Thus, if $f$ satisfies (1.2.16),

$$C_n^{p/n} \left(\int_{\mathbb{R}^n} f(\mathbf{y})^{p^*} d\mathbf{y}\right)^{1-p/n} \le \left(\frac{p}{n-p}\right)^p \int_{\mathbb{R}^n} |\nabla f(\mathbf{x})|^p d\mathbf{x}$$

which gives (1.3.6) with constant $C \le \frac{p}{(n-p)} C_n^{-1/n}$. □

The reverse implication involves the weak space $L^{q,\infty}(\mathbb{R}^n)$ defined as the space of Lebesgue measurable functions $f$ on $\mathbb{R}^n$ which are such that

$$\sup_{t>0} t^q \mu_f(t) < \infty, \tag{1.4.1}$$

where $\mu_f(t) = |\{\mathbf{x} \in \mathbb{R}^n : |f(\mathbf{x})| > t\}|$, is the distribution function of $f$. From the proof of Proposition 3.4.2 in [49], it is readily seen that for all $0 < q < \infty$,

$$\{\sup_{t>0} t^q \mu_f(t)\}^{1/q} = \sup_{t>0} t^{1/q} f^*(t) =: \|f\|_{q,\infty};$$

for $1 < q \leq \infty$, $\|f\|_{q,\infty}$ is a norm on $L^{q,\infty}(\mathbb{R}^n)$; see [49], Lemma 3.4.6 and Theorem 3.4.7. Since, for any $t \geq 0$,

$$\int_{\mathbb{R}^n} |f(\mathbf{x})|^q d\mathbf{x} \geq t^q \int_{\mu_f(t)} \chi_{\mu_f(t)}(\mathbf{x}) d\mathbf{x} = t^q \mu_f(t),$$

it follows that we have the continuous embedding $L^q(\mathbb{R}^n) \hookrightarrow L^{q,\infty}(\mathbb{R}^n)$, with

$$\|f\|_{q,\infty} \leq \|f\|_q. \tag{1.4.2}$$

We need the following weak form of Hölder's inequality which is proved in [21]

**Lemma 1.4.2** *Let* $1 < p < \infty$ *and* $p' = p/(p-1)$. *If* $f \in L^{p,\infty}(\mathbb{R}^n)$ *and* $g \in L^{p',\infty}(\mathbb{R}^n)$ *then* $fg \in L^{1,\infty}(\mathbb{R}^n)$ *and*

$$\|fg\|_{1,\infty} \leq p'^{1/p'} p^{1/p} \|f\|_{p,\infty} \|g\|_{p',\infty}$$

*Proof* Let $\varepsilon > 0, t > 0$ be arbitrary and set

$$\begin{aligned} A &= \{\mathbf{x} \in \mathbb{R}^n : \varepsilon|f(\mathbf{x})| > t^{1/p}\}, \\ B &= \{\mathbf{x} \in \mathbb{R}^n : \varepsilon^{-1}|g(\mathbf{x})| > t^{1/p'}\}, \\ E &= \{\mathbf{x} \in \mathbb{R}^n : |f(\mathbf{x})g(\mathbf{x})| > t\}. \end{aligned}$$

Since

$$|f(\mathbf{x})g(\mathbf{x})| \leq p^{-1}(\varepsilon|f(\mathbf{x})|)^p + p'^{-1}(\varepsilon^{-1}|g(\mathbf{x})|)^{1/p'},$$

we have $E \subseteq A \cup B$ and this implies that $t|E| \leq t|A| + t|B|$. On substituting $s := \varepsilon^{-1} t^{1/p}$, $r := \varepsilon t^{1/p'}$, it follows that

$$\begin{aligned} t|E| &\leq \varepsilon^p s^p |\{\mathbf{x} : |f(\mathbf{x})| > s\} + \varepsilon^{-p'} r^{p'} |\{\mathbf{x} : |g(\mathbf{x})| > r\} \\ &\leq \varepsilon^p \|f\|_{p,\infty}^p + \varepsilon^{-p'} \|g\|_{p',\infty}^{p'}. \end{aligned}$$

The minimum value of the right-hand side is

$$\left[(p'/p)^{1/p'} + (p/p')^{1/p}\right] \|f\|_{p,\infty} \|g\|_{p',\infty} = p^{1/p} p'^{1/p'} \|f\|_{p,\infty} \|g\|_{p',\infty},$$

attained when $\varepsilon = (p' \|g\|_{p',\infty}^{p'} / p \|f\|_{p,\infty}^p)^{1/pp'}$. □

We are now ready to prove

**Proposition 1.4.3** *For $1 \leq p < n$, Sobolev's inequality (1.3.6) with best possible constant C, implies the weak Hardy inequality*

$$\|f/|\cdot|\|_{p,\infty} \leq \left(\frac{n-p}{p}\right)^{\frac{1}{n}} \left(\frac{n}{n-p}\right)^{\frac{1}{p}} \omega_n C \|\nabla f\|_p, \quad (f \in C_0^\infty(\mathbb{R}^n)).$$

*Proof* It is readily seen that

$$\left\|\frac{f}{|\cdot|}\right\|_{p,\infty} = \left\|\left(\frac{f}{|\cdot|}\right)^p\right\|_{1,\infty}$$

and by Lemma 1.4.2 with $1 < q < \infty$,

$$\begin{aligned}\left\|\left(\frac{f}{|\cdot|}\right)^p\right\|_{1,\infty} &\leq q^{1/q} q'^{1/q'} \|f^p\|_{q,\infty} \|\frac{1}{|\cdot|^p}\|_{q',\infty} \\ &= q^{1/q} q'^{1/q'} \|f\|_{pq,\infty}^p \|\frac{1}{|\cdot|}\|_{pq',\infty}^p.\end{aligned}$$

We now choose $q = p^*/p$, and so $q' = n/p$, to get

$$\begin{aligned}\left\|\left(\frac{f}{|\cdot|}\right)^p\right\|_{1,\infty} &\leq \left(\frac{n}{p}\right)^{\frac{1}{n}} \left(\frac{p^*}{p}\right)^{\frac{1}{p^*}} \|f\|_{p^*,\infty} \|\frac{1}{|\cdot|}\|_{n,\infty} \\ &= \left(\frac{n}{p}\right)^{\frac{1}{n}} \left(\frac{p^*}{p}\right)^{\frac{1}{p^*}} \omega_n \|f\|_{p^*,\infty},\end{aligned}$$

since $\|\frac{1}{|\cdot|}\|_{n,\infty} = \omega_n$. The proposition follows from (1.4.2). □

## 1.5 The CLR Inequality

In [39, 110, 131], Cwikel, Lieb and Rosenbljum independently proved an inequality for the number of negative eigenvalues of the self-adjoint operator $-\Delta - V$ in $L^2(\mathbb{R}^n)$, which has important implications in semi-classical spectral analysis, in which the transition between classical and quantum mechanics is studied. This inequality is usually referred to as the **CLR inequality**, with the authors' names listed alphabetically and in the reverse chronological order of discovery. We proceed to give a presentation of the background theory, but leave the reader to consult the references given later for the very different original proofs of the inequality. Brief reminders about notions and results from operator theory, and especially on quadratic forms, will be given within the background theory, but for a comprehensive account, [48] may be consulted. The inner product and norm on $L^2(\mathbb{R}^n)$ will be denoted by $(\cdot,\cdot)$ and $\|\cdot\|$ respectively.

### 1.5.1 Background Theory

We shall assume that the potential $V$ is a real-valued function which satisfies the conditions

$$V \geq 0, \quad V \in L^{n/2}(\mathbb{R}^n), \quad n \geq 3; \tag{1.5.1}$$

it is in fact sufficient to assume that $V \in L^1_{loc}(\mathbb{R}^n)$ with its positive part $V_+ := \frac{1}{2}(|V| - V) \in L^{n/2}(\mathbb{R}^n)$. The operator $-\Delta - V$ is defined in the *form sense*, which we now recall. Consider the quadratic form

$$t[u,\varphi] := \int_{\mathbb{R}^n} \{\nabla u(\mathbf{x}) \cdot \nabla \bar{\varphi}(\mathbf{x}) - V(\mathbf{x})u(\mathbf{x})\bar{\varphi}(\mathbf{x})\}\, d\mathbf{x}, \quad u,\varphi \in C_0^\infty(\mathbb{R}^n), \tag{1.5.2}$$

and let $(\cdot,\cdot)$ and $\|\cdot\|$ denote the $L^2(\mathbb{R}^n)$ inner product and norm respectively. We shall prove in the lemma below that $t$ is bounded below and closable, i.e., $t[u] := t[u,u] \geq -k\|u\|^2$ for some constant $k$ and all $u \in C_0^\infty(\mathbb{R}^n)$, and the completion $\mathcal{Q}$ of $C_0^\infty(\mathbb{R}^n)$ with respect to the norm

$$\|u\|_{\mathcal{Q}} := \left\{t[u] + (k+1)\|u\|^2\right\}^{1/2}$$

is continuously embedded in $L^2(\mathbb{R}^n)$, and therefore can be identified with a subspace of $L^2(\mathbb{R}^n)$. Note that for all $k \geq 0$, the norms $\|\cdot\|_{\mathcal{Q}}$ are equivalent and hence the corresponding spaces $\mathcal{Q}$ are isomorphic. Let $E$ denote the embedding $\mathcal{Q} \hookrightarrow L^2(\mathbb{R}^n)$. The *adjoint* $E^*$ of $E$ is defined by

$$(E^*v)(u) = (v, Eu), \quad u \in \mathcal{Q}, \ v \in L^2(\mathbb{R}^n),$$

and since

$$|(v, Eu)| \leq \|v\|\|Eu\| \leq \|E\|\|v\|\|u\|_{\mathcal{Q}},$$

$E^*$ therefore maps $L^2(\mathbb{R}^n)$ linearly into the space of bounded, conjugate linear functionals on $\mathcal{Q}$; this space is called the *adjoint* of $\mathcal{Q}$ and denoted by $\mathcal{Q}^*$. We therefore have the triplet of spaces

$$\mathcal{Q} \hookrightarrow L^2(\mathbb{R}^n) \hookrightarrow \mathcal{Q}^*,$$

with respective embeddings $E$ and $E^*$ which are continuous, injective and have dense ranges. If $t$ is closable and $\{u_m\}, \{v_m\} \subset \mathcal{Q}$, are such that $u_m \to u,\ v_m \to v$ in $\mathcal{Q}$, then the *closure* $\bar{t}$ of $t$ is defined on $\mathcal{Q}$ by

$$\bar{t}[u, v] = \lim_{m\to\infty} t[u_m, v_m].$$

Moreover, $t$ is said to be *closed* if $t = \bar{t}$, and to have *domain* $\mathcal{Q}$.

Let $t$ now stand for the closure of the quadratic form defined by (1.5.2); $t$ is therefore a closed quadratic form with domain $\mathcal{Q}$. It then follows from Kato's representation theorems ([83], Sect. VI.2) that there exists a self-adjoint operator $T$ with the following properties:

(i) the domain $D(T)$ of $T$ lies in $\mathcal{Q}$ and

$$t[u, \varphi] = (Tu, \varphi), \quad \text{for } u \in D(T), \quad \varphi \in \mathcal{Q};$$

(ii) $D(T)$ is a *core* of $t$, i.e. $D(T)$ is dense in $\mathcal{Q}$;
(iii) $D((T + k)^{1/2}) = \mathcal{Q}$ and $t[u] = \|(T + k)^{1/2}u\|^2$ for $u \in D((T + k)^{1/2})$.

The space $\mathcal{Q}$ is called the *form domain* of $T$. Given a symmetric operator $T_0$ in our Hilbert space $L^2(\mathbb{R}^n)$ which is bounded below, the form $t_0[\cdot]$ defined by $t_0[u, v] = (T_0u, v)$ has a closure $t[\cdot]$ and in this case the self-adjoint operator $T$ in Kato's theorem is the **Friedrichs extension** of $T_0$.

In (i), a strictly separate identification of the spaces $\mathcal{Q}$ and $L^2(\mathbb{R}^n)$ would require us to write, for $u \in D(T), \varphi \in \mathcal{Q}$

$$t[u, \varphi] = (Tu, E\varphi) = (E^*Tu, \varphi) := (E^*Tu)(\varphi).$$

For $u \in C_0^\infty(\mathbb{R}^n)$, by the Hölder and Sobolev inequalities, we have

$$\begin{aligned}\|V^{1/2}u\|^2 &= \int_{\mathbb{R}^n} V|u|^2 d\mathbf{x} \\ &\le \left(\int_{\mathbb{R}^n} V^{n/2} d\mathbf{x}\right)^{2/n} \left(\int_{\mathbb{R}^n} |u|^{2n/(n-2)} d\mathbf{x}\right)^{(n-2)/n} \\ &\le \gamma_n \|V\|_{n/2} \|\nabla u\|^2\end{aligned}$$

where $\gamma_n$ is the norm of the Sobolev embedding $H^1(\mathbb{R}^n) \hookrightarrow L^2(\mathbb{R}^n)$. It will be shown in the proposition below that $H^1(\mathbb{R}^n)$ and $\mathcal{Q}$ are isomorphic. Hence $(V^{1/2}u, V^{1/2}u)$ is a bounded quadratic form on $\mathcal{Q} \times \mathcal{Q}$, and so there exists a bounded linear operator $\hat{V} : \mathcal{Q} \to \mathcal{Q}^*$ such that

$$(\hat{V}u, v) = (V^{1/2}u, V^{1/2}v), \quad u, v \in \mathcal{Q}.$$

Multiplication by $V$ is said to be *compact relative to the form* $t_0[u] = \|\nabla u\|^2$ if $\hat{V} : \mathcal{Q} \to \mathcal{Q}^*$ is compact, where the norm of $\mathcal{Q}$ is now given by $\|u\|_{\mathcal{Q}} = \left(t_0[u] + \|u\|^2\right)^{1/2} = \|u\|_{H^1}$. Setting $T_0 : -\Delta$ and $\hat{T}_0 := E^* T_0$, we have

$$t[u, v] = (\hat{T}u, v), \quad \hat{T} = \hat{T}_0 - \hat{V},$$

for $u \in D(T) \subset \mathcal{Q}$ and $v \in \mathcal{Q}$. Thus $\hat{T} : D(T) \subset \mathcal{Q} \to \mathcal{Q}^*$.

To prepare the ground for the CLR inequality we begin with the following proposition. We refer to [48] for the information required on the essential spectrum.

**Proposition 1.5.1** *Suppose that the conditions (1.5.1) on $V$ are satisfied. Then the quadratic form $t$ is bounded below and closable on $C_0^\infty(\mathbb{R}^n)$, and multiplication by $V$ is compact relative to the form $t_0$. Then $T$ has essential spectrum $[0, \infty)$, and in $(-\infty, 0)$, the spectrum of $T$ consists only of isolated eigenvalues of finite multiplicity.*

*Proof* Since $C_0^\infty(\mathbb{R}^n)$ is dense in $L^{n/2}(\mathbb{R}^n)$, given any $\varepsilon > 0$, there exists $U \in C_0^\infty(\mathbb{R}^n)$ with support $\Omega_\varepsilon$ say, such that

$$\|V - U\|_{n/2} < \varepsilon \quad \text{and} \quad \sup_{\mathbf{x} \in \Omega_\varepsilon} |U(\mathbf{x})| < k_\varepsilon$$

for some $k_\varepsilon > 0$. Then, for all $\varphi \in C_0^\infty(\mathbb{R}^n)$, on using the Hölder and Sobolev inequalities,

$$\begin{aligned} \|V^{1/2}\varphi\|^2 &\leq \|V - U\|_{n/2}\|\varphi\|^2_{2n/(n-2)} + \|U\|_\infty \|\varphi\|^2_{\Omega_\varepsilon} \\ &\leq \varepsilon\|\nabla\varphi\|^2 + k_\varepsilon\|\varphi\|^2_{\Omega_\varepsilon}. \end{aligned} \tag{1.5.3}$$

Hence

$$t[\varphi] \geq (1 - \varepsilon)\|\nabla\varphi\|^2 - k_\varepsilon\|\varphi\|^2,$$

and on choosing $\varepsilon < 1$, we see that $t$ is bounded below. Furthermore

$$t[\varphi] \leq (1 + \varepsilon)\|\nabla\varphi\|^2 + k_\varepsilon\|\varphi\|^2.$$

Thus $\|\cdot\|_{\mathcal{Q}}$ is equivalent to the $H^1(\mathbb{R}^n)$ norm $(\|\nabla \cdot\|^2 + \|\cdot\|^2)^{1/2}$ for any choice of $\varepsilon \in (0, 1)$.

Let $\{\varphi_m\}$ be a bounded sequence in $\mathcal{Q}$ which converges weakly to zero. Then, from (1.5.3), for arbitrary $\varepsilon > 0$,

$$\begin{aligned} (\hat{V}\varphi_m, \varphi_m) &= \|V^{1/2}\varphi_m\|^2 \\ &\leq \varepsilon\|\varphi_m\|^2_{H^1} + k_\varepsilon\|\varphi_m\|^2_{\Omega_\varepsilon}. \end{aligned}$$

Since the embedding $\mathcal{Q} \hookrightarrow L^2(\Omega_\varepsilon)$ is compact,(as $\mathcal{Q}$ is isomorphic to $H^1(\mathbb{R}^n)$), $\|\varphi_m\|_{\Omega_\varepsilon} \to 0$ as $m \to \infty$ and consequently, with $\|\varphi_m\|_{\mathcal{Q}} \le K$ say,

$$\limsup_{m\to\infty} \|V^{1/2}\varphi_m\| \le K^2\varepsilon.$$

Hence, since $\varepsilon$ is arbitrary, we have that $\|V^{1/2}\varphi_m\| \to 0$ as $m \to \infty$. This implies that

$$\begin{aligned}
\|\hat{V}\varphi_m\|_{\mathcal{Q}^*} &= \sup_{\|u\|_{\mathcal{Q}}=1} |(\hat{V}\varphi_m, u)| \\
&= \sup_{\|u\|_{\mathcal{Q}}=1} (V^{1/2}\varphi_m, V^{1/2}u) \\
&\le \sup_{\|u\|_{H^1}=1} \|V^{1/2}\varphi_m\|\|V^{1/2}u\| \\
&\le \sup_{\|u\|_{H^1}=1} \|V\|_{n/2}\|u\|_{2n/(n-2)}\|V^{1/2}\varphi_m\| \\
&\le C\|V\|_{n/2}\|V^{1/2}\varphi_m\| \\
&\to 0
\end{aligned}$$

as $m \to \infty$. Thus $\hat{V} : \mathcal{Q} \to \mathcal{Q}^*$ is compact. We now show that this property of $V$ ensures that $T$ and $-\Delta$ have the same essential spectra, namely $[0, \infty)$, and hence that the spectrum of $T$ in $(-\infty, 0)$ consists only of isolated eigenvalues of finite multiplicity.

From (1.5.3), and for arbitrary $\varepsilon > 0$ and $u \in \mathcal{Q}$,

$$\|\hat{V}u\|^2_{\mathcal{Q}^*} \le \varepsilon\|\nabla u\|^2 + k_\varepsilon\|u\|^2.$$

Let $b^2 = k_\varepsilon/\varepsilon$ and define on $\mathcal{Q}$ the equivalent norm

$$\|u\|_{\mathcal{Q}} = \|(T_0^{1/2} + ib)u\| = \left(\|\nabla u\|^2 + b^2\|u\|^2\right)^{1/2}.$$

Then

$$\|\hat{V}u\|^2_{\mathcal{Q}^*} \le \varepsilon\|u\|^2_{\mathcal{Q}}. \tag{1.5.4}$$

For $z \in \mathbb{R}$,

$$\begin{aligned}
\hat{T} - izE^* &= \hat{T}_0 - izE^* - \hat{V} \\
&= \{I_{\mathcal{Q}^*} - \hat{V}(\hat{T}_0 - izE^*)^{-1}\}(\hat{T}_0 - izE^*),
\end{aligned} \tag{1.5.5}$$

where $I_{\mathcal{Q}^*}$ is the identity on $\mathcal{Q}^*$. We shall now prove that $(\hat{T}-izE^*)^{-1} \in \mathcal{B}(\mathcal{Q}, \mathcal{Q}^*)$, the space of bounded linear operators mapping $\mathcal{Q}$ into $\mathcal{Q}^*$, and

$$\|(\hat{T} - izE^*)^{-1}\|_{\mathcal{B}(\mathcal{Q},\mathcal{Q})} \leq 1 + b^2/|z|. \tag{1.5.6}$$

Since $T_0$ is self-adjoint, we know that, with $I$ the identity on $L^2(\mathbb{R}^n)$, the range of $T_0 - izI$, $\mathcal{R}(T_0 - izI)$, coincides with $L^2(\mathbb{R}^n)$, and so $\mathcal{R}(\hat{T}_0 - izE^*)$ is dense in $\mathcal{Q}^*$. Let $\varphi \in \mathcal{R}(T_0 - izI)$, $\theta = (T_0^{1/2} - ibI)^{-1}\varphi$ and $|z| \geq b^2$. Then since $T_0$ and $T_0^{1/2}$ commute,

$$\begin{aligned}
\|(T_0 - izI)^{-1}\varphi\|_{\mathcal{Q}}^2 &= \|(T_0^{1/2} + ibI)(T_0 - izI)^{-1}\varphi\|^2 \\
&= \|(T_0 + b^2I)(T_0^{1/2} - ibI)^{-1}(T_0 - izI)^{-1}\varphi\|^2 \\
&= \|(T_0 + b^2I)(T_0 - izI)^{-1}\theta\|^2 \\
&= \|T_0(T_0 - izI)^{-1}\theta\|^2 + 2b^2\left(T_0(T_0 - izI)^{-1}\theta, (T_0 - izI)^{-1}\theta\right) \\
&\quad + b^4\|(T_0 - izI)^{-1}\theta\|^2 \\
&\leq \|(T_0 - izI)(T_0 - izI)^{-1}\theta\|^2 \\
&\quad + 2b^2\|T_0(T_0 - izI)^{-1}\theta\|\|(T_0 - izI)^{-1}\theta\|,
\end{aligned}$$

on using $b^2 \leq |z|$ and the identity

$$\|(T_0 - izI)u\|^2 = \|T_0u\|^2 + z^2\|u\|^2.$$

Hence, $\|T_0u\|^2 \leq \|(T_0 - izI)u\|^2$, $\|u\| \leq |z|^{-1}\|(T_0 - izI)u\|$ and

$$\begin{aligned}
\|(T_0 - izI)^{-1}\varphi\|_{\mathcal{Q}}^2 &\leq \|\theta\|^2 + 2b^2\|\theta\|\|(T_0 - izI)^{-1}\theta\| \\
&\leq \left(1 + \frac{2b^2}{|z|}\right)\|\theta\|^2 \\
&\leq \left(1 + \frac{b^2}{|z|}\right)^2\|\theta\|^2.
\end{aligned} \tag{1.5.7}$$

Furthermore, if $\varphi^* := E^*\varphi$,

$$\begin{aligned}
\|\varphi^*\|_{\mathcal{Q}^*} &= \sup_{\|\psi\|_{\mathcal{Q}} \leq 1} |(\varphi^*, \psi)| \\
&= \sup_{\|\psi\|_{\mathcal{Q}} \leq 1} |(\varphi, E\psi)| \\
&= \sup_{\|\psi\|_{\mathcal{Q}} \leq 1} \left|([T_0^{1/2} - ibI]^{-1}\varphi, [T_0^{1/2} - ibI]\psi)\right|
\end{aligned}$$

$$= \|(T_0^{1/2} - ibI)^{-1}\varphi\|$$
$$= \|\theta\|. \tag{1.5.8}$$

Thus $\|\theta\| = \|E^*\varphi\|_{\mathcal{Q}^*} \le \|\varphi\|$ and

$$\|(T_0 - izI)^{-1}\| \le 1 + \frac{b^2}{|z|}. \tag{1.5.9}$$

As $E^*$ is injective, $\hat{T}_0 - izE^* = E^*(T_0 - izI)$ has an inverse, and for $\varphi^* = E^*\varphi \in \mathcal{R}(\hat{T} - izE^*)$ and $\psi \in \mathcal{Q}$,

$$\begin{aligned}\left|\left((\hat{T}_0 - izE^*)^{-1}\varphi^*, \psi\right)_{\mathcal{Q}}\right| &= \left|\left((T_0 - izI)^{-1}\varphi, \psi\right)_{\mathcal{Q}}\right| \\ &\le \|(T_0 - izI)^{-1}\varphi\|_{\mathcal{Q}}\|\psi\|_{\mathcal{Q}}.\end{aligned}$$

Therefore, we have from (1.5.7) and (1.5.8),

$$\begin{aligned}\|(\hat{T}_0 - izE^*)^{-1}\varphi^*\|_{\mathcal{Q}} &\le \|(T_0 - izI)^{-1}\varphi\|_{\mathcal{Q}} \\ &\le \left(1 + \frac{b^2}{|z|}\right)\|\varphi^*\|_{\mathcal{Q}^*}.\end{aligned}$$

Since $\mathcal{R}(\hat{T}_0 - izE^*)$ is dense in $\mathcal{Q}^*$, $(\hat{T}_0 - izE^*)^{-1}$ extends by continuity to a map in $\mathcal{B}(\mathcal{Q}^*, \mathcal{Q})$, and (1.5.6) follows.

From (1.5.4) and (1.5.6),

$$\|\hat{V}(\hat{T}_0 - izE^*)^{-1}\|_{\mathcal{Q}^*} \le \varepsilon\left(1 + \frac{b^2}{|z|}\right). \tag{1.5.10}$$

Hence in (1.5.5), for $\varepsilon < 1$ and $|z|$ large enough,

$$\{I_{\mathcal{Q}^*} - \hat{V}(\hat{T}_0 - izE^*)^{-1}\}^{-1} \in \mathcal{B}(\mathcal{Q}^*) := \mathcal{B}(\mathcal{Q}^*, \mathcal{Q}^*),$$

and consequently $(\hat{T} - izE^*)^{-1} \in \mathcal{B}(\mathcal{Q}^*, \mathcal{Q})$. Moreover $(T - izI)^{-1} = (\hat{T} - izE^*)^{-1}E^* \in \mathcal{B}(L^2(\mathbb{R}^n), \mathcal{Q}) \subset \mathcal{B}(L^2(\mathbb{R}^n))$. It now follows from (1.5.5) that

$$(T - izI)^{-1} - (T_0 - izI)^{-1} = (\hat{T} - izE^*)^{-1}\hat{V}(T_0 - izI)^{-1}, \tag{1.5.11}$$

where $(T_0 - izI)^{-1}$ is understood as a map from $L^2(\mathbb{R}^n)$ to $\mathcal{Q}$. Since we have shown that $\hat{V} : \mathcal{Q} \to \mathcal{Q}^*$ is compact, the right-hand side of (1.5.11) is compact in $L^2(\mathbb{R}^n)$ and hence by Weyl's Theorem (see [48], Theorem IX.2.1), $T$ and $T_0$ have the same essential spectrum. As this is $[0, \infty)$ for $T_0$, the proof is complete. □

The CLR inequality can now be given.

**Theorem 1.5.2** *Let V satisfy (1.5.1). Then the number $N(T)$ of negative eigenvalues of the self-adjoint operator T defined in Proposition 1.5.1, counting multiplicities, satisfies the inequality*

$$N(T) \leq C_n \int_{\mathbb{R}^n} V^{n/2}(\mathbf{x})d\mathbf{x}, \quad n \geq 3, \tag{1.5.12}$$

*for some constant $C_n$.*

We refer the reader to [39, 110, 131], for the original proofs of Theorem 1.5.2, which use very different techniques. Later proofs were obtained by Li and Yau in [109] and Conlon [38]. The optimal value of the constant $C_n$ in (1.5.13) is not known; the best known estimate is that obtained by Lieb in [110].

## 1.5.2 Comparison of the CLR and Sobolev Inequalities

The inequality (1.5.12) is sharp in the following sense. On replacing $V$ by $\alpha V$, where $\alpha > 0$ is a large *coupling* constant, and denoting the corresponding operator $-\Delta - \alpha V$ by $T_\alpha$, (1.5.12) becomes

$$N(T_\alpha) \leq C_n \alpha^{n/2} \int_{\mathbb{R}^n} V^{n/2}(\mathbf{x})d\mathbf{x}. \tag{1.5.13}$$

But $N(T_\alpha)$ is known to satisfy the asymptotic formula

$$\lim_{\alpha \to \infty} \alpha^{-n/2} N(T_\alpha) = c_n \int_{\mathbb{R}^n} V^{n/2}(\mathbf{x})d\mathbf{x}, \quad c_n = (2\sqrt{n})^{-n}[\Gamma(1+n/2)]^{-1};$$

see [112], Sect. 4.1.1. Thus (1.5.13) is sharp both in the power of $\alpha$ and in the function class of $V$.

Theorem 1.5.2 has the following special case of Sobolev's embedding theorem, Theorem 1.3.3, as converse.

**Theorem 1.5.3** *Suppose that Theorem 1.5.2 is satisfied. Then*

$$\|u\|^2_{2n/(n-2)} \leq C_n^{2/n} \|\nabla u\|^2, \quad \text{for all } u \in H^1(\mathbb{R}^n).$$

*Proof* Suppose that (1.5.12) is satisfied. Then

$$C_n \int_{\mathbb{R}^n} V^{n/2}(\mathbf{x})d\mathbf{x} < 1$$

implies that $N(T) = 0$; thus $T$ is non-negative and for all $u \in H^1(\mathbb{R}^n)$,

$$\int_{\mathbb{R}^n} |\nabla u|^2 d\mathbf{x} \geq \int_{\mathbb{R}^n} V|u|^2 d\mathbf{x}. \tag{1.5.14}$$

Let $W$ be an arbitrary member of $L^{n/2}(\mathbb{R}^n)$ with $\|W\|_{L^{n/2}(\mathbb{R}^n)} = 1$, and set

$$V = (\widetilde{C}_n)^{-2/n}|W|, \quad \text{for any } \widetilde{C}_n > C_n.$$

Then

$$C_n \int_{\mathbb{R}^n} V^{n/2} d\mathbf{x} = (C_n/\widetilde{C}_n) \int_{\mathbb{R}^n} |W|^{n/2} d\mathbf{x} < 1$$

and we infer from (1.5.14) that

$$\left|\int_{\mathbb{R}^n} W|u|^2 d\mathbf{x}\right| \leq \int_{\mathbb{R}^n} |W||u|^2 d\mathbf{x} \leq \widetilde{C}_n^{2/n} \int_{\mathbb{R}^n} |\nabla u|^2 d\mathbf{x}, \quad u \in H^1(\mathbb{R}^n).$$

From this it follows that $|u|^2$ belongs to $L^{n/(n-2)}(\mathbb{R}^n)$, the dual of $L^{n/2}(\mathbb{R}^n)$, and

$$\begin{aligned} \left(\int_{\mathbb{R}^n} |u|^{2n/(n-2)} d\mathbf{x}\right)^{(n-2)/n} &= \||u|^2\|_{L^{n/(n-2)}(\mathbb{R}^n)} \\ &= \sup_{\|W\|_{n/2}=1} \left|\int_{\mathbb{R}^n} W|u|^2 d\mathbf{x}\right| \\ &\leq \widetilde{C}_n^{2/n} \int_{\mathbb{R}^n} |\nabla u|^2 d\mathbf{x}. \end{aligned}$$

Since $\widetilde{C}_n > C_n$ is arbitrary, the theorem follows. □

The proof of Li and Yau in [109] is of particular interest to us as it only uses the Sobolev inequality (1.3.6) and the fact that the kernel of the heat operator $\exp(t\Delta)$ is positive; recall that (see [111])

$$(\exp(t\Delta)f)(\mathbf{x}) = \left(\frac{1}{4\pi t}\right)^{n/2} \int_{\mathbb{R}^n} \exp\left(-\frac{|\mathbf{x}-\mathbf{y}|^2}{4t}\right) f(\mathbf{y}) d\mathbf{y}.$$

Therefore Li and Yau's result and Theorem 1.5.3 imply that the Sobolev and CLR inequalities in $\mathbb{R}^n$ are equivalent in view of the positivity of the heat operator. In [103], Levin and Solomyak derive an abstract version of Li and Yau's proof, in which the quadratic form is given by $t[u] = q[u] - \int_\Omega V|u|^2 d\sigma$, where $q$ is a general quadratic form associated with a *Markov* generator and $(\Omega, \sigma)$ is a measure space with sigma-finite measure. Such a quadratic form $q$ has an abstract description given by the following Beurling-Deny conditions in which $q$ is bounded below and closed, and its domain is denoted by $\mathcal{H}_1(q)$:

(a) $q[u+iv] = q[u] + q[v]$ for real $u, v \in \mathcal{H}_1(q)$;
(b) if $u \in \mathcal{H}_1(q)$ is real, then $|u| \in \mathcal{H}_1(q)$ and $q[|u|] \leq q[u]$;
(c) if $0 \leq u \in \mathcal{H}_1(q)$, then $u \wedge 1 := \min\{u, 1\} \in \mathcal{H}_1(q)$ and $q[u \wedge 1] \leq q[u]$.

These conditions are satisfied by our quadratic form $t_0[u] = \|\nabla u\|^2$ in $L^2(\mathbb{R}^n)$ in view of Theorem 1.3.8. The approach of Levin and Solomyak in [103] applies *inter alia* to analogous discrete problems on graphs. Their theory will be of relevance to our discussion in Chap. 4, Sect. 4.3, below.

## 1.6 The Uncertainty Principle and Heisenberg's Inequality

In quantum mechanics, the state of a system consisting of a single particle in $\mathbb{R}^3$ is described by a wave function $\psi \in L^2(\mathbb{R}^3)$ satisfying

$$\int_{\mathbb{R}^3} |\psi(\mathbf{x})|^2 d\mathbf{x} = 1.$$

The function $|\psi|^2$ is interpreted as the probability density of the position of the particle; the probability that the particle is in a set $N$ is given by

$$\int_{\mathbb{R}^3} |\psi(\mathbf{x})|^2 \chi_N d\mathbf{x},$$

where $\chi_N$ is the characteristic function of $N$. On taking Planck's constant to be normalised, i.e., $\hbar = 1$, the momentum of the particle is defined to be $-i\nabla\psi(\mathbf{x})$. In view of (1.3.13), the operator $-i\nabla$ is unitarily equivalent to multiplication by $\mathbf{p}$, which justifies the standard use of $\mathbf{p}$ to represent the momentum; $\mathbf{p}$ is also the standard notation for momentum in classical mechanics. From Plancherel's theorem,

$$\int_{\mathbb{R}^3} |\hat{\psi}(\mathbf{p})|^2 d\mathbf{p} = \int_{\mathbb{R}^3} |\psi(\mathbf{x})|^2 d\mathbf{x} = 1$$

and $|\hat{\psi}(\mathbf{p})|^2$ is interpreted as the probability density of the particle's momentum.

**Heisenberg's uncertainty principle** asserts that the position $\mathbf{x}$ and momentum $\mathbf{p}$ can not be determined simultaneously. The position $\mathbf{x}$ and momentum $\mathbf{p} = -i\nabla$ are now linear operators, and the readily verified commutator identity

$$[\mathbf{p}\cdot\mathbf{a}, \mathbf{x}\cdot\mathbf{b}] := (\mathbf{p}\cdot\mathbf{a})(\mathbf{x}\cdot\mathbf{b}) - (\mathbf{x}\cdot\mathbf{b})(\mathbf{p}\cdot\mathbf{a}) = -i(\mathbf{a}\cdot\mathbf{b}), \quad (\mathbf{a}, \mathbf{b} \in \mathbb{C}^3) \qquad (1.6.1)$$

implies intuitively that $|\mathbf{x}|$ and $|\mathbf{p}|$ can not be simultaneously small. The principle is enshrined in the Hardy inequality

$$\begin{aligned}\int_{\mathbb{R}^3} |\mathbf{p}\hat{\psi}(\mathbf{p})|^2 d\mathbf{p} &= \int_{\mathbb{R}^3} |\nabla\psi(\mathbf{x})|^2 d\mathbf{x} \\ &\geq \frac{1}{4}\int_{\mathbb{R}^3} \frac{|\psi(\mathbf{x})|^2}{|\mathbf{x}|^2} d\mathbf{x},\end{aligned} \tag{1.6.2}$$

for it implies that if the particle is localized at the origin (i.e., the wave function is supported in a neighbourhood of the origin), $|\mathbf{x}|$ and $|\mathbf{p}|$ cannot both be small.

On choosing $\mathbf{a} = \mathbf{b} = (\delta_{1j}, \delta_{2j}, \delta_{3j})$ in (1.6.1), where $\delta_{ij}$ is the Kronecker delta, we have for $\psi \in C_0^\infty(\mathbb{R}^3)$,

$$(-i\partial_j)(x_j\psi(\mathbf{x})) - x_j(-i\partial_j\psi(\mathbf{x})) = -i\psi(\mathbf{x}),$$

for $j = 1, 2, 3$, and integration by parts gives

$$\begin{aligned}-i\int_{\mathbb{R}^3} |\psi(\mathbf{x})|^2 d\mathbf{x} &= \int_{\mathbb{R}^3} \overline{\psi}(\mathbf{x})\left[(-i\partial_j)(x_j\psi(\mathbf{x})) - x_j(-i\partial_j\psi(\mathbf{x}))\right] d\mathbf{x} \\ &= 2i\mathrm{Re}\int_{\mathbb{R}^3} \left[(\partial_j\overline{\psi(\mathbf{x})})(x_j\psi(\mathbf{x}))\right] d\mathbf{x}.\end{aligned}$$

Hence,

$$\begin{aligned}3\int_{\mathbb{R}^3} |\psi(\mathbf{x})|^2 d\mathbf{x} &\leq 2\sum_{j=1}^{3}\int_{\mathbb{R}^3} |\partial_j\psi(\mathbf{x})||x_j\psi(\mathbf{x})| d\mathbf{x} \\ &\leq 2\left(\int_{\mathbb{R}^3} |\mathbf{x}|^2|\psi(\mathbf{x})|^2 d\mathbf{x}\right)^{\frac{1}{2}}\left(\int_{\mathbb{R}^3} |\nabla\psi(\mathbf{x})|^2 d\mathbf{x}\right)^{\frac{1}{2}} \\ &= 2\left(\int_{\mathbb{R}^3} |\mathbf{x}|^2|\psi(\mathbf{x})|^2 d\mathbf{x}\right)^{\frac{1}{2}}\left(\int_{\mathbb{R}^3} |\mathbf{p}\hat{\psi}(\mathbf{p})|^2 d\mathbf{p}\right)^{\frac{1}{2}}\end{aligned}$$

Thus, if $\int_{\mathbb{R}^3} |\psi(\mathbf{x})|^2 d\mathbf{x} = 1$, the uncertainty principle takes the form of **Heisenberg's inequality**

$$\left(\int_{\mathbb{R}^3} |\mathbf{x}|^2|\psi(\mathbf{x})|^2 d\mathbf{x}\right)^{\frac{1}{2}}\left(\int_{\mathbb{R}^3} |\mathbf{p}\hat{\psi}(\mathbf{p})|^2 d\mathbf{p}\right)^{\frac{1}{2}} \geq 3/2. \tag{1.6.3}$$

It's analogue in $\mathbb{R}^n$, $n \geq 3$, is

$$\left(\int_{\mathbb{R}^n} |\mathbf{x}|^2|\psi(\mathbf{x})|^2 d\mathbf{x}\right)^{\frac{1}{2}}\left(\int_{\mathbb{R}^n} |\mathbf{p}\hat{\psi}(\mathbf{p})|^2 d\mathbf{p}\right)^{\frac{1}{2}} \geq n/2, \quad \int_{\mathbb{R}^n} |\psi(\mathbf{x})|^2 d\mathbf{x} = 1. \tag{1.6.4}$$

Heisenberg's inequality (1.6.4) is a consequence of Hardy's inequality (1.2.16) in $L^2(\mathbb{R}^n)$, but with a smaller constant. For, by the Cauchy-Schwarz inequality,

$$\left(\int_{\mathbb{R}^n} |f(\mathbf{x})|^2 d\mathbf{x}\right)^2 \leq \left(\int_{\mathbb{R}^n} |\mathbf{x}|^2 |f(\mathbf{x})|^2 d\mathbf{x}\right) \left(\int_{\mathbb{R}^n} \frac{1}{|\mathbf{x}|^2} |f(\mathbf{x})|^2 d\mathbf{x}\right),$$

and on substituting (1.2.16), this yields, for $n \geq 3$,

$$\left(\int_{\mathbb{R}^n} |\mathbf{x}|^2 |f(\mathbf{x})|^2 d\mathbf{x}\right)^{\frac{1}{2}} \left(\int_{\mathbb{R}^n} |\nabla f(\mathbf{x})|^2 d\mathbf{x}\right)^{\frac{1}{2}} \geq \left(\frac{n-2}{2}\right) \int_{\mathbb{R}^n} |f(\mathbf{x})|^2 d\mathbf{x}. \qquad (1.6.5)$$

## 1.7 Relativistic Hardy-Type Inequalities

Hardy's inequality (1.2.16) with $p = 2$ and $n \geq 3$, is associated with the Dirichlet Laplacian $-\Delta$, and can be expressed in terms of the $L^2(\mathbb{R}^n)$ inner-product and norm as

$$(-\Delta u, u) \geq \left(\frac{n-2}{2}\right)^2 \left\| \frac{u}{|\cdot|} \right\|^2,$$

first on $C_0^\infty(\mathbb{R}^n)$, and then by extension to $D_n^1 := D_0^{1,2}(\mathbb{R}^n)$, which contains $H_n^1 := H^{1,2}(\mathbb{R}^n)$ as a proper subspace; $H_n^1$ is the form domain of $-\Delta$ and is the domain of the square root, $\sqrt{-\Delta}$. There is a relativistic analogue due to Kato in which the Laplacian is replaced by the pseudo differential operator $\sqrt{-\Delta}$, whose definition is motivated by (1.3.13):

$$\left(\mathbb{F}\sqrt{-\Delta} f\right)(\mathbf{p}) = |\mathbf{p}|(\mathbb{F}f)(\mathbf{p}), \qquad (1.7.1)$$

where $\mathbb{F}$ is the Fourier transform. Its domain as a self-adjoint operator in $L^2(\mathbb{R}^n)$ is

$$\mathcal{D}(\sqrt{-\Delta}) = \{f : \hat{f}, \ |\mathbf{p}|^{1/2}\hat{f} \in L^2(\mathbb{R}^n)\}, \quad \hat{f} = \mathbb{F}f :$$

equivalently, $\mathcal{D}(\sqrt{-\Delta})$ is the completion of the Schwarz space $\mathcal{S}(\mathbb{R}^n)$ with respect to the norm

$$\|u\|_{H^{1/2}(\mathbb{R}^n)} := \left\{\|\hat{u}\|^2_{L^2(\mathbb{R}^n)} + \||\cdot|^{1/2}\hat{u}\|^2_{L^2(\mathbb{R}^n)}\right\}^{1/2}.$$

In the statement of **Kato's inequality** that follows, we make use of Plancherel's theorem to give

$$\int_{\mathbb{R}^n} |\mathbf{p}||\hat{f}(\mathbf{p})|^2 d\mathbf{p} = \int_{\mathbb{R}^n} \overline{f(\mathbf{x})}\sqrt{-\Delta} f(\mathbf{x}) d\mathbf{x}. \tag{1.7.2}$$

Note that the Fourier transform in [111] is defined as $\int_{\mathbb{R}^n} e^{-2\pi i(\mathbf{p}\cdot\mathbf{x})} f(\mathbf{x}) d\mathbf{x}$, which accounts for the differences between some identities in this book and their analogues in [111].

**Theorem 1.7.1** *For all $f \in \mathcal{S}(\mathbb{R}^n)$, $n \geq 2$,*

$$\int_{\mathbb{R}^n} \frac{|f(\mathbf{x})|^2}{|\mathbf{x}|} d\mathbf{x} \leq c_n^2 \int_{\mathbb{R}^n} |\mathbf{p}||\hat{f}(\mathbf{p})|^2 d\mathbf{p}, \tag{1.7.3}$$

$$= c_n^2 \int_{\mathbb{R}^n} \overline{f(\mathbf{x})}\sqrt{-\Delta} f(\mathbf{x}) d\mathbf{x}, \tag{1.7.4}$$

*where the best possible value of the constant $c_n$ is*

$$c_n = \frac{\Gamma(\frac{n-1}{4})}{\sqrt{2}\Gamma(\frac{n+1}{4})};$$

*thus, in particular $c_3 = \sqrt{\pi/2}$. The inequality is strict for non-trivial functions $f$.*

Theorem 8.4 in [111] gives the following Sobolev inequality corresponding to (1.7.3).

**Theorem 1.7.2** *Let $n \geq 2$ and $q = 2n/(n-1)$. Then, for all $f \in \mathcal{S}(\mathbb{R}^n)$,*

$$\|f\|_q^2 \leq C_n \int_{\mathbb{R}^n} |\mathbf{p}||\hat{f}(\mathbf{p})|^2 d\mathbf{p}, \tag{1.7.5}$$

*where the best possible constant is*

$$C_n = \left\{(n-1)2^{\frac{1}{n}}\pi^{\frac{3n+1}{2n}}\right\}^{-1} \Gamma\left(\frac{n+1}{2}\right)^{\frac{1}{n}}.$$

*There is equality if and only if $f$ is a constant multiple of a function of the form $[\mu^2 + (\mathbf{x} - \boldsymbol{a})^2]^{-(n-1)/2}$, with $\mu > 0$ and $\boldsymbol{a} \in \mathbb{R}^n$ arbitrary.*

Kato's inequality is a special case of the following general inequality obtained by Herbst in [77], which also contains the Hardy inequality.

**Theorem 1.7.3** *Let $\alpha > 0$ and $n\alpha^{-1} > q > 1$. Then the operator $|\mathbf{x}|^{-\alpha}|\mathbf{p}|^{-\alpha}$ defines a bounded linear operator $C_\alpha$ from $L^q(\mathbb{R}^n)$ into itself with norm*

$$\|C_\alpha : L^q(\mathbb{R}^n) \to L^q(\mathbb{R}^n)\| = \gamma(n,\alpha) := \frac{\Gamma(\frac{1}{2}[nq^{-1}-\alpha])\Gamma(\frac{1}{2}n(q')^{-1})}{2^\alpha\Gamma(\frac{1}{2}[n(q')^{-1}+\alpha])\Gamma(\frac{1}{2}nq^{-1})}, \tag{1.7.6}$$

*where $q' = q/(q-1)$. If $q \geq n\alpha^{-1}$ or $q = 1$, then $C_\alpha$ is unbounded.*

In the case $q = 2$, (1.7.6) implies that

$$\int_{\mathbb{R}^n} \frac{1}{|\mathbf{x}|^{2\alpha}}|f(\mathbf{x})|^2 d\mathbf{x} \leq \gamma^2(n,\alpha)\int_{\mathbb{R}^n} |\mathbf{p}|^{2\alpha}|\hat{f}(\mathbf{p})|^2 d\mathbf{p}, \tag{1.7.7}$$

for all functions $f$ for which the right-hand side is finite. This is Hardy's inequality when $\alpha = 1$ and Kato's inequality when $\alpha = 1/2$.

We establish Kato's inequality in the case $n = 3$, and follow the proof given in [15], Theorem 2.2.4. Two preliminary lemmas, and some auxiliary results from Fourier theory are required. The first lemma involves the normalised spherical harmonics $Y_{l,m}$ and the Legendre function of the second kind, namely

$$Q_l(z) = \frac{1}{2}\int_{-1}^{1}\frac{P_l(t)}{z-t}dt, \tag{1.7.8}$$

where the $P_l$ are the Legendre polynomials; [150] may be consulted for all the properties of Legendre polynomials that we use. We recall that in spherical polar co-ordinates $\mathbf{x} = (x_1, x_2, x_3) = r\boldsymbol{\omega},\ \boldsymbol{\omega} \in \mathbb{S}^2$,

$$x_1 = r\sin\theta\cos\varphi,\ \ x_2 = r\sin\theta\sin\varphi,\ \ x_3 = r\cos\varphi,$$

the normalised spherical harmonics are given in terms of the *associated Legendre polynomials*

$$P_l^k(x) = \frac{(-1)^k}{2^l l!}(1-x^2)^{k/2}\frac{d^{k+l}}{dx^{k+l}}(x^2-1)^l$$

by

$$Y_{l,k}(\theta,\varphi) = \sqrt{\frac{(2l+1)(l-k)!}{4\pi(l+k)!}}e^{ik\varphi}P_l^k(\cos\theta),\quad k > 0,$$
$$Y_{l,-k}(\theta,\varphi) = (-1)^k\overline{Y_{l,k}(\theta,\varphi)};$$

we adopt the convention that $Y_{l,k} = 0$ for $|k| > l$. From our perspective, the most important role played by the normalised spherical harmonics is that they form an orthonormal basis for $L^2(\mathbb{S}^2)$.

**Lemma 1.7.4** *Let* $\mathbf{p} = p\boldsymbol{\omega}_{\mathbf{p}},\ \mathbf{p}' = p'\boldsymbol{\omega}_{\mathbf{p}'} \in \mathbb{R}^3$. *Then*

$$\int_{\mathbb{S}^2}\int_{\mathbb{S}^2} \frac{1}{|\mathbf{p}-\mathbf{p}'|^2} Y_{l',m'}(\omega_{\mathbf{p}'})Y_{l,m}(\omega_{\mathbf{p}})d\boldsymbol{\omega}_{\mathbf{p}'}d\boldsymbol{\omega}_{\mathbf{p}} = \frac{2\pi}{pp'}Q_l\left(\frac{p'^2+p^2}{2pp'}\right)\delta_{ll'}\delta_{mm'}, \tag{1.7.9}$$

*where* $\delta$ *denotes the Kronecker delta.*

*Proof* We set $z = (p^2 + p'^2)/2pp'$ and $\mathbf{p}\cdot\mathbf{p}' = pp'\cos\gamma$. Then, from [150], Chap. XV, we have that

$$\begin{aligned}\frac{1}{|\mathbf{p}-\mathbf{p}'|^2} &= \frac{1}{2pp'(z-\cos\gamma)}\\ &= \frac{1}{2pp'}\sum_{l''=0}^{\infty}(2l''+1)Q_{l''}(z)P_{l''}(\cos\gamma)\\ &= \frac{4\pi}{2pp'}\sum_{l''=0}^{\infty}Q_{l''}(z)\sum_{m''=-l''}^{l''}\overline{Y}_{l'',m''}(\boldsymbol{\omega}_{\mathbf{p}})Y_{l'',m''}(\boldsymbol{\omega}_{\mathbf{p}'}).\end{aligned} \tag{1.7.10}$$

The lemma follows from the orthonormality of the spherical harmonics, □

The next lemma is a consequence of the generalisation of Hilbert's double series inequality established in [75], Chap. IX, Sect. 319.

**Lemma 1.7.5** *For* $u, v \in L^2(\mathbb{R}_+; xdx)$ *and* $l \in \mathbb{N}_0$,

$$\begin{aligned}I_l &= \int_0^\infty\int_0^\infty \overline{u}(x)v(y)Q_l\left(\frac{1}{2}\left[\frac{x}{y}+\frac{y}{x}\right]\right)dxdy\\ &\le C_l\left(\int_0^\infty x|u(x)|^2dx\right)\left(\int_0^\infty y|v(y)|^2dy\right),\end{aligned} \tag{1.7.11}$$

*where*

$$C_l = \int_0^\infty Q_l\left(\frac{1}{2}\left[x+\frac{1}{x}\right]\right)x^{-1}dx$$

*is sharp. In particular,*

$$C_l = \begin{cases}\pi^2/2, & \text{if } l=0,\\ 2, & \text{if } l=1,\end{cases} \tag{1.7.12}$$

*and* $C_l \le 2$ *for* $l > 2$.

*Proof* Since the functions

$$K(x, y) := x^{-1/2}y^{-1/2}Q_l\left(\frac{1}{2}\left[\frac{x}{y} + \frac{y}{x}\right]\right)$$

are homogeneous of degree $-1$, that is, $K(\lambda x, \lambda y) = \lambda^{-1}K(x, y)$, the inequality ( 1.7.11) follows immediately from Hilbert's inequality, and the exhibited constant $C_l$ is sharp.

To prove (1.7.12), we use the result from [150], Chap. XV, Sect. 32, that for $t > 1$, the Legendre functions $Q_l(t)$ have the integral representation

$$Q_l(t) = \int_{t+\sqrt{t^2-1}}^{\infty} \frac{z^{-l-1}}{\sqrt{1-2tz+z^2}}dz, \tag{1.7.13}$$

to infer that for $t > 1$,

$$Q_0(t) \geq Q_1(t) \geq \cdots \geq Q_l(t) \geq 0. \tag{1.7.14}$$

Thus $C_l \leq C_0$ if $l \in \mathbb{N}_0$, and $C_l \leq C_1$ if $l \in \mathbb{N}$. Furthermore

$$\begin{aligned} C_0 &= \int_0^{\infty} Q_0\left(\frac{1}{2}\left[x + \frac{1}{x}\right]\right)x^{-1}dx = \int_0^{\infty} \ln\left|\frac{x+1}{x-1}\right|\frac{dx}{x} \\ &= 2\int_0^1 \ln\left|\frac{x+1}{x-1}\right|\frac{dx}{x} \\ &= 4\int_0^1 \left(\sum_{k=0}^{\infty}\frac{x^{2k}}{2k+1}\right)dx \\ &= 4\sum_{k=0}^{\infty}\frac{1}{(2k+1)^2} \\ &= \frac{\pi^2}{2}, \end{aligned} \tag{1.7.15}$$

and

$$\begin{aligned} C_1 &= \int_0^{\infty} Q_1\left(\frac{1}{2}\left[x + \frac{1}{x}\right]\right)x^{-1}dx \\ &= 2\int_0^1 Q_1\left(\frac{1}{2}\left[x + \frac{1}{x}\right]\right)x^{-1}dx \\ &= 2\int_0^1 \left\{\frac{1}{2}\left(x + \frac{1}{x}\right)\ln\left|\frac{x+1}{x-1}\right| - 1\right\}\frac{dx}{x} \end{aligned}$$

$$= 2 \lim_{\varepsilon \to 0_+, \delta \to 1_-} \left[ \frac{1}{2} \left( x + \frac{1}{x} \right) \ln \left| \frac{x+1}{x-1} \right| \right]_{\varepsilon}^{\delta}$$
$$= 2, \tag{1.7.16}$$

as asserted in (1.7.12). The lemma is therefore proved. □

Finally we need the following result which follows from Corollary 5.10 in [111] on using (1.3.10), with an adjustment for the difference between our Fourier transform and that in [111]. For $f \in \mathcal{S}(\mathbb{R}^n)$ and

$$\hat{g}(\mathbf{p}) = \left( \sqrt{\frac{2}{\pi}} |\cdot|^{-2} * \hat{f} \right)(\mathbf{p})$$
$$= \frac{1}{(2\pi)^{3/2}} \sqrt{\frac{2}{\pi}} \int_{\mathbb{R}^3} |\mathbf{p} - \mathbf{p}'|^{-2} \hat{f}(\mathbf{p}') d\mathbf{p}',$$

we have

$$g(\mathbf{x}) = \frac{1}{|\mathbf{x}|} f(\mathbf{x}). \tag{1.7.17}$$

At a formal level, $\left( \sqrt{\frac{2}{\pi}} |\cdot|^{-2} \right)(\mathbf{x}) = \mathbb{F}\left( |\cdot|^{-1} \right)(\mathbf{x})$, and this would imply (1.7.17) by (1.3.12). However, this has to be justified since $|\cdot|^{-1} \notin L^1(\mathbb{R}^3)$. Corollary 5.10 in [111] provides us with a way to sidestep this problem.

We are now ready to prove Kato's inequality.

**Proof of Theorem 1.7.1** By Parseval's formula (1.3.12),and (1.7.17),

$$I = \frac{1}{(2\pi)^{3/2}} \sqrt{\frac{2}{\pi}} \int_{\mathbb{R}^3} \int_{\mathbb{R}^3} |\mathbf{p} - \mathbf{p}'|^{-2} \hat{f}(\mathbf{p}') \overline{\hat{f}}(\mathbf{p}) d\mathbf{p}' d\mathbf{p}$$
$$= \int_{\mathbb{R}^3} \left\{ \left( \sqrt{\frac{2}{\pi}} |\cdot|^{-2} \right) * \hat{f} \right\} (\mathbf{p}) \overline{\hat{f}}(\mathbf{p}) d\mathbf{p}$$
$$= \int_{\mathbb{R}^3} \mathbb{F}^{-1} \left\{ \left( \sqrt{\frac{2}{\pi}} |\cdot|^{-2} \right) * \hat{f} \right\} (\mathbf{x}) \overline{\mathbb{F}^{-1}\left(\hat{f}\right)(\mathbf{x})} d\mathbf{x}$$
$$= \int_{\mathbb{R}^3} \frac{|f(\mathbf{x})|^2}{|\mathbf{x}|} d\mathbf{x}. \tag{1.7.18}$$

Since the spherical harmonics $\{Y_{l,m}\}$ form an orthonormal basis of $L^2(\mathbb{S}^2)$, then, in terms of polar co-ordinates $\mathbf{p} = p\boldsymbol{\omega}_{\mathbf{p}}$, we can write

$$\hat{f}(\mathbf{p}) = \sum_{l,m} c_{l,m}(p) Y_{l,m}(\boldsymbol{\omega}_{\mathbf{p}}), \quad c_{l,m}(p) = \int_{\mathbb{S}^2} \hat{f}(p\boldsymbol{\omega}_{\mathbf{p}}) Y_{l,m}(\boldsymbol{\omega}_{\mathbf{p}}) d\boldsymbol{\omega}_{\mathbf{p}},$$

where the summation is over $m = -l, -l+1, \cdots, l-1, l,\ l \in \mathbb{N}_0$. On substituting in the integral $I$ of (1.7.18), and using Lemma 1.7.4, we obtain

$$\begin{aligned}
I &= \frac{1}{2\pi^2} \sum_{l,m} \sum_{l',m'} \int_0^\infty \int_0^\infty p^2 p'^2 c_{l',m'}(p') \overline{c_{l,m}(p)} dp' dp \\
&\times \int_{\mathbb{S}^2} \int_{\mathbb{S}^2} \frac{1}{|\mathbf{p} - \mathbf{p}'|^2} Y_{l',m'}(\boldsymbol{\omega}_{\mathbf{p}'}) Y_{l,m}(\boldsymbol{\omega}_{\mathbf{p}}) d\boldsymbol{\omega}_{\mathbf{p}'} d\boldsymbol{\omega}_{\mathbf{p}} \\
&= \frac{1}{\pi} \sum_{l,m} \int_0^\infty \int_0^\infty pp' Q_l \left( \frac{p^2 + p'^2}{2pp'} \right) c_{l,m}(p') \overline{c_{l,m}(p)} dp' dp \\
&= \frac{1}{\pi} \int_0^\infty \int_0^\infty Q_l \left( \frac{1}{2} \left[ \frac{p}{p'} + \frac{p'}{p} \right] \right) \left( p' c_{l'm'}(p') \right) \left( p \overline{c_{l,m}(p)} \right) dp' dp.
\end{aligned} \tag{1.7.19}$$

It now follows from Lemma 1.7.5 and (1.7.14) that

$$\begin{aligned}
I &\leq \frac{\pi}{2} \sum_{l,m} \int_0^\infty p^3 |c_{l,m}(p)|^2 dp \\
&= \frac{\pi}{2} \int_{\mathbb{R}^3} |\mathbf{p}| |\hat{f}(\mathbf{p})|^2 d\mathbf{p},
\end{aligned}$$

with sharp constant $\pi/2$. Theorem 1.7.1 is therefore proved.□

The operator $\sqrt{-\Delta}$ is the "massless" case of the so-called *quasi-relativistic* operator $\sqrt{-\Delta + m^2}$ which has been used as a model for a free relativistic spin zero particle of mass $m$; see [77, 147, 148]. The massless case of the Dirac operator is

$$\boldsymbol{\alpha} \cdot (-i\nabla) = -i \sum_{j=1}^{3} \alpha_j \partial_j, \quad \partial_j := \frac{\partial}{\partial x_j}, \tag{1.7.20}$$

where the $\alpha_j$ are the Dirac matrices,which in the standard representation, are given by

$$\alpha_j = \begin{pmatrix} 0 & \sigma_j \\ \sigma_j & 0 \end{pmatrix}, \quad j = 1, 2, 3, \tag{1.7.21}$$

and the $\sigma_j$ are the Pauli matrices

$$\sigma_1 = \begin{pmatrix} 0 & 1 \\ 1 & 0 \end{pmatrix} \quad \sigma_2 = \begin{pmatrix} 0 & -i \\ i & 0 \end{pmatrix} \quad \sigma_3 = \begin{pmatrix} 1 & 0 \\ 0 & -1 \end{pmatrix};$$

the Pauli matrices are Hermitian and satisfy

$$\sigma_j\sigma_k + \sigma_k\sigma_j = 2I_2\delta_{jk}, \tag{1.7.22}$$

where $I_2$ is the unit $2 \times 2$ matrix, and $\delta_{jk}$ is the Kronecker delta. The massless Dirac operator is therefore determined by the **Weyl-Dirac** operator

$$\mathbb{D}_0 := \boldsymbol{\sigma} \cdot (-i\nabla) \tag{1.7.23}$$

which acts on $\mathbb{C}^2$-valued functions, whereas the massless Dirac operator acts on $\mathbb{C}^4$-valued functions. In [46], the following theorem featuring a Hardy-type inequality involving the Weyl-Dirac operator is established. We use the notation $H^1(\mathbb{R}^3;\mathbb{C}^2)$, $L^2(\mathbb{R}^3;\mathbb{C}^2)$, $C_0^\infty(\mathbb{R}^3;\mathbb{C}^2)$ to denote the spaces of $\mathbb{C}^2$-valued functions whose components lie in $H^{1,2}(\mathbb{R}^3)$, $L^2(\mathbb{R}^3)$, $C_0^\infty(\mathbb{R}^3)$, respectively.

**Theorem 1.7.6** *For all $\varphi \in H^1(\mathbb{R}^3;\mathbb{C}^2)$,*

$$\int_{\mathbb{R}^3}\left(\frac{|(\boldsymbol{\sigma}\cdot\nabla)\varphi|^2}{1+|\mathbf{x}|^{-1}} + |\varphi|^2\right)d\mathbf{x} \geq \int_{\mathbb{R}^3}\frac{|\varphi(\mathbf{x})|^2}{|\mathbf{x}|}d\mathbf{x}. \tag{1.7.24}$$

*On replacing $\varphi(\mathbf{x})$ by $\varepsilon^{-1}\varphi(\varepsilon^{-1}\mathbf{x})$ and allowing $\varepsilon \to 0$, (1.7.24) yields*

$$\int_{\mathbb{R}^3}|\mathbf{x}||(\boldsymbol{\sigma}\cdot\nabla)f(\mathbf{x})|^2 d\mathbf{x} \geq \int_{\mathbb{R}^3}\frac{|f(\mathbf{x})|^2}{|\mathbf{x}|}d\mathbf{x}, \quad f \in C_0^\infty(\mathbb{R}^3;\mathbb{C}^2). \tag{1.7.25}$$

*Proof* We shall follow the analytic proof given in [47]. The following facts will be needed:

**(i)** If $h$ is a radial function which is differentiable in $\mathbb{R}_+ = (0,\infty)$, then

$$[(\boldsymbol{\sigma}\cdot\nabla),(\boldsymbol{\sigma}\cdot\mathbf{x})h] = |\mathbf{x}|h' + 2(1+\boldsymbol{\sigma}\cdot L)h + h, \tag{1.7.26}$$

where $[\cdots]$ is the commutator, and $L = -i\mathbf{x}\wedge\nabla$.

**(ii)** $L$ is the *orbital angular momentum operator*; it acts only on the angular variables.

**(iii)** $1+\boldsymbol{\sigma}\cdot L$ is a self-adjoint operator in $L^2(\mathbb{R}^3)$ whose spectrum is the discrete set $\{\pm 1, \pm 2, \cdots\}$.

**(iv)** We denote by $X_\pm$, the positive and negative spectral subspaces of $1+\boldsymbol{\sigma}\cdot L$, and by $P_\pm$, the associated projections.

For $\varphi \in H^1(\mathbb{R}^3,\mathbb{C}^2)$, let $\varphi_\pm := P_\pm\varphi$. Then, from (1.7.26) and **(iii)**,

$$([(\boldsymbol{\sigma}\cdot\nabla),(\boldsymbol{\sigma}\cdot\mathbf{x})h]\,\varphi_+,\varphi_+) \geq \int_{\mathbb{R}^3}\left(3h(\mathbf{x}) + h'(\mathbf{x})|\mathbf{x}|\right)|\varphi_+(\mathbf{x})|^2 d\mathbf{x} \tag{1.7.27}$$

and

$$([(\sigma\cdot\nabla),(\sigma\cdot\mathbf{x})h]\varphi_-,\varphi_-)\le\int_{\mathbb{R}^3}\left(-h(\mathbf{x})+h'(\mathbf{x})|\mathbf{x}|\right)|\varphi_-(\mathbf{x})|^2d\mathbf{x}. \tag{1.7.28}$$

We now use in (1.7.27) the fact that $(\sigma\cdot\nabla)$ is a skew-symmetric operator in $L^2(\mathbb{R}^3)$ since the Pauli matrices are Hermitian, and then apply the Cauchy-Schwarz inequality. For any positive function $g$ (to be chosen later), we have that

$$\begin{aligned}
I &:= \int_{\mathbb{R}^3}(3h+h'|\mathbf{x}|)|\varphi_+|^2d\mathbf{x}\\
&\le -((\sigma\cdot\nabla)\varphi_+,(\sigma\cdot\mathbf{x})h\varphi_+)-((\sigma\cdot\mathbf{x})h\varphi_+,(\sigma\cdot\nabla)\varphi_+)\\
&\le 2\|g^{-1/2}(\sigma\cdot\mathbf{x})h\varphi_+\|\|g^{1/2}(\sigma\cdot\nabla)\varphi_+\|\\
&\le \int_{\mathbb{R}^3}g|(\sigma\cdot\nabla)\varphi_+|^2d\mathbf{x}+\int_{\mathbb{R}^3}\frac{1}{g}|(\sigma\cdot\mathbf{x})h\varphi_+|^2d\mathbf{x}\\
&\le \int_{\mathbb{R}^3}g|(\sigma\cdot\nabla)\varphi_+|^2d\mathbf{x}+\int_{\mathbb{R}^3}\frac{1}{g}|\mathbf{x}|^2h^2|\varphi_+|^2d\mathbf{x};
\end{aligned} \tag{1.7.29}$$

the last inequality follows since $|(\sigma\cdot\mathbf{x})|^2=\sum_{j,k}x_jx_k\sigma_j\sigma_k=|\mathbf{x}|^2$. Similarly, from (1.7.28),

$$\int_{\mathbb{R}^3}(h-h'|\mathbf{x}|)|\varphi_-|^2d\mathbf{x}\le\int_{\mathbb{R}^3}g|(\sigma\cdot\nabla)\varphi_-|^2d\mathbf{x}+\int_{\mathbb{R}^3}\frac{1}{g}|(\sigma\cdot\mathbf{x})h\varphi_-|^2d\mathbf{x}. \tag{1.7.30}$$

We now choose $h(\mathbf{x})=\frac{1}{|\mathbf{x}|}$ and $g(\mathbf{x})=\frac{|\mathbf{x}|}{1+|\mathbf{x}|}$. Then $3h+h'|\mathbf{x}|=h-h'|\mathbf{x}|=\frac{2}{|\mathbf{x}|}$, and hence from (1.7.29) and (1.7.30),

$$\int_{\mathbb{R}^3}\frac{1}{|\mathbf{x}|}|\varphi_+|^2d\mathbf{x}\le\int_{\mathbb{R}^3}\frac{1}{1+|\mathbf{x}|^{-1}}|(\sigma\cdot\nabla)\varphi_+|^2d\mathbf{x}+\int_{\mathbb{R}^3}|\varphi_+|^2d\mathbf{x}, \tag{1.7.31}$$

and

$$\int_{\mathbb{R}^3}\frac{1}{|\mathbf{x}|}|\varphi_-|^2d\mathbf{x}\le\int_{\mathbb{R}^3}\frac{1}{1+|\mathbf{x}|^{-1}}|(\sigma\cdot\nabla)\varphi_-|^2d\mathbf{x}+\int_{\mathbb{R}^3}|\varphi_-|^2d\mathbf{x}. \tag{1.7.32}$$

Since $\varphi=\varphi_++\varphi_-$ and the subspaces $X_+$ and $X_-$ are orthogonal, it follows that

$$\int_{\mathbb{R}^3}|\varphi|^2d\mathbf{x}=\int_{\mathbb{R}^3}|\varphi_+|^2d\mathbf{x}+\int_{\mathbb{R}^3}|\varphi_-|^2d\mathbf{x}.$$

The proof will be completed by Lemma 5 in [47], which asserts that

$$P_-(\sigma\cdot\nabla)^2P_+=P_+(\sigma\cdot\nabla)^2P_-\ \text{ in }\ H^1(\mathbb{R}^3,\mathbb{C}^2). \tag{1.7.33}$$

To verify this, the crucial point is that $(\boldsymbol{\sigma} \cdot \nabla)$ anticommutes with $1 + \boldsymbol{\sigma} \cdot L$, and this is shown by direct computation. Therefore $(\boldsymbol{\sigma} \cdot \nabla)^2$ commutes with $1 + \boldsymbol{\sigma} \cdot L$. Now let $\Phi_\pm \in X_\pm$ be eigenvectors of $1 + \boldsymbol{\sigma} \cdot L$ corresponding to eigenvalues $\lambda_\pm$, $\lambda_- < 0 < \lambda_+$. Then

$$\begin{aligned}
((\boldsymbol{\sigma} \cdot \nabla)\Phi_-, (\boldsymbol{\sigma} \cdot \nabla)\Phi_+) &= -\frac{1}{\lambda_+}\left(\Phi_-, (\boldsymbol{\sigma} \cdot \nabla)^2[1 + \boldsymbol{\sigma} \cdot L]\Phi_+\right) \\
&= -\frac{1}{\lambda_+}\left(\Phi_-, [1 + \boldsymbol{\sigma} \cdot L](\boldsymbol{\sigma} \cdot \nabla)^2\Phi_+\right) \\
&= -\frac{1}{\lambda_+}\left([1 + \boldsymbol{\sigma} \cdot L]\Phi_-, (\boldsymbol{\sigma} \cdot \nabla)^2\Phi_+\right) \\
&= -\frac{\lambda_-}{\lambda_+}\left(\Phi_-, (\boldsymbol{\sigma} \cdot \nabla)^2\Phi_+\right) \\
&= \frac{\lambda_-}{\lambda_+}((\boldsymbol{\sigma} \cdot \nabla)\Phi_-, (\boldsymbol{\sigma} \cdot \nabla)\Phi_+)
\end{aligned}$$

which is only possible if $((\boldsymbol{\sigma} \cdot \nabla)\Phi_-, (\boldsymbol{\sigma} \cdot \nabla)\Phi_+) = 0$, or equivalently, (1.7.33) holds. This gives

$$\begin{aligned}
\|(\boldsymbol{\sigma} \cdot \nabla)(\varphi_+ + \varphi_-)\|^2 &= \|(\boldsymbol{\sigma} \cdot \nabla)\varphi_+\|^2 + \|(\boldsymbol{\sigma} \cdot \nabla)\varphi_-\|^2 \\
&\quad + 2\mathrm{Re}\left[((\boldsymbol{\sigma} \cdot \nabla)\varphi_+, (\boldsymbol{\sigma} \cdot \nabla)\varphi_-)\right] \\
&= \|(\boldsymbol{\sigma} \cdot \nabla)\varphi_+\|^2 + \|(\boldsymbol{\sigma} \cdot \nabla)\varphi_-\|^2.
\end{aligned}$$

The theorem therefore follows from (1.7.31) and (1.7.32). □

*Remark 1.7.7* The case $\sigma = -1, n = 3, p = 2$ of (1.2.20) gives

$$\int_{\mathbb{R}^3} \frac{|u(\mathbf{x})|^2}{|\mathbf{x}|} d\mathbf{x} \leq \int_{\mathbb{R}^3} |\mathbf{x}||\nabla u(\mathbf{x})|^2 d\mathbf{x}, \quad u \in C_0^\infty(\mathbb{R}^3). \tag{1.7.34}$$

However, one shouldn't be misled into thinking that (1.7.25) is a consequence of (1.7.34); it is not, for $\int_{\mathbb{R}^3} |\mathbf{x}||(\boldsymbol{\sigma} \cdot \nabla)u(\mathbf{x})|^2 d\mathbf{x} \neq \int_{\mathbb{R}^3} |\mathbf{x}||\nabla u(\mathbf{x})|^2 d\mathbf{x}$ when $u \in C_0^\infty(\mathbb{R}^3; \mathbb{C}^2)$! The inequalities (1.7.34) and (1.7.25) are sharp. Indeed, it is shown in [46] that the powers of $|\mathbf{x}|$ and the constants in (1.7.24) are optimal.

Theorem 1.7.6 is a special case of a more general inequality with weights, obtained in [51]. In [2], Adimurthi and Tintarev established the following result in $L^2(\mathbb{R}^n)$ for all $n \geq 2$. Let $2m = 2^{n/2}$, when $n$ is even, and $2m = 2^{(n+1)/2}$ when $n$ is odd, and let $\sigma_j, j = 1, 2, \cdots, m$ be Hermitian $m \times m$ matrices satisfying

$$\sigma_i\sigma_j + \sigma_j\sigma_i = 2\delta_{ij}, \quad i, j = 1, 2, \cdots, m;$$

when $n = 3$, the $\sigma_j, j = 1, 2, 3$, are the Pauli matrices. The result in [2] is

**Theorem 1.7.8** *Let $b \in \mathbb{R}$ and $n \geq 2$. Then for all $f \in C_0^\infty(\mathbb{R}^n \setminus \{0\}; \mathbb{C}^m)$,*

$$\int_{\mathbb{R}^n} |\mathbf{x}|^{-b} |(\boldsymbol{\sigma} \cdot \nabla) f(\mathbf{x})|^2 d\mathbf{x} \geq c_{n,b} \int_{\mathbb{R}^n} |\mathbf{x}|^{-b-2} |f(\mathbf{x})|^2 d\mathbf{x}, \tag{1.7.35}$$

*where*

$$c_{n,b} = \min_{k \in \mathbb{Z} \setminus \{1,2,\cdots,n-2\}} \left( k - \frac{n-2-b}{2} \right)^2 \tag{1.7.36}$$

*is the best possible constant. In particular, $c_{3,-1} = 1$ and so (1.7.25) is recovered.*

Note that the maximum value of $c_{n,b}$ is $[(n-2)/2]^2$ which is attained when $b = 0$ and (1.7.35) becomes Hardy's inequality. Also, $c_{2,0} = 0$, when the Hardy inequality is known to be invalid.

In [2], Theorem 1.2, the following Sobolev-type inequality is derived.

**Theorem 1.7.9** *Let $b \in \mathbb{R}, n > 2$ and $2^* = 2n/(n-2)$. Suppose that in (1.7.36), $c_{n,b} \neq 0$. Then there exists a positive constant $C$ which depends only on $n$ and $b$, such that for all $f \in C_0^\infty(\mathbb{R}^n \setminus \{0\}; \mathbb{C}^m)$*

$$\int_{\mathbb{R}^n} |\mathbf{x}|^{-b} |(\boldsymbol{\sigma} \cdot \nabla) f(\mathbf{x})|^2 d\mathbf{x} \geq C \left( \int_{\mathbb{R}^n} |\mathbf{x}|^{\beta} |f(\mathbf{x})|^{2^*} d\mathbf{x} \right)^{2/2^*}, \tag{1.7.37}$$

*where $\beta = bn/(n-2)$.*

# Chapter 2
# Boundary Curvatures and the Distance Function

## 2.1 Introduction

Let $\Omega$ be an open subset of $\mathbb{R}^n$, $n \geq 2$, with non-empty boundary, and set

$$\delta(\mathbf{x}) := \inf\{|\mathbf{x} - \mathbf{y}| : \mathbf{y} \in \mathbb{R}^n \setminus \Omega\}$$

for the distance of $\mathbf{x} \in \Omega$ to the boundary $\partial\Omega$ of $\Omega$. Our main objective in this chapter is to gather information about the regularity properties of $\delta$. This is of intrinsic interest for the way it relates to the geometry of $\Omega$ and its boundary. However we have an additional motive in that it prepares the ground for the study in subsequent chapters of inequalities of the form

$$\int_\Omega |\nabla f(\mathbf{x})|^p d\mathbf{x} \geq \left(\frac{p-1}{p}\right)^p \int_\Omega \{1 + a(\delta, \partial\Omega)(\mathbf{x})\} \frac{|f(\mathbf{x})|^p}{\delta(\mathbf{x})^p} d\mathbf{x}, \quad f \in C_0^\infty(\Omega);$$

the case when $a(\delta, \partial\Omega) = 0$ is the Hardy inequality for $\Omega$. To give a flavour to what follows, we note that the subset of $\Omega$ which we introduce below and call the *skeleton*, is such that it is precisely the set of points in $\Omega$ at which $\delta$ ceases to be differentiable. If $\partial\Omega$ is assumed to belong to the class $C^2$ then $\delta \in C^2$ outside the skeleton, and its Laplacian is given by an explicit formula involving the principal curvatures at $\partial\Omega$. This is obviously of value for the analysis of inequalities like the above. For the properties of $\delta$ on the skeleton, and another related set called the *ridge*, to be defined below, we follow the treatment of Evans and Harris in [53]. However there are other earlier works on these topics, notably those of Bunt [33], Motzkin [120] and Federer [58], and these will be cited where appropriate. Furthermore, ideas from Balinsky, Evans and Lewis in [20], and Lewis et al. in [107] will be used in the inequalities which feature the boundary curvatures. The Appendix on boundary curvatures and the distance function in [68] is also an important reference.



A.A. Balinsky et al., *The Analysis and Geometry of Hardy's Inequality*,
Universitext, DOI 10.1007/978-3-319-22870-9_2

## 2.2 The Ridge and Skeleton of $\Omega$

Suppose that $\Omega$ does not contain a half-space. We shall call the set $N(\mathbf{x}) := \{\mathbf{y} \in \partial\Omega : \delta(\mathbf{x}) = |\mathbf{x} - \mathbf{y}|\}$ the **near set** of $\mathbf{x}$ on $\partial\Omega$: when $N(\mathbf{x}) = \{\mathbf{y}\}$ we usually write $\mathbf{y} = N(\mathbf{x})$. The following lemma follows easily from the definition.

**Lemma 2.2.1**

(1) *For each* $\mathbf{x} \in \Omega$, *the set of near points* $N(\mathbf{x})$ *is compact.*
(2) *If* $B$ *is a bounded subset of* $\Omega$, *then* $\cup_{\mathbf{x}\in B} N(\mathbf{x})$ *is bounded.*
(3) *Let* $(\mathbf{x}_i)$ *be a sequence in* $\Omega$ *which converges to* $\mathbf{x} \in \Omega$. *If* $\mathbf{y}_i \in N(\mathbf{x}_i)$ *for all* $i$ *and* $\mathbf{y}_i \to \mathbf{y}$, *then* $\mathbf{y} \in N(\mathbf{x})$.

**Lemma 2.2.2** *Let* $\mathbf{x} \in \Omega$, $\mathbf{y} \in N(\mathbf{x})$ *and* $\mathbf{u} = t\mathbf{x} + (1-t)\mathbf{y}$, *where* $0 < t < 1$. *Then* $N(\mathbf{u}) = \mathbf{y}$.

*Proof* Suppose to the contrary that there exists a $\mathbf{y}' \in N(\mathbf{u})$ which also lies in the ball $B(\mathbf{u}, |\mathbf{y} - \mathbf{u}|)$ centre $\mathbf{u}$ and radius $|\mathbf{y} - \mathbf{u}|$. Then

$$\begin{aligned} |\mathbf{x} - \mathbf{y}'| &\le |\mathbf{x} - \mathbf{u}| + |\mathbf{u} - \mathbf{y}'| \\ &< |\mathbf{x} - \mathbf{u}| + |\mathbf{y} - \mathbf{u}| \\ &= |\mathbf{x} - \mathbf{y}| \end{aligned}$$

which contradicts the fact that $\mathbf{y} \in N(\mathbf{x})$. □

An immediate consequence of the last lemma is

**Corollary 2.2.3** *For* $\mathbf{x} \in \Omega$ *and* $\mathbf{y} \in N(\mathbf{x})$, *let*

$$\lambda : \sup\{t \in (0, \infty) : \mathbf{y} \in N(\mathbf{y} + t[\mathbf{x} - \mathbf{y}])\}.$$

*Then, for all* $t \in (0, \lambda)$, $N(\mathbf{y} + t[\mathbf{x} - \mathbf{y}]) = \mathbf{y}$.

This leads to the following notions introduced in [53]; see also [49], Sect. 5.1.1.

**Definition 2.2.4** For $\mathbf{x} \in \Omega$, $\mathbf{y} \in N(\mathbf{x})$ and $\lambda$ defined in Corollary 2.2.3, the point $p(\mathbf{x}) := \mathbf{y} + \lambda(\mathbf{x} - \mathbf{y})$ is called the **ridge point** of $\mathbf{x}$ in $\Omega$ and the set $\mathcal{R}(\Omega) = \{p(\mathbf{x}) : \mathbf{x} \in \Omega\}$ is called the **ridge** of $\Omega$.

In [65], the set of centres of maximal open balls contained in $\Omega$, denoted by $\mathcal{R}_C(\Omega)$, is called the **central set** of $\Omega$, and the following is proved.

**Lemma 2.2.5** *The ridge* $\mathcal{R}(\Omega)$ *and central set* $\mathcal{R}_C(\Omega)$ *of a proper open subset* $\Omega$ *of* $\mathbb{R}^n$ *coincide.*

*Proof* Following the proof in [65], Proposition 3A, we show that $\mathcal{R}_C(\Omega)$ is the set of points in $\Omega$ not lying in any open interval $(\mathbf{x}, \mathbf{y})$, where $\mathbf{x} \in \Omega$ and $\mathbf{y} \in N(\mathbf{x})$.

Set $U_{\mathbf{x}} := \{\mathbf{u} : |\mathbf{u} - \mathbf{x}| < \delta(\mathbf{x}, \partial\Omega)\}$, the largest open ball centre $\mathbf{x}$ which lies in $\Omega$. Then $\mathcal{R}_C(\Omega) = \{\mathbf{w} : \mathbf{w} \in \Omega, U_{\mathbf{w}} \not\subseteq U_{\mathbf{x}}$ for every $\mathbf{x} \in \Omega\}$. If $\mathbf{w} \notin \mathcal{R}(\Omega)$, then by Corollary 2.2.3, $\mathbf{w} \in (\mathbf{x}, \mathbf{y})$ where $\mathbf{x} \in \Omega$ and $\mathbf{y} \in N(\mathbf{x})$. Therefore $U_{\mathbf{w}} \subseteq U_{\mathbf{x}}$ and so $\mathbf{w} \notin \mathcal{R}_C(\Omega)$.

Suppose that $\mathbf{w} \notin \mathcal{R}_C(\Omega)$. Then there exists $\mathbf{x} \in \Omega$ such that $U_{\mathbf{w}} \subset U_{\mathbf{x}}$. If $\mathbf{y} \in N(\mathbf{w})$ then the spheres $\partial U_{\mathbf{w}}$, $\partial U_{\mathbf{x}}$ must be tangent at $\mathbf{y}$ and $\mathbf{y} \in N(\mathbf{x})$. Thus $\mathbf{w} \in (\mathbf{x}, \mathbf{y})$ and $\mathbf{w} \notin \mathcal{R}(\Omega)$. □

Note that the point of assuming that $\Omega$ contains no half space is to ensure that $p(\mathbf{x})$ is defined for all $\mathbf{x} \in \Omega$; if $\Omega$ does contain a half-space, then for some $\mathbf{x} \in \Omega$, $\lambda = \infty$ in Corollary 2.2.3 and we put $p(\mathbf{x}) = \infty$. It follows from Lemma 2.2.2 that if $\mathbf{x} \notin \mathcal{R}(\Omega)$ then card $N(\mathbf{x}) = 1$, i.e., $N(\mathbf{x})$ is a unique point. The converse is not true as is easily seen from the example of an ellipse; in that case the ridge is the straight line along the major axis joining the centres of curvatures $A, B$ of the points $C, D$, say, of intersection of the major axis and $\partial\Omega$, whereas $N(A) = \{C\}$ and $N(B) = \{D\}$.

**Definition 2.2.6** The **skeleton** of $\Omega$ is the set

$$\mathcal{S}(\Omega) := \{\mathbf{x} \in \Omega : \text{card } N(\mathbf{x}) > 1\}. \tag{2.2.1}$$

The significance of the skeleton $\mathcal{S}(\Omega)$ was exposed by the work of Federer in [58], but even earlier by Bunt in [33] and Motzkin in [120]. Sets $\Omega$ for which $\mathcal{S}(\Omega) = \emptyset$ are of particular interest in arbitrary metric spaces and are sometimes called *Chebyshev sets*. The proof of the next theorem follows that in [53], but the result was established by Motzkin in [120].

**Theorem 2.2.7** *The function $\delta$ is differentiable at $\mathbf{x} \in \Omega$ if and only if the cardinality of $N(\mathbf{x}) = 1$, i.e. $N(\mathbf{x})$ contains only one element. If $\delta$ is differentiable at $\mathbf{x}$ then $\nabla\delta(\mathbf{x}) = (\mathbf{x} - \mathbf{y})/|\mathbf{x} - \mathbf{y}|$, where $\mathbf{y} = N(\mathbf{x})$. Also $\nabla\delta$ is continuous on its domain of definition.*

*Proof* Suppose that $N(\mathbf{x}) = \{\mathbf{y}\}$ and that $\mathbf{y} + \mathbf{k} \in N(\mathbf{x} + \mathbf{h})$. Then

$$\begin{aligned}\delta(\mathbf{x} + \mathbf{h})^2 - \delta(\mathbf{x})^2 &= |\mathbf{y} + \mathbf{k} - \mathbf{x} - \mathbf{h}|^2 - |\mathbf{y} - \mathbf{x}|^2 \\ &= 2(\mathbf{x} - \mathbf{y}) \cdot \mathbf{h} + 2(\mathbf{y} - \mathbf{x}) \cdot \mathbf{k} - 2\mathbf{h} \cdot \mathbf{k} + |\mathbf{h}|^2 + |\mathbf{k}|^2 \\ &= 2(\mathbf{x} - \mathbf{y}) \cdot \mathbf{h} + \eta,\end{aligned}$$

where $\eta = 2(\mathbf{y} - \mathbf{x}) \cdot \mathbf{k} - 2\mathbf{h} \cdot \mathbf{k} + |\mathbf{h}|^2 + |\mathbf{k}|^2$. But $|\mathbf{y} - \mathbf{x}|^2 \le |\mathbf{y} + \mathbf{k} - \mathbf{x}|^2$ and $|\mathbf{y} + \mathbf{k} - \mathbf{x} - \mathbf{h}|^2 \le |\mathbf{y} - \mathbf{x} - \mathbf{h}|^2$, so that

$$0 \le -2(\mathbf{x} - \mathbf{y}) \cdot \mathbf{k} + |\mathbf{k}|^2$$

and

$$0 \le 2(\mathbf{x} - \mathbf{y} + \mathbf{h}) \cdot \mathbf{k} - |\mathbf{k}|^2.$$

Therefore

$$-2\mathbf{h}\cdot\mathbf{k}+|\mathbf{h}|^2 \le \eta \le |\mathbf{h}|^2,$$

and as $|\mathbf{h}| \to 0$, $\eta = o(|\mathbf{h}|)$, since $|\mathbf{k}| \to 0$ by Lemma 2.2.1(3). Thus $\delta^2$ is differentiable at $\mathbf{x}$ with gradient $2(\mathbf{x}-\mathbf{y})$ and so $\delta$ is differentiable with

$$\nabla\delta(\mathbf{x}) = \frac{(\mathbf{x}-\mathbf{y})}{|\mathbf{x}-\mathbf{y}|}.$$

Conversely, suppose that $\delta$ is differentiable at $\mathbf{x}$ and that $\mathbf{y} \in N(\mathbf{x})$. Let $\mathbf{u} = t\mathbf{x}+(1-t)\mathbf{y}$, where $0 < t < 1$. Then $\mathbf{y} \in N(\mathbf{u})$ by Lemma 2.2.2, and as $|\mathbf{u}-\mathbf{x}| \to 0$, we have

$$\begin{aligned}-|\mathbf{u}-\mathbf{x}| &= |\mathbf{u}-\mathbf{y}| - |\mathbf{x}-\mathbf{y}| = \delta(\mathbf{u}) - \delta(\mathbf{x})\\ &= \nabla\delta(\mathbf{x})\cdot(\mathbf{u}-\mathbf{x}) + o(|\mathbf{u}-\mathbf{x}|).\end{aligned}$$

On dividing through by $1-t$ and letting $t \to 1$, we obtain

$$-|\mathbf{y}-\mathbf{x}| = \nabla\delta(\mathbf{x})\cdot(\mathbf{y}-\mathbf{x}).$$

Now $|\delta(\mathbf{x}+\mathbf{h})-\delta(\mathbf{x})| \le |\mathbf{h}|$ so that $|\nabla\delta(\mathbf{x})| \le 1$. It follows that

$$\nabla\delta(\mathbf{x}) = \frac{(\mathbf{x}-\mathbf{y})}{|\mathbf{x}-\mathbf{y}|}.$$

and hence

$$\mathbf{y} = \mathbf{x} - \delta(\mathbf{x})\nabla\delta(\mathbf{x});$$

$\mathbf{y}$ is therefore unique. The continuity of $\nabla\delta$ on its domain of definition follows from Lemma 2.2.1(3). □

A consequence of the last theorem is that $\mathcal{S}(\Omega)$ is the set of points in $\Omega$ at which $\delta$ is not differentiable. It is readily shown that $\delta$ is uniformly Lipschitz. For if $\mathbf{x}, \mathbf{y} \in \Omega$, choose $\mathbf{z} \in \partial\Omega$ such that $\delta(\mathbf{y}) = |\mathbf{y}-\mathbf{z}|$. Then

$$\delta(\mathbf{x}) \le |\mathbf{x}-\mathbf{z}| \le |\mathbf{x}-\mathbf{y}| + \delta(\mathbf{y})$$

together with the inequality obtained by interchanging $\mathbf{x}$ and $\mathbf{y}$, yield

$$|\delta(\mathbf{x})-\delta(\mathbf{y})| \le |\mathbf{x}-\mathbf{y}|.$$

It follows from *Rademacher's theorem* that $\mathcal{S}(\Omega)$ is of zero Lebesgue measure. Whether or not $\mathcal{R}(\Omega)$ is of zero Lebesgue measure is not known in general, but it is proved in [65], Proposition $3N$, that it is if $\Omega$ is a proper open subset of $\mathbb{R}^2$.

Another important subset of $\Omega$ which is relevant to our needs is $\Sigma(\Omega) := \Omega \setminus G(\Omega)$, where $G(\Omega)$ is the *good set* defined by Li and Nirenberg in [108] as *the largest open subset of* $\Omega$ *such that every point* $\mathbf{x} \in G(\Omega)$ *has a unique near point.*

**Lemma 2.2.8**

$$(1).\quad \mathcal{S}(\Omega) \subseteq \mathcal{R}(\Omega) \subseteq \overline{\mathcal{S}(\Omega)}, \tag{2.2.2}$$

$$(2).\quad \Sigma(\Omega) = \overline{\mathcal{R}(\Omega)} = \overline{\mathcal{S}(\Omega)}, \tag{2.2.3}$$

*where the closures are relative to* $\Omega$.

*Proof*

(1). This is proved in [65]. Let $\mathbf{x} \in \mathcal{S}(\Omega)$ and $\mathbf{y} \in N(x)$. If $\mathbf{x} \notin \mathcal{R}(\Omega)$, then by Corollary 2.2.3, for some $r > 1$, $\mathbf{y} = N(\mathbf{y} + t[\mathbf{x} - \mathbf{y}])$ for all $t < r$. But this implies that $\mathbf{y}$ is the unique member of $N(\mathbf{x})$, contrary to $\mathbf{x} \in \mathcal{S}(\Omega)$. Hence $\mathcal{S}(\Omega) \subseteq \mathcal{R}(\Omega)$.

To prove that $\mathcal{R}(\Omega) \subseteq \overline{\mathcal{S}(\Omega)}$, we shall show that assuming the existence of an $\mathbf{x}_0 \in \mathcal{R}(\Omega)\setminus\overline{\mathcal{S}(\Omega)}$ leads to a contradiction. Let $f : \mathbf{x} \mapsto N(\mathbf{x}) : \mathbb{R}^n\setminus\overline{\mathcal{S}(\Omega)} \to \mathbb{R}^n \setminus \Omega$, which by Theorem 2.2.7 is continuous. Let $\mathbf{x}_0 \in \mathcal{R}(\Omega) \setminus \overline{\mathcal{S}(\Omega)}$, set $\gamma = \delta(\mathbf{x}_0, \partial\Omega)$, and let $\varepsilon > 0$ be such that the open ball $B_\varepsilon(\mathbf{x}_0)$, centre $\mathbf{x}_0$ and radius $\varepsilon$, lies in $\Omega \setminus \mathcal{S}(\Omega)$ and $\delta(f(\mathbf{z}), f(\mathbf{x}_0)) \le \gamma$ for all $\mathbf{z} \in B_\varepsilon(\mathbf{x}_0)$. Set $E := \{\mathbf{z} : \delta(\mathbf{z}, \mathbf{x}_0) = \varepsilon\}$. Then $\mathbf{x}_0 \notin t\mathbf{z} + (1-t)f(\mathbf{z})$ for any $\mathbf{z} \in E,\ t \in [0, 1]$. There is therefore a homotopy in $\mathbb{R}^n \setminus \{\mathbf{x}_0\}$ between the identity function on $E$ and $f_E$, the restriction of $f$ to $E$. On projecting this homotopy onto $E$ from the centre $\mathbf{x}_0$, we have a homotopy in $E$ between the constant function on $E$ and a function taking values in the contraction $K$ of $f[E]$ back onto $E$. But as $f[E] \subseteq B_\gamma(f(\mathbf{x}_0))$, $K \neq E$, which is impossible by Theorems 3-4a in [93], Sect. 59.IV. Therefore there is no such point $\mathbf{x}_0$ and $\mathcal{R}(\Omega) \subseteq \overline{\mathcal{S}(\Omega)}$.

(2). Since $S(\Omega)$ is the set containing all $\mathbf{x} \in \Omega$ with non-unique near points, $S(\Omega) \subseteq \Sigma(\Omega)$ implying that $\overline{\mathcal{S}(\Omega)} \subseteq \Sigma(\Omega)$. Since $\mathcal{S}(\Omega) \subseteq \mathcal{R}(\Omega) \subseteq \overline{\mathcal{S}(\Omega)}$, then $\overline{\mathcal{R}(\Omega)} = \overline{\mathcal{S}(\Omega)} \subseteq \Sigma(\Omega)$. The set $\Omega \setminus \overline{\mathcal{R}(\Omega)}$ is an open set containing only points with unique near points. Therefore $\Omega \setminus \overline{\mathcal{R}(\Omega)} \subseteq G(\Omega)$ since $G(\Omega)$ is the largest such set. We now have that $\mathbf{x} \notin \overline{\mathcal{R}(\Omega)}$ implies that $\mathbf{x} \notin \Sigma(\Omega)$ or, equivalently, $\Sigma(\Omega) \subseteq \overline{\mathcal{R}(\Omega)}$ which completes the proof. □

Finally, we show that

$$\mathcal{R}(\mathbb{R}^n \setminus \overline{\Omega}) = \mathcal{S}(\mathbb{R}^n \setminus \overline{\Omega}) = \emptyset$$

if and only if $\overline{\Omega}$ is convex. The result is attributed to independent work of Bunt [33] and Motzkin [120], but it is usually called Motzkin's Theorem.

**Theorem 2.2.9** *Let $\overline{\Omega}$ be a subset of $\mathbb{R}^n$ and let $\delta(\cdot)$ be the distance function in $\overline{\Omega}^c := \mathbb{R}^n \setminus \overline{\Omega}$. The following are equivalent:*

(i) *$\overline{\Omega}$ is convex.*
(ii) *$\delta(\mathbf{x})$ is differentiable at every point in $\overline{\Omega}^c$.*
(iii) *For every $\mathbf{x} \in \overline{\Omega}^c$ there is a unique point in $\overline{\Omega}$ at minimal distance from $\mathbf{x}$, i.e., $\mathcal{S}(\overline{\Omega}^c) = \emptyset$.*

*Proof* The equivalence of (ii) and (iii) is shown in Theorem 2.2.7. It will suffice to show the equivalence of (i) and (iii).

Suppose that $\overline{\Omega}$ is convex. Let $\mathbf{z} \in N(\mathbf{x}) \subset \overline{\Omega}$ for some $\mathbf{x} \in \overline{\Omega}^c$. Since $\overline{\Omega}$ is convex, then for any $y \in \overline{\Omega}$ and any $\varepsilon \in [0, 1]$, $\mathbf{z} + \varepsilon(\mathbf{y} - \mathbf{z}) \in \overline{\Omega}$ which implies that

$$|\mathbf{x} - \mathbf{z}|^2 \leq |\mathbf{x} - (\mathbf{z} + \varepsilon(\mathbf{y} - \mathbf{z}))|^2 = |\mathbf{x} - \mathbf{z}|^2 - 2\varepsilon < \mathbf{x} - \mathbf{z}, \mathbf{y} - \mathbf{z} > +\varepsilon^2|\mathbf{y} - \mathbf{z}|^2,$$

where $< \cdot, \cdot >$ denotes the scaler product. By letting $\varepsilon \to 0$, we see that $< \mathbf{x} - \mathbf{z}, \mathbf{y} - \mathbf{z} > \leq 0$, and on letting $\varepsilon = 1$ in the expression above, we have that

$$|\mathbf{x} - \mathbf{y}|^2 = |\mathbf{x} - \mathbf{z}|^2 - 2(\mathbf{x} - \mathbf{z}, \mathbf{y} - \mathbf{z}) + |\mathbf{y} - \mathbf{z}|^2 > |\mathbf{x} - \mathbf{z}|^2$$

indicating that (i) implies (iii).

Our proof of the reverse implication is based on that of Theorem 2.1.30 in [80]. Assume that $\overline{\Omega}$ is not convex. It will suffice to show that there is an open ball $B$ with $B \cap \overline{\Omega} = \emptyset$ such that $\overline{B} \cap \overline{\Omega}$ contains more than one point. Since $\overline{\Omega}$ is not convex there exist distinct points $\mathbf{x}_1, \ \mathbf{x}_2 \in \overline{\Omega}$ such that the open segment between $\mathbf{x}_1$ and $\mathbf{x}_2$ is contained in $\overline{\Omega}^c$. We may assume that the midpoint of this segment is the origin so that $\mathbf{x}_2 = -\mathbf{x}_1$. Choose $\rho > 0$ so that $\overline{B(0, \rho)} \cap \overline{\Omega} = \emptyset$; thus $\overline{B(0, \rho)}$ is at a positive distance from $\overline{\Omega}$. Let $S$ be the set of points $(\omega, r)$ in $\mathbb{R}^{n+1}$ which are such that the family of balls $\{B(\omega, r)\}$ satisfy

$$B(\omega, r) \supset B(0, \rho), \qquad B(\omega, r) \cap \overline{\Omega} = \emptyset.$$

Then

$$r \geq |\omega| + \rho, \qquad |\omega \pm \mathbf{x}_1|^2 \geq r^2;$$

hence

$$(|\omega| + \rho)^2 \leq r^2 \leq \frac{1}{2}(|\omega + \mathbf{x}_1|^2 + |\omega - \mathbf{x}_1|^2) = |\omega|^2 + |\mathbf{x}_1|^2.$$

From this we infer that

$$|\omega| \le \frac{|\mathbf{x}_1|^2 - \rho^2}{2\rho}, \qquad r \le \frac{|\mathbf{x}_1|^2 + \rho^2}{2\rho}. \tag{2.2.4}$$

It follows that $S$ is bounded, and since it is clearly closed, it is a compact subset of $\mathbb{R}^{n+1}$. Let $(\omega_0, r_0) \in S$ be such that $r_0$ is maximal amongst the points $(\omega, r)$ in $S$; this implies that $\overline{B(\omega_0, r_0)}$ must intersect the boundary of $\Omega$. Let $\mathbf{y}_1 \in \overline{B(\omega_0, r_0)} \cap \overline{\Omega}$ and suppose it is unique. We shall prove that this leads to a contradiction.

Let $\theta$ be any vector with $\langle \theta, \omega_0 - \mathbf{y}_1 \rangle > 0$. Then for small $\varepsilon > 0$, we have that

$$\overline{B(\omega_0 + \varepsilon\theta, r_0)} \cap \overline{\Omega} = \varnothing. \tag{2.2.5}$$

We claim that $B(\omega_0 + \varepsilon\theta, r_0) \supsetneq B(0, \rho)$. Otherwise, $(\omega_0 + \varepsilon\theta, r_0) \in S$, which, in view of (2.2.5), contradicts the maximality of $r_0$. Thus $B(\omega_0 + \varepsilon\theta, r_0)$ must intersect $B(0, \rho)$, and so there exists a point $\mathbf{y}_\varepsilon \in \partial B(\omega_0 + \varepsilon\theta, r_0) \cap \partial B(0, \rho)$. On allowing $\varepsilon \to 0$, $\mathbf{y}_\varepsilon \to \mathbf{y}_2 \in \partial B(\omega_0, r_0) \cap \partial B(0, \rho)$. Since $r_0 > \rho$, $\mathbf{y}_2$ is unique, and as $\overline{B(0, \rho)} \cap \overline{\Omega} = \varnothing$ then $\mathbf{y}_2 \neq \mathbf{y}_1$ and the segment between $\mathbf{y}_1$ and $\mathbf{y}_2$ lies in $\overline{B(\omega_0, r_0)}$. For small $\varepsilon > 0$, $B(\omega_0 + \varepsilon(\mathbf{y}_2 - \mathbf{y}_1), r_0) \supset B(0, \rho)$ and

$$\overline{B(\omega_0 + \varepsilon(\mathbf{y}_2 - \mathbf{y}_1), r_0)} \cap \overline{\Omega} = \varnothing. \tag{2.2.6}$$

Hence $(\omega_0 + \varepsilon(\mathbf{y}_2 - \mathbf{y}_1), r_0) \in S$, and (2.2.6) contradicts that $r_0$ is maximal. □

**Theorem 2.2.10** *The functions $p$ and $\delta \circ p$ are continuous on $\Omega$ if and only if $\mathcal{R}(\Omega)$ is closed relative to $\Omega$.*

*Proof* If $p$ is continuous, then so is the map $\mathbf{x} \to p(\mathbf{x}) - \mathbf{x}$ and hence $\mathcal{R}(\Omega) = \{\mathbf{x} \in \Omega : p(\mathbf{x}) - \mathbf{x} = 0\}$ is closed.

Our proof of the converse implication, uses the fact that $r := \delta \circ p$ is upper semi-continuous. To prove this, we show that if $\{\mathbf{x}_n\}$ is a sequence of points in $\Omega$ which converges to $\mathbf{x} \in \Omega$ and is such that $\{r(\mathbf{x}_n)\}$ tends to a limit or infinity, then $\lim_{n\to\infty} r(\mathbf{x}_n) \le r(\mathbf{x})$. Let $\mathbf{y}_n \in N(\mathbf{x}_n)$ and

$$\lambda_n := \min\left\{\frac{r(\mathbf{x}_n)}{\delta(\mathbf{x}_n)}, \frac{2r(\mathbf{x})}{\delta(\mathbf{x})}\right\}$$

so that $1 \le \lambda_n \le 2r(\mathbf{x})/\delta(\mathbf{x})$. Then $p(\mathbf{x}_n) = \mathbf{y}_n + \frac{r(\mathbf{x}_n)}{\delta(\mathbf{x}_n)}(\mathbf{x}_n - \mathbf{y}_n)$. By Lemma 2.2.1(2), $\{\mathbf{y}_n\}$ is bounded and hence there exists a subsequence $\{\mathbf{y}_{n(k)}\}$ converging to $\mathbf{y}$ say, and such that $\{\lambda_{n(k)}\}$ converges to some $\lambda \ge 1$. Since

$$\mathbf{y}_n \in N(\mathbf{y}_n + \lambda_n(\mathbf{x}_n - \mathbf{y}_n)),$$

it follows from Lemma 2.2.1(3) that $\mathbf{y} \in N(\mathbf{y} + \lambda(\mathbf{x} - \mathbf{y}))$ and hence $\lambda \leq r(\mathbf{x})/\delta(\mathbf{x})$. Therefore

$$\lambda = \lim_{n\to\infty} \{r(\mathbf{x}_n)/\delta(\mathbf{x}_n)\}$$

and so $\lim_{n\to\infty} r(\mathbf{x}_n) = \lambda\delta(\mathbf{x}) \leq r(\mathbf{x})$, as asserted.

Suppose now that $\mathcal{R}(\Omega)$ is closed. Since $r$ is upper semi-continuous, it is bounded above on compact subsets of $\Omega$ and so is $|p|$. Therefore to prove that $p$, and consequently $r$, is continuous on $\Omega$, it is sufficient to show that if $\{\mathbf{x}_n\}$ is a sequence in $\Omega$ which converges to $\mathbf{x}$, and $\{p(\mathbf{x}_n)\}$ converges to $\mathbf{z}$ say, then $\mathbf{z} = p(\mathbf{x})$. Since $\mathcal{R}(\Omega)$ is assumed to be closed, it follows that $\mathbf{z}$ lies on $\mathcal{R}(\Omega)$. If $\mathbf{y} \in N(\mathbf{x})$ is the limit of $\{\mathbf{y}_n\}, \mathbf{y}_n \in N(\mathbf{x}_n)$, then the straight line through $\mathbf{y}$ and $\mathbf{x}$ meets $\mathcal{R}(\Omega)$ at $\mathbf{z}$. This implies that $\mathbf{z} = p(\mathbf{x})$ and the continuity of $p$ and $r$ is established. □

*Remark 2.2.11* Let $\mathbf{n}(\mathbf{y})$ denote the unit inward normal at $\mathbf{y} \in \partial\Omega$, and suppose that $\mathbf{n}(\mathbf{y})$ exists for all $\mathbf{y} \in \partial\Omega$. Then we can write

$$p(\mathbf{x}) = \mathbf{y} + \bar{s}(\mathbf{y})\mathbf{n}(\mathbf{y}) =: m(\mathbf{y}), \quad \mathbf{y} \in N(\mathbf{x}), \tag{2.2.7}$$

where

$$\bar{s}(\mathbf{y}) = \sup\{t > 0 : \mathbf{y} \in N(\mathbf{y} + t\mathbf{n}(\mathbf{y}))\} = \delta \circ p(\mathbf{x}); \tag{2.2.8}$$

we set $m(\mathbf{y}) = \mathbf{y}$ if $\mathbf{y} \in \overline{\mathcal{R}(\Omega)} \cap \partial\Omega$. If $\mathcal{R}(\Omega)$ is closed, then

$$\bar{s}(\mathbf{y}) = \sup\{t > 0 : \mathbf{y} + t\mathbf{n}(\mathbf{y}) \in G(\Omega)\}. \tag{2.2.9}$$

Note that the function $m(\cdot)$ is defined on $\partial\Omega$. In the terminology of [108], $m(\mathbf{y})$ is a **cut point** and the set $\{m(\mathbf{y}) : \mathbf{y} \in \partial\Omega\}$ is called the **cut locus**. It is shown in [108], Corollary 4.11, that if $\Omega$ has a $C^{2,1}$ boundary, $\Sigma(\Omega)$ is the cut locus, and as we saw in (2.2.3), this is $\overline{\mathcal{R}(\Omega)}$.

The following lemma clarifies the connections between notions and terminology in [53, 108].

**Lemma 2.2.12** *Let* $\mathbf{n}(\cdot)$ *be continuous on* $\partial\Omega$. *Then the following are equivalent:*

(1) *$p$ and $\delta \circ p$ are continuous on $\Omega$;*
(2) *$m$ and $\bar{s}$ are continuous on $\partial\Omega$;*
(3) *$\mathcal{R}(\Omega)$ is closed and equal to $\Sigma(\Omega)$.*

*Proof* Since the equivalence of (1) and (3) follows from Lemma 2.2.8 and Theorem 2.2.10, it is sufficient to prove that (1) and (2) are equivalent.

Suppose that $p$ is continuous, and hence $\mathcal{R}(\Omega)$ is closed by Theorem 2.2.10. Let $\mathbf{y}_i \in \partial\Omega, \mathbf{y}_i \to \mathbf{y}, \mathbf{y} \notin \mathcal{R}(\Omega)$. Then there exist $\mathbf{x}_i \in G$ such that $\mathbf{y}_i = N(\mathbf{x}_i)$. Since the sequence $(\mathbf{x}_i)$ is bounded, it contains a subsequence $(\mathbf{x}_{k(i)})$ which converges to some point $\mathbf{x}$ say. Hence by Lemma 2.2.1, $(\mathbf{y}_{k(i)})$ converges to $\mathbf{y} \in N(\mathbf{x})$. Therefore

$$m(\mathbf{y}_{k(i)}) = p(\mathbf{x}_{k(i)}) \to p(\mathbf{x}) = m(\mathbf{y}),$$

and

$$\bar{s}(\mathbf{y}_{k(i)}) = \delta \circ p(\mathbf{x}_{k(i)}) \to (\delta \circ p)(\mathbf{x}) = \bar{s}(\mathbf{y}).$$

Conversely, suppose that $m$ and $\bar{s}$ are continuous on $\partial\Omega$ and let $\mathbf{x}_i \to \mathbf{x}$ in $\Omega$. Let $\mathbf{y}_i \in N(\mathbf{x}_i)$. Then the sequence $(\mathbf{y}_i)$ is bounded and so contains a convergent subsequence $(\mathbf{y}_{k(i)})$ whose limit must be $\mathbf{y} \in N(\mathbf{x})$, by Lemma 2.2.1. Thus

$$(\delta \circ p)(\mathbf{x}_{k(i)}) = \bar{s}(\mathbf{y}_{k(i)}) \to \bar{s}(\mathbf{y}) = (\delta \circ p)(\mathbf{x})$$

and

$$p(\mathbf{x}_{k(i)}) = m(\mathbf{y}_{k(i)}) \to m(\mathbf{y}) = p(\mathbf{x}).$$

□

*Remark 2.2.13*

(1) It is shown in [65] that if $\Omega$ is connected, the sets $\mathcal{S}(\Omega)$ and $\mathcal{R}(\Omega)$ are connected.
(2) For any proper open subset $\Omega$ of $\mathbb{R}^2$, it is proved in [65] that $\mathcal{R}(\Omega)$ has zero two-dimensional Lebesgue measure.
(3) In [115], p.10, an example is given of a convex open subset $\Omega$ of $\mathbb{R}^2$ with a $C^{1,1}$ boundary, which is such that $\overline{\mathcal{S}(\Omega)}$ has non-zero Lebesgue measure. Thus, in view of the previous remark and Lemma 2.2.8, $\mathcal{R}(\Omega)$ is not closed in this example.
(4) Let $\Omega$ be bounded and with a $C^{2,1}$ boundary; see Definition 1.3.7. Then it is proved in [81, 108] that its cut locus $\Sigma(\Omega)$ is arcwise connected, and its $(n-1)$-dimensional Hausdorff measure is finite, thus implying that $\Sigma(\Omega)$ has zero $n$-dimensional Lebesgue measure. Furthermore, $m, \ \bar{s} \in C^{0,1}_{loc}(\partial\Omega)$, which in view of Lemmas 2.2.8 and 2.2.12, means that the cut locus $\Sigma(\Omega)$ of $\Omega$ coincides with $\mathcal{R}(\Omega)$.

For further details and properties of $\mathcal{S}(\Omega)$, $\mathcal{R}(\Omega)$, and $\Sigma(\Omega)$ we refer the reader to [49, 53], Chap. 5, and [108].

Suppose that $\mathbf{n}$ and the functions $m, \bar{s}$ defined in (2.2.7) and (2.2.8) lie in $C(\partial\Omega)$; note Remark 2.2.13(4) above. We then have from Theorem 2.2.10 and Remark 2.2.11(1) that $\mathcal{R}(\Omega)$ is closed in $\Omega$ and is the cut locus $\Sigma(\Omega)$.

For

$$\Lambda(\mathbf{x}) := \bar{s}(N(\mathbf{x})) = \delta \circ p(\mathbf{x}),$$

define the **normalized distance function** by

$$h(\mathbf{x}) := \frac{\delta(\mathbf{x})}{\Lambda(\mathbf{x})}, \qquad \mathbf{x} \in G(\Omega),$$

with $h(\mathbf{x}) = 1$ on $\Sigma(\Omega)$ and 0 on $\partial\Omega$. We then have the following lemma proved in [107].

**Lemma 2.2.14** *Let $\mathcal{R}(\Omega)$ be closed and $\mathbf{n} \in C(\partial\Omega)$. Then, for $\mathbf{x} \in G(\Omega), 0 < h(\mathbf{x}) < 1$, and for each $\bar{\mathbf{x}} \in \Sigma$*

$$\lim_{\mathbf{x}\to\bar{\mathbf{x}},\ \mathbf{x}\in G} h(\mathbf{x}) = 1. \tag{2.2.10}$$

*Hence $h \in C(\Omega)$.*

*Proof* The fact that $h(\mathbf{x}) \in (0, 1)$ for $\mathbf{x} \in G(\Omega)$ follows since $\delta(\mathbf{x}) < \Lambda(\mathbf{x})$ in $G(\Omega)$. As $\Sigma(\Omega) = \{m(\mathbf{y}) : \mathbf{y} \in \partial\Omega\}$, for each $\bar{\mathbf{x}} \in \Sigma(\Omega)$, there exists $\mathbf{y} \in \partial\Omega$ such that

$$\bar{\mathbf{x}} = m(\mathbf{y}) = \mathbf{y} + \bar{s}(\mathbf{y})\mathbf{n}(\mathbf{y}). \tag{2.2.11}$$

Let

$$\{\mathbf{x}_i\} \subset G(\Omega) \qquad \text{for} \quad \mathbf{x}_i \to \bar{\mathbf{x}} \in \Sigma(\Omega), \tag{2.2.12}$$

and let $\mathbf{y}_i = N(\mathbf{x}_i)$; thus $|\mathbf{x}_i - \mathbf{y}_i| = \delta(\mathbf{x}_i)$. Then $\Lambda(\mathbf{x}_i) > |\mathbf{x}_i - \mathbf{y}_i| = \delta(\mathbf{x}_i)$, which implies that

$$\liminf_{i\to\infty} \Lambda(\mathbf{x}_i) \geq \delta(\bar{\mathbf{x}}). \tag{2.2.13}$$

It will suffice to show that, for $\{\mathbf{x}_i\}$ given in (2.2.12),

$$\limsup_{i\to\infty} \Lambda(\mathbf{x}_i) \leq \delta(\bar{\mathbf{x}}). \tag{2.2.14}$$

Suppose that (2.2.14) does not hold. Then, there is an $\alpha > 0$ such that

$$\Lambda(\mathbf{x}_i) > \delta(\bar{\mathbf{x}}) + \alpha$$

for $i$ sufficiently large and hence, since $\Lambda$ is continuous on $\Omega$,

$$\Lambda(\bar{\mathbf{x}}) \geq \delta(\bar{\mathbf{x}}) + \alpha. \tag{2.2.15}$$

Since $|\mathbf{y}_i - \mathbf{x}_i| = \delta(\mathbf{x}_i)$ converges to $\delta(\mathbf{x})$, the sequence $\{\mathbf{y}_i\}$ is bounded and hence contains a subsequence (still called $\{\mathbf{y}_i\}$) which converges to a limit $\hat{\mathbf{y}}$ say, in $\partial(\Omega)$. Also $s_i := \delta(\mathbf{x}_i) \to \delta(\bar{\mathbf{x}})$. Furthermore,

$$\mathbf{x}_i = \mathbf{y}_i + s_i\mathbf{n}(\mathbf{y}_i),$$

implies that

$$\bar{\mathbf{x}} = \hat{\mathbf{y}} + \delta(\bar{\mathbf{x}})\mathbf{n}(\hat{\mathbf{y}}).$$

Since $\delta(\bar{\mathbf{x}}) < \Lambda(\bar{\mathbf{x}}) - \alpha$ by (2.2.15), it follows that $\bar{\mathbf{x}} \in G(\Omega)$, contrary to assumption. Consequently (2.2.14) must hold, and the lemma is proved. □

## 2.3 The Distance Function for a Convex Domain

In the next chapter use will be made of the result we shall now prove, that for a convex domain $\Omega$, $-\Delta\delta$ is a non-negative Radon measure on $\Omega$. This means that there exists a non-negative Radon measure $\mu$ on $\Omega$ such that

$$-\int_{\Omega} \delta(\mathbf{x})\Delta\varphi(\mathbf{x})d\mathbf{x} = \int_{\Omega} \varphi(\mathbf{x})d\mu(\mathbf{x}), \quad \text{for all} \quad \varphi \in C_0^{\infty}(\Omega). \tag{2.3.1}$$

A Radon measure is a measure which is *locally finite* and *inner regular*, these properties being defined respectively by

(1) every point has a neighbourhood of finite measure,
(2) for any measurable set $A$, $\mu(A) = \sup\{\mu(K) : K \subseteq A, K \text{ compact}\}$.

The proof given in Theorem 2.3.2 below is taken from [52], Theorem 2, p. 239. The result will be seen to be a consequence of the fact that $\delta$ is a concave function of $\Omega$, and the Riesz representation theorem, which we now recall. Let $C_0(\Omega)$ be the set of continuous functions on $\Omega$ which are compactly supported in $\Omega$, and let $L : C_0(\Omega) \to [0, \infty)$ be a non-negative linear functional with the property that, for any compact subset $K$ of $\Omega$, there exists a constant $M_K$ such that

$$|L(f)| \leq M_K \|f\|_{\infty}, \quad \|f\|_{\infty} := \sup_{\mathbf{x} \in \Omega} |f(\mathbf{x})|,$$

for all $f \in C_0(\Omega)$ with support in $K$. Then the Riesz representation theorem asserts that there exists a non-negative Radon measure $\mu$ such that

$$L(f) = \int_{\Omega} f(\mathbf{x})d\mu(\mathbf{x})$$

for all $f \in C_0(\Omega)$. In fact we shall need the Riesz theorem in the following form:

**Corollary 2.3.1** *Let $L$ be a linear functional on $C_0^{\infty}(\Omega)$ which is non-negative on $C_0^{\infty}(\Omega)$:*

$$L(\varphi) \geq 0 \quad \text{for all} \quad \varphi \in C_0^{\infty}(\Omega), \ \varphi \geq 0. \tag{2.3.2}$$

*Then there exists a non-negative Radon measure $\mu$ on $\Omega$ such that*

$$L(\varphi) = \int_{\Omega} \varphi(\mathbf{x})d\mu(\mathbf{x}), \quad \text{for all} \quad \varphi \in C_0^\infty(\Omega). \tag{2.3.3}$$

*Proof* Let $\varphi \in C_0^\infty(\Omega)$ have support in a compact subset $K$ of $\Omega$, and let $\boldsymbol{\zeta} \in C_0^\infty(\Omega)$ be such that $\boldsymbol{\zeta} = 1$ on $K$ and $0 \leq \boldsymbol{\zeta} \leq 1$. Set $g = \|\varphi\|_\infty \zeta - \varphi$. Then $g \geq 0$ and (2.3.2) implies that

$$0 \leq L(g) = \|\varphi\|_\infty L(\zeta) - L(\varphi)$$

and so

$$L(\varphi) \leq C\|\varphi\|_\infty,$$

where $C = L(\boldsymbol{\zeta})$. We now contend that $L$ extends to a linear functional on $C_0(\Omega)$ satisfying the hypothesis of the Riesz representation theorem. This follows since any non-negative function $f \in C_0(\Omega)$ is the uniform limit of a sequence of non-negative functions in $C_0^\infty(\Omega)$. To be specific, let $f_\varepsilon$ be a regularisation of $f$ defined by a mollifier $\rho$, a $C_0^\infty(\mathbb{R}^n)$ function supported in the unit ball $B(0, 1)$. Thus if $\mathbf{x} \in \Omega$ and $\varepsilon < \delta(\mathbf{x})$,

$$f_\varepsilon(\mathbf{x}) := \int_{B(0,1)} \rho(\mathbf{z})f(\mathbf{x} - \varepsilon\mathbf{z})d\mathbf{z}. \tag{2.3.4}$$

If $K \subset \Omega$ contains the support of $f$ and $\varepsilon < \text{dist}(K, \partial\Omega)$, then $f_\varepsilon \in C_0^\infty(\Omega)$ and as $\varepsilon \to 0, f_\varepsilon$ converges uniformly on $K$ to $f$. If $f \geq 0$ so are $f_\varepsilon \geq 0$ and hence our needs are satisfied. The corollary therefore follows. □

**Theorem 2.3.2** *Let $\Omega$ be a convex domain in $\mathbb{R}^n$. Then $\delta$ is concave on $\Omega$ and $-\Delta\delta$ is a non-negative Radon measure in the sense of (2.3.1).*

*Proof* To prove that $\delta$ is concave on $\Omega$, we repeat the argument in [23], Example 2. Let $\mathbf{x}, \mathbf{y} \in \Omega$ and for $\lambda \in (0, 1)$, set $\mathbf{z} = \lambda\mathbf{x} + (1 - \lambda)\mathbf{y}$. Since $\Omega$ is assumed to be convex, $\mathbf{z} \in \Omega$. Let $\mathbf{z}_0 \in N(\mathbf{z})$ and denote by $T(\mathbf{z}_0)$, the tangent plane to $\partial\Omega$ at $\mathbf{z}_0$; $T(\mathbf{z}_0)$ is therefore orthogonal to the vector $\mathbf{z} - \mathbf{z}_0$. Let $\mathbf{x}_0, \mathbf{y}_0$, be the projections of $\mathbf{x}, \mathbf{y}$, respectively on $T(\mathbf{z}_0)$. It follows from the convexity of $\Omega$ and a simple similarity argument that

$$\delta(\mathbf{z}) = |\mathbf{z} - \mathbf{z}_0| = \lambda|\mathbf{x} - \mathbf{x}_0| + (1 - \lambda)|\mathbf{y} - \mathbf{y}_0| \geq \lambda\delta(\mathbf{x}) + (1 - \lambda)\delta(\mathbf{y})$$

which proves that $\delta$ is concave.

Let $f(\mathbf{x}) = -\delta(\mathbf{x}); f$ is therefore convex on $\Omega$. As $\varepsilon \to 0$, the regularisations $f_\varepsilon$ of $f$ defined by (2.3.4) converge uniformly to $f$ on any open subset $\Omega'$ of $\Omega$ which is such that $\overline{\Omega'} \subset \Omega$, and is convex since, for sufficiently small $\varepsilon$,

$$
\begin{aligned}
f_\varepsilon(\lambda\mathbf{x} + (1-\lambda)\mathbf{y}) &= \int_{B(0,1)} \rho(\mathbf{z}) f(\lambda[\mathbf{x} - \varepsilon\mathbf{z}] + (1-\lambda)[\mathbf{y} - \varepsilon\mathbf{z}]) d\mathbf{z}. \\
&\leq \lambda f_\varepsilon(\mathbf{x}) + (1-\lambda) f_\varepsilon(\mathbf{y}).
\end{aligned}
$$

Also $f_\varepsilon \in C^\infty(\Omega')$. It follows that $D^2 f_\varepsilon \geq 0$, where $D^2 f_\varepsilon = (\partial_i\partial_j f_\varepsilon)_{1\leq i,j\leq n}$ is the Hessian of $f_\varepsilon$. Thus $D^2 f_\varepsilon$ has non-negative eigenvalues and its trace $\Delta f_\varepsilon$ is non-negative. We therefore have for all $\varphi \in C_0^\infty(\Omega)$ with $\varphi \geq 0$,

$$
\int_\Omega f_\varepsilon(\mathbf{x})\Delta\varphi(\mathbf{x}) d\mathbf{x} = \int_\Omega \varphi(\mathbf{x})\Delta f_\varepsilon(\mathbf{x}) d\mathbf{x} \geq 0.
$$

On allowing $\varepsilon \to 0$, we conclude that

$$
L(\varphi) := \int_\Omega f(\mathbf{x})\Delta\varphi d\mathbf{x} \geq 0.
$$

The hypothesis of Corollary 2.3.1 is therefore satisfied and the theorem is proved. □

The following result is established in [35].

**Proposition 2.3.3** *For any proper open subset* $\Omega$ *of* $\mathbb{R}^n$,

$$
(n-1) - \delta\Delta\delta \geq 0 \tag{2.3.5}
$$

*in the distributional sense*

*Proof* The proof uses the fact that $A(\mathbf{x}) := |\mathbf{x}|^2 - \delta^2(\mathbf{x})$ defines a convex function on $\mathbb{R}^n$. To see this, let $\mathbf{x} \in \mathbb{R}^n$ and $\mathbf{y} \in \partial\Omega$ such that $\delta(\mathbf{x}) = |\mathbf{x} - \mathbf{y}|$. Then, for all $\mathbf{z} \in \mathbb{R}^n$,

$$
\begin{aligned}
&A(\mathbf{x} + \mathbf{z}) + A(\mathbf{x} - \mathbf{z}) - 2A(\mathbf{x}) \\
&= 2|\mathbf{z}|^2 - \left\{\delta^2(\mathbf{x} + \mathbf{z}) + \delta^2(\mathbf{x} - \mathbf{z}) - 2\delta^2(\mathbf{x})\right\} \\
&\geq 2|\mathbf{z}|^2 - \left\{|\mathbf{x} + \mathbf{z} - \mathbf{y}|^2 + |\mathbf{x} - \mathbf{z} - \mathbf{y}|^2 - 2|\mathbf{x} - \mathbf{y}|^2\right\} = 0.
\end{aligned}
$$

Since $A$ is continuous it is therefore convex. We therefore infer, as in the proof of the previous theorem, that for all $\varphi \in C_0^\infty(\Omega)$, $\varphi \geq 0$,

$$
\int_\Omega A(\mathbf{x})\Delta\varphi(\mathbf{x}) d\mathbf{x} \geq 0,
$$

and this gives

$$0 \leq \int_\Omega \{2n - 2\mathrm{div}(\delta\nabla\delta)\}\,\varphi(\mathbf{x})d\mathbf{x} = 2\int_\Omega (n-1-\delta\Delta\delta)\varphi(\mathbf{x})d\mathbf{x}$$

which yields the proposition. □

The above proposition is shown to have the following interesting consequence in [127], Lemma 2.2 and Theorem 2.7.

**Theorem 2.3.4** *Let $\Omega$ be a proper open subset of $\mathbb{R}^n$. Then for all $u \in C_0^\infty(\Omega)$ and $s \geq 1$,*

$$\int_\Omega \frac{|\nabla u|}{\delta^{s-1}}d\mathbf{x} \geq (s-1)\int_\Omega \frac{|u|}{\delta^s}d\mathbf{x} + \int_\Omega \frac{|u|}{\delta^{s-1}}(-\Delta\delta)d\mathbf{x} \tag{2.3.6}$$

*where $\Delta\delta$ is meant in the distributional sense. Therefore, if $-\Delta\delta \geq 0$,*

$$\int_\Omega \frac{|\nabla u|}{\delta^{s-1}}d\mathbf{x} \geq (s-1)\int_\Omega \frac{|u|}{\delta^s}d\mathbf{x}. \tag{2.3.7}$$

*The constant $(s-1)$ in (2.3.7) is sharp.*

*If $\Omega$ is bounded and $-\Delta\delta \geq 0$, all the constants in (2.3.6) are sharp and equality holds for $u_\varepsilon = \delta(\mathbf{x})^{s-1+\varepsilon}$, $\varepsilon > 0$; this function lies in the weighted Sobolev space $W_0^{1,1}(\Omega;\delta^{1-s})$, which is the completion of $C_0^\infty(\Omega)$ with respect to the norm $\int_\Omega(|\nabla u(\mathbf{x})| + |u(\mathbf{x})|)\delta(\mathbf{x})^{1-s}d\mathbf{x}$.*

*Proof* For any vector field $\mathbf{V}$ on $\Omega$, we have for all $u \in C_0^\infty(\Omega)$, on integration by parts,

$$\int_\Omega \mathrm{div}\mathbf{V}|u|d\mathbf{x} = -\int_\Omega \mathbf{V}\cdot\nabla|u|d\mathbf{x}$$

and hence since $|\nabla|u|| \leq |\nabla u|$ a.e. on $\Omega$,

$$\int_\Omega \mathrm{div}\mathbf{V}|u|d\mathbf{x} \leq \int_\Omega |\mathbf{V}||\nabla u|d\mathbf{x}.$$

The inequality (2.3.6) follows on choosing $\mathbf{V} = -\delta^{1-s}\nabla\delta$. To prove that the constant is sharp in (2.3.7) we pick $\mathbf{y} \in \partial\Omega$ to be the centre of a ball $B_\mu(\mathbf{y})$ of small radius $\mu$, and define the family of functions $u_\varepsilon(\mathbf{x}) := \varphi(\mathbf{x})(\delta(\mathbf{x}))^{s-1+\varepsilon}$, $\varepsilon > 0$, where

$\varphi \in C_0^\infty(B_\mu(\mathbf{y}))$, $0 \le \varphi \le 1$, and $\varphi = 1$ in $B_{\mu/2}(\mathbf{y})$. Then

$$I := \frac{\int_\Omega \frac{|\nabla u_\varepsilon|}{\delta^{s-1}} d\mathbf{x}}{\int_\Omega \frac{|u_\varepsilon|}{\delta^s} d\mathbf{x}} \le s - 1 + \varepsilon + \frac{\int_\Omega |\nabla\varphi| \delta^\varepsilon d\mathbf{x}}{\int_\Omega \varphi \delta^{-1+\varepsilon} d\mathbf{x}}$$
$$\le s - 1 + \varepsilon + \frac{C}{\int_{\Omega \cap B_{\mu/2}(\mathbf{y})} \delta^{-1+\varepsilon} d\mathbf{x}},$$

where $C$ is a constant independent of $\varepsilon$. Since $\delta(\mathbf{x}) \le |\mathbf{x} - \mathbf{y}|$, the integral in the last inequality diverges as $\varepsilon \to 0$. Hence $I \le s - 1 + o(1)$ as $\varepsilon \to 0$; it follows that $(s-1)$ is sharp. The fact that $u_\varepsilon$ gives equality is readily verified. □

Under the hypothesis of Theorem 2.3.4 but with $s \ge n$, it is proved in [10] that

$$\int_\Omega \frac{|\nabla u|}{\delta^{s-1}} d\mathbf{x} \ge (s - n) \int_\Omega \frac{|u|}{\delta^s} d\mathbf{x},$$

the proof involving a covering of $\Omega$ by cubes. An elementary proof of this inequality is given in [127], Theorem 2.3.

## 2.4 Domains with $C^2$ Boundaries

In the following lemma, it is assumed that $\Omega$ has a $C^2$ boundary. This means that locally, after a rotation of coordinates, $\partial\Omega$ is the graph of a $C^2$ function. To be specific (see also Sect. 1.2.4), for any $\mathbf{y} \in \partial\Omega$, let $\mathbf{n}(\mathbf{y})$, $T(\mathbf{y})$ denote respectively the unit inward normal to $\partial\Omega$ at $\mathbf{y}$ and the tangent plane to $\Omega$ at $\mathbf{y}$. The $\partial\Omega$ is said to be of class $C^2$ if, given any point $\mathbf{y}_0 \in \partial\Omega$ there exists a neighbourhood $\mathcal{N}(\mathbf{y}_0)$ in which $\partial\Omega$ is given in terms of local coordinates by $x_n = \varphi(x_1, x_2, \cdots, x_{n-1})$, $\varphi \in C^2(T(\mathbf{y}_0) \cap \mathcal{N}(\mathbf{y}_0))$, $x_n$ lies in the direction of $\mathbf{n}(\mathbf{y}_0)$ and with $\mathbf{x}' = (x_1, x_2, \cdots, x_{n-1})$, we have

$$\mathbf{D}\varphi(\mathbf{y}_0') \equiv (D_1, D_2, \cdots, D_{n-1})\varphi(\mathbf{y}_0')$$
$$\equiv [(\partial/\partial x_1, \partial/\partial x_2, \cdots, \partial/\partial x_{n-1})\varphi](\mathbf{y}_0') = 0.$$

The **principal curvatures** $\kappa_1, \cdots, \kappa_{n-1}$ of $\partial\Omega$ at $\mathbf{y}_0$ are the eigenvalues of the Hessian matrix

$$[\mathbf{D}^2\varphi(\mathbf{y}_0')] = (D_i D_j \varphi(\mathbf{y}_0'))_{i,j=1,\cdots,n-1}$$

and the corresponding eigenvectors are called the **principal directions**. By a rotation of coordinates, we can assume that the coordinate axes lie in the principal directions at $\mathbf{y}_0$. Then with respect to this coordinate system we have that $\mathbf{y} \in \mathcal{N}(\mathbf{y}_0)$

can be expressed as

$$\mathbf{y} = \boldsymbol{\gamma}(s_1, s_2 \cdots s_{n-1}), \boldsymbol{\gamma} = (\gamma_1, \cdots, \gamma_n, ), \tag{2.4.1}$$

where $\gamma_j \in C^2(T(\mathbf{y}_0) \cap \mathcal{N}(\mathbf{y}_0))$. At $\mathbf{y}_0$

$$\frac{\partial \boldsymbol{\gamma}}{\partial s_i} =: \mathbf{v}_i = (v_i^1, \cdots, v_i^n), \quad i = 1, 2, \cdots, n-1,$$

are unit vectors in the direction of the principal directions and we have

$$\begin{aligned} &\langle \mathbf{v}_i, \mathbf{v}_j \rangle = \delta_{i,j}, \quad \langle \mathbf{v}_i, \mathbf{n} \rangle = 0, \\ &\frac{\partial \mathbf{n}(\mathbf{s}')}{\partial s_i} = \kappa_i(\mathbf{s}') \frac{\partial \boldsymbol{\gamma}(\mathbf{s}')}{\partial s_i} = \kappa_i(\mathbf{s}') \mathbf{v}_i(\mathbf{s}'), \end{aligned} \tag{2.4.2}$$

where the angle brackets denotes scalar product and $\mathbf{n}(\mathbf{s}')$ is the unit inward normal vector at $\boldsymbol{\gamma}(s_1, s_2 \cdots s_{n-1})$. In (2.4.2), the signs of the principal curvatures are determined by the direction of the normal $\mathbf{n}$. If $\Omega$ is convex, the principal curvatures of $\partial\Omega$ are non-positive, while if the domain under consideration is $\overline{\Omega}^c = \mathbb{R}^n \setminus \overline{\Omega}$, the principal curvatures are non-negative.

**Lemma 2.4.1** *Let $\Omega$ be a domain in $\mathbb{R}^n$, $n \geq 2$, with a $C^2$ boundary. Let $\kappa_j(\mathbf{y})$, $j = 1, \ldots, n-1$, be the principal curvatures at $\mathbf{y} \in \partial\Omega$ with respect to the unit inward normal. Then for $\mathbf{x} \in G(\Omega) = \Omega \setminus \overline{\mathcal{R}(\Omega)}$ and $\mathbf{y} = N(\mathbf{x})$,*

$$1 + \delta(\mathbf{x})\kappa_j(\mathbf{y}) > 0, \quad j = 1, \ldots, n-1.$$

*Proof* If $\kappa_j(\mathbf{y}) \geq 0$, the inequality is trivial. Suppose $\kappa_j(\mathbf{y}) < 0$ for some $\mathbf{y} \in \partial\Omega$ which is the unique near point of $\mathbf{x} \in G(\Omega)$. Let $B_\delta(\mathbf{x})$ be the ball centered at $\mathbf{x}$ with radius $\delta(\mathbf{x})$ satisfying

$$\{\mathbf{y}\} = \overline{B_\delta(\mathbf{x})} \cap \mathbb{R}^n \setminus \Omega.$$

Recall that the principal radius $r_i$ is the radius of the osculating circle and for $\mathbf{y} \in \partial\Omega$, $r_i = 1/|\kappa_i(\mathbf{y})|$. Since $\partial\Omega$ is $C^2$, then $\delta(\mathbf{x}) \leq r_i$, for otherwise, $\overline{B_\delta(\mathbf{x})}$ would enclose the osculating circle and would intersect $\partial\Omega$ more than once, contradicting the fact that $\mathbf{y}$ is the unique nearest point of $\mathbf{x}$. Therefore,

$$1 + \delta(\mathbf{x})\kappa_i(\mathbf{y}) \geq 0 \tag{2.4.3}$$

for any $\mathbf{x} \in G(\Omega)$.

Since $\mathbf{x}$ is in the open set $G(\Omega)$, there is an open neighborhood $O_\varepsilon(\mathbf{x})$ centered at $\mathbf{x}$ with radius $\varepsilon > 0$ also contained in $G(\Omega)$. From the definition of a ridge in

Definition 2.2.4, there is a $\lambda$ such that for all $t \in (0, \lambda)$,

$$\{\mathbf{y}\} = N(\mathbf{y} + t[\mathbf{x} - \mathbf{y}]) = N(\mathbf{y} + t\delta(\mathbf{x})\mathbf{n})$$

where $\mathbf{n}$ is the unit inward normal at $\mathbf{y}$. Clearly, $\lambda \geq 1$ and, in fact, $\lambda > 1$ since otherwise $\mathbf{x} \notin G$. Consequently, there is an $s \in (0, \varepsilon)$ such that

$$\mathbf{x}_s := \mathbf{y} + (\delta(\mathbf{x}) + s)\mathbf{n} \in O_\varepsilon(\mathbf{x}) \subset G(\Omega)$$

and $\{\mathbf{y}\} = N(\mathbf{x}_s)$, i.e., $\delta(\mathbf{x}_s) = s + \delta(\mathbf{x})$. Since $\mathbf{x}_s \in G(\Omega)$, we may apply (2.4.3) to conclude that

$$1 + \delta(\mathbf{x})\kappa_i(\mathbf{y}) > 1 + (\delta(\mathbf{x}) + s)\kappa_i(\mathbf{y}) \geq 0$$

which completes the proof. □

**Lemma 2.4.2** *Let $\Omega$ be a domain in $\mathbb{R}^n$, $n \geq 2$, with $C^2$ boundary, and $\delta(\mathbf{x}) := \operatorname{dist}(\mathbf{x}, \partial\Omega)$. Then $\delta \in C^2(G(\Omega))$, $G(\Omega) = \Omega \setminus \overline{\mathcal{R}(\Omega)}$, and for $g(\mathbf{x}) = g(\delta(\mathbf{x}))$, $g \in C^2(\mathbb{R}^+)$,*

$$\Delta_{\mathbf{x}} g(\mathbf{x}) = \frac{\partial^2 g}{\partial \delta^2}(\mathbf{x}) + \sum_{i=1}^{n-1} \left( \frac{\kappa_i(\mathbf{y})}{1 + \delta(\mathbf{x})\kappa_i(\mathbf{y})} \right) \frac{\partial g}{\partial \delta}(\mathbf{x}), \quad \mathbf{x} \in \Omega \setminus \overline{\mathcal{R}(\Omega)}, \tag{2.4.4}$$

*where the $\kappa_i(\mathbf{y})$ are the principal curvatures of $\partial\Omega$ at the unique near point $\mathbf{y}$ of $\mathbf{x}$. In particular,*

$$\Delta_{\mathbf{x}} \delta(\mathbf{x}) = \tilde{\kappa}(\mathbf{y}) := \sum_{i=1}^{n-1} \left( \frac{\kappa_i(\mathbf{y})}{1 + \delta(\mathbf{x})\kappa_i(\mathbf{y})} \right), \quad \mathbf{x} \in \Omega \setminus \overline{\mathcal{R}(\Omega)}, \ \mathbf{y} = N(\mathbf{x}). \tag{2.4.5}$$

*Proof* For $\mathbf{x}_0 \in G(\Omega)$, let $\mathbf{y}_0 = N(\mathbf{x}_0)$, and consider the coordinate system in (2.4.1) for points in $\mathcal{N}(\mathbf{y}_0)$. We define a mapping $\Gamma$ from $\mathcal{U} = T(\mathbf{y}_0) \cap \mathcal{N}(\mathbf{y}_0) \times \mathbb{R}$ into $\mathbb{R}^n$ by

$$\mathbf{x} = \Gamma(\boldsymbol{\gamma}(\mathbf{s}'), s_n) = \boldsymbol{\gamma}(s') + s_n \mathbf{n}(s'), \tag{2.4.6}$$

where $\mathbf{x} \in G(\Omega)$ is such that $N(\mathbf{x}) = \mathbf{y} = \boldsymbol{\gamma}(s')$ and $s_n = \delta(\mathbf{x})$. Then, for $i = 1, 2, \cdots, n$,

$$\frac{\partial x_i}{\partial s_j} = \frac{\partial \gamma_i}{\partial s_j} + s_n \frac{\partial n^i}{\partial s_j}, \quad j = 1, 2, \cdots n-1, \quad \frac{\partial x_i}{\partial s_n} = n^i, \tag{2.4.7}$$

and so, by (2.4.2), at $\mathbf{y}_0 = \boldsymbol{\gamma}(\mathbf{s}_0')$,

$$\frac{\partial \mathbf{x}}{\partial s_j} = (1 + \delta\kappa_j)\mathbf{v}_j, \; j = 1, 2, \cdots, n-1, \quad \frac{\partial \mathbf{x}}{\partial s_n} = \mathbf{n}. \tag{2.4.8}$$

It follows that the Jacobian matrix of $\Gamma$ at $(\boldsymbol{\gamma}(\mathbf{s}_0'), \delta(\mathbf{x}_0))$, where $\mathbf{y}_0' = \boldsymbol{\gamma}(\mathbf{s}_0')$ and $\mathbf{y}_0 = \mathcal{N}(\mathbf{x}_0)$, is

$$\begin{pmatrix} (1+\delta\kappa_1)v_1^1 & \cdots & (1+\delta\kappa_{n-1})v_{n-1}^1 & n^1 \\ \vdots & & \vdots & \vdots \\ (1+\delta\kappa_1)v_1^n & \cdots & (1+\delta\kappa_{n-1})v_{n-1}^n & n^n \end{pmatrix} \tag{2.4.9}$$

This has the inverse

$$\begin{pmatrix} (1+\delta\kappa_1)^{-1}v_1^1 & \cdots & (1+\delta\kappa_1)^{-1}v_1^n \\ \vdots & & \vdots \\ (1+\delta\kappa_{n-1})^{-1}v_{n-1}^1 & \cdots & (1+\delta\kappa_{n-1})^{-1}v_{n-1}^n \\ n^1 & \cdots & n^n \end{pmatrix} \tag{2.4.10}$$

and hence the Jacobian (2.4.9) has a non-zero determinant. Since $\Gamma \in C^1(\mathcal{U})$, it follows from the inverse mapping theorem that for some neighbourhood $\mathcal{M} = \mathcal{M}(\mathbf{x}_0)$ of $\mathbf{x}_0$, the inverse map is in $C^1(\mathcal{M})$. From

$$\mathbf{x} = \mathbf{y} + \delta(\mathbf{x})\mathbf{n}(\mathbf{y}) = \boldsymbol{\gamma}(\mathbf{s}') + \delta(\mathbf{x})\mathbf{n}(\mathbf{s}') \tag{2.4.11}$$

we have that $\delta(\mathbf{x}) = (\mathbf{x} - \boldsymbol{\gamma}(s')) \cdot \mathbf{n}(s')$ and so $D\delta(\mathbf{x}) = \mathbf{n}(\mathbf{s}')$. Thus $\delta \in C^2(\mathcal{M})$ and consequently $\delta \in C^2(G(\Omega))$.

Since $(\partial\mathbf{s}/\partial\mathbf{x}) = (\partial\mathbf{x}/\partial\mathbf{s})^{-1}$, it follows that $(\partial\mathbf{s}/\partial\mathbf{x})$ at $\mathbf{y}_0$ is the matrix (2.4.10), and for $j = 1, 2, \cdots n$, $\; i = 1, 2, \cdots n-1$,, this yields at $\mathbf{y}_0$

$$\frac{\partial s_i}{\partial x_j} = [1 + \delta\kappa_i]^{-1} v_i^j, \quad \frac{\partial s_n}{\partial x_j} = n^j. \tag{2.4.12}$$

As $\delta(\mathbf{x}) = s_n$, we have on employing the usual summation convention,

$$\frac{\partial^2 \delta}{\partial x_j^2} = \frac{\partial n^j}{\partial s_i}\frac{\partial s_i}{\partial x_j},$$

which at $\mathbf{y}_0$ is

$$\sum_{i=1}^{n-1} [1 + \delta\kappa_i]^{-1} v_i^j \frac{\partial n^j}{\partial s_i} + \frac{\partial n^j}{\partial \delta} n^j.$$

Consequently, at $\mathbf{y}_0$,

$$\Delta\delta = \sum_{j=1}^{n}\left\{\sum_{i=1}^{n-1}[1+\delta\kappa_i]^{-1}v_i^j\frac{\partial n^j}{\partial s_i} + n^j\frac{\partial n^j}{\partial\delta}\right\}$$
$$= \sum_{i=1}^{n-1}[1+\delta\kappa_i]^{-1}\langle\mathbf{v}_i, \frac{\partial\mathbf{n}}{\partial s_i}\rangle + \langle\mathbf{n}, \frac{\partial\mathbf{n}}{\partial\delta}\rangle$$
$$= \sum_{i=1}^{n-1}\kappa_i[1+\delta\kappa_i]^{-1} \tag{2.4.13}$$

by (2.4.2). From the Chain Rule, we have

$$\frac{\partial g}{\partial x^j} = \frac{\partial g}{\partial s_i}\frac{\partial s_i}{\partial x_j} + \frac{\partial g}{\partial\delta}\frac{\partial\delta}{\partial x_j} = \frac{\partial g}{\partial\delta}\frac{\partial\delta}{\partial x_j}$$

and

$$\Delta_{\mathbf{x}}g = \frac{\partial}{\partial s_k}\left[\frac{\partial g}{\partial\delta}\frac{\partial\delta}{\partial x_j}\right]\frac{\partial s_k}{\partial x_j}$$
$$= \left[\frac{\partial^2 g}{\partial s_k\partial\delta}n^j + \frac{\partial g}{\partial\delta}\frac{\partial n^j}{\partial s_k}\right]\frac{\partial s_k}{\partial x_j}.$$

Hence at $\mathbf{y}_0$, we get from (2.4.12)

$$\Delta_{\mathbf{x}}g = \frac{\partial^2 g}{\partial\delta^2} + \frac{\partial g}{\partial\delta}\sum_{j=1}^{n}\sum_{k=1}^{n-1}\kappa_k v_k^j[1+\delta\kappa_k]^{-1}v_k^j$$
$$= \frac{\partial^2 g}{\partial\delta^2} + \frac{\partial g}{\partial\delta}\sum_{k=1}^{n-1}\kappa_k[1+\delta\kappa_k]^{-1}.$$

The lemma is therefore proved. □

**Corollary 2.4.3** *Let $\Omega$ be a convex domain in $\mathbb{R}^n, n \geq 2$, with a $C^2$-boundary. Then $\delta$ is superharmonic ( i.e. $\Delta\delta \leq 0$) in $G(\Omega) = \Omega \setminus \overline{\mathcal{R}(\Omega)}$ and subharmonic (i.e. $\Delta\delta \geq 0$) in $\mathbb{R}^n \setminus \overline{\Omega}$.*

*Proof* We noted in the paragraph following (2.4.2) that if $\Omega$ is convex, then the principal curvatures of $\partial\Omega$ are non-positive in $\Omega$ and those of the boundary of $\mathbb{R}^n \setminus \overline{\Omega}$ are non-negative in $\mathbb{R}^n \setminus \overline{\Omega}$. The corollary is therefore a consequence of Lemma 2.4.1, (2.4.5) and Lemma 2.2.8. It was first proved in [8] by a different method. □

*Remark 2.4.4* It follows from (2.4.12) that $[\kappa_i/(1+\delta\kappa_i)](\mathbf{y}), i = 1, 2, \cdots, n-1$, are the principal curvatures of the level surface of $\delta$ through $\mathbf{x}$ at $\mathbf{x}$.

*Remark 2.4.5* From Lemmas 2.4.1 and 2.4.2 it follows that the inequality (2.3.5) is strict in $G(\Omega)$.

*Remark 2.4.6* If $\Omega$ is a domain with a $C^2$ boundary, there exists $\varepsilon > 0$ such that $\Omega_\varepsilon := \{\mathbf{x} \in \Omega : \delta(\mathbf{x}) < \varepsilon\} \subset G(\Omega)$. For $\Omega$ satisfies a *uniform sphere condition*: this means that for each point $\mathbf{y}_0 \in \partial\Omega$, there exists a ball $B$, depending on $\mathbf{y}_0$, which is such that $\overline{B} \cap (\mathbb{R}^n \setminus \Omega) = \mathbf{y}_0$, and the radii of the balls $B$ are bounded from below by a positive constant, $\varepsilon$, say. Moreover, any $\mathbf{x} \in \Omega_\varepsilon$ has a unique near point $\mathbf{y}$ on $\partial\Omega$, with $\mathbf{x} = \mathbf{y} + \mathbf{n}(\mathbf{y})\delta(\mathbf{x})$. Hence, $\Omega_\varepsilon \subset G(\Omega)$.

## 2.5 Mean Curvature

The **mean curvature** of $\partial\Omega$ at $\mathbf{y}$ is defined to be

$$H(\mathbf{y}) := \frac{1}{n-1}\sum_{j=1}^{n-1}\kappa_j(\mathbf{y}), \qquad \mathbf{y} \in \partial\Omega,$$

where we adopt the convention that the standard unit sphere $\mathbb{S}^{n-1} \subset \mathbb{R}^n$ has mean curvature $-1$ everywhere. From Lemma 2.4.1, we have that

$$1 + \delta(\mathbf{x})H(\mathbf{y}) > 0, \quad \mathbf{y} = N(\mathbf{x}), \quad \mathbf{x} \in G(\Omega). \tag{2.5.1}$$

As noted in the paragraph after (2.4.2), if $\Omega$ is convex and has a $C^2$ boundary, then the principal curvatures satisfy $\kappa_i \leq 0$, $i = 1, 2, \cdots, n-1$. It is well-known (see [143], Chap. 13) that $\Omega$ is strictly convex if and only if $\kappa_j < 0$ for each $j = 1, 2, \cdots, n-1$. A weaker property is now introduced.

**Definition 2.5.1** A domain $\Omega \subset \mathbb{R}^n$ with a $C^2$ boundary $\partial\Omega$ is said to be **mean convex** (with respect to the inward normal) if

$$H(\mathbf{y}) < 0, \qquad \mathbf{y} \in \partial\Omega,$$

and **weakly mean convex** if

$$H(\mathbf{y}) \leq 0, \qquad \mathbf{y} \in \partial\Omega.$$

If $H \equiv 0$ on $\partial\Omega$ then $\partial\Omega$ is said to be a **minimal surface**.

*Example 2.5.2* Let $\Omega$ be the ring torus in $\mathbb{R}^3$ with minor radius $r$ and major radius $R \geq 2r$. This is the "doughnut-shaped" domain generated by rotating a disc of radius $r$ about a co-planar axis at a distance $R$ from the center of the disc. The ridge of $\Omega$ is clearly

$$\mathcal{R}(\Omega) = \{\mathbf{x} : \rho(\mathbf{x}) = 0\},$$

where $\rho(\mathbf{x})$ is the distance from the point $\mathbf{x}$ in $\Omega$ to the center of the cross-section and $\delta(\mathbf{x}) = r - \rho(\mathbf{x})$.

For $\mathbf{x} = (x_1, x_2, x_3) \in \Omega \setminus \mathcal{R}(\Omega)$, let $\mathbf{y} = (y_1, y_2, y_3) = N(\mathbf{x})$ have the parametric co-ordinates

$$\begin{aligned} y_1 &= (R + r\cos s^2)\cos s^1 \\ y_2 &= (R + r\cos s^2)\sin s^1 \ , \\ y_3 &= r\sin s^2 \end{aligned}$$

where $s^1, s^2 \in (-\pi, \pi]$. The principal curvatures at $\mathbf{y} \in \partial\Omega$ are

$$\kappa_1 = -\frac{1}{r}, \qquad \kappa_2 = -\frac{\cos s^2}{R + r\cos s^2},$$

e.g., see Kreyszig [90]. Hence

$$H(\mathbf{y}) = -\frac{R + 2r\cos s^2}{2r(R + r\cos s^2)} \leq -\frac{R - 2r}{2r(R - r)}. \tag{2.5.2}$$

Therefore $\Omega$ is mean convex if $R > 2r$ and weakly mean convex if $R = 2r$. This is a classic example of a domain which is mean (or weakly mean) convex, but not convex.

It follows from (2.4.5) that

$$\begin{aligned} \Delta\delta(\mathbf{x}) = \sum_{i=1}^{2}\left(\frac{\kappa_i}{1 + \delta\kappa_i}\right)(\mathbf{y}) &= -\frac{R + 2(r - \delta)\cos s^2}{(r - \delta)(R + (r - \delta)\cos s^2)} \\ &= -\frac{\sqrt{x_1^2 + x_2^2} + (r - \delta)\cos s^2}{(r - \delta)\sqrt{x_1^2 + x_2^2}} \leq 0, \end{aligned}$$

since $R + r\cos s^2 = \sqrt{x_1^2 + x_2^2} + \delta(\mathbf{x})\cos s^2$ and $R \geq 2r$. The fact that $\Delta\delta \leq 0$ was proved by Armitage and Kuran [8].

An interesting result is obtained in [107], Proposition 2.6, on the relationship between $\Delta\delta(\mathbf{x})$ in $\Omega$ and the mean curvature of the boundary, by the use of an inequality involving the *elementary symmetric functions* of a vector $\lambda \in \mathbb{R}^n$. Recall that the $k^{\text{th}}$ elementary symmetric function of the vector $\lambda \in \mathbb{R}^n$ is given by

$$\sigma_k(\lambda) := \sum_{1 \le i_1 < \cdots < i_k \le n} \lambda_{i_1} \cdots \lambda_{i_k}, \qquad \lambda = (\lambda_1, \ldots, \lambda_n) \in \mathbb{R}^n,$$

for $k = 0, 1, \ldots, n$, with $\sigma_0 = 1$, $\sigma_1(\lambda) = \lambda_1 + \cdots + \lambda_n$, and $\sigma_n(\lambda) = \lambda_1 \cdots \lambda_n$. The *elementary symmetric means* are defined as

$$M_k(\lambda) := \sigma_k(\lambda) \Big/ \binom{n}{k}, \qquad k = 0, 1, \ldots, n.$$

The following inequality is due to Newton [123] and MacLaurin [114]. If $\lambda = (\lambda_1, \ldots, \lambda_n)$ with each $\lambda_i \ge 0$, then

$$M_k(\lambda)^2 \ge M_{k-1}(\lambda) M_{k+1}(\lambda), \qquad 1 \le k \le n-1, \tag{2.5.3}$$

with equality if, and only if $\lambda_1 = \cdots = \lambda_n$, and consequently, for $\lambda_k > 0$, $k = 1, \ldots, n$,

$$\frac{\sigma_{n-1}(\lambda)}{\sigma_n(\lambda)} \ge \cdots \ge c(n,k) \frac{\sigma_{k-1}(\lambda)}{\sigma_k(\lambda)} \ge \cdots \ge n^2 \frac{1}{\sigma_1(\lambda)} \tag{2.5.4}$$

where $c(n,k) := \frac{n(n-k+1)}{k}$. The equalities hold if, and only if $\lambda_1 = \cdots = \lambda_n$.

**Proposition 2.5.3** *Let $\Omega$ be a domain in $\mathbb{R}^n$, $n \ge 2$, with a $C^2$ boundary. Let $\kappa_j(\mathbf{y})$, $j = 1, \ldots, n-1$, be the principal curvatures at $\mathbf{y} \in \partial\Omega$ with respect to the unit inward normal, and let $H(\mathbf{y})$ be the mean curvature at $\mathbf{y}$. Then for all $\mathbf{x} \in G(\Omega)$ and $\mathbf{y} = N(\mathbf{x}) \in \partial\Omega$,*

$$\Delta\delta(\mathbf{x}) = \tilde{\kappa}(\mathbf{y}) := \sum_{i=1}^{n-1} \left( \frac{\kappa_i(\mathbf{y})}{1 + \delta(\mathbf{x})\kappa_i(\mathbf{y})} \right) \le \frac{(n-1)H(\mathbf{y})}{1 + \delta(\mathbf{x})H(\mathbf{y})}, \tag{2.5.5}$$

*where $1 + \delta(\mathbf{x})H(\mathbf{y}) > 0$ by (2.5.1). Equality holds if, and only if $\kappa_1 = \cdots = \kappa_{n-1}$.*

*Proof* Let

$$\lambda_i := 1 + \delta(\mathbf{x})\kappa_i(\mathbf{y}), \qquad i = 1, \ldots, n-1,$$

which is positive-valued according to Lemma 2.4.1. Then

$$\delta(\mathbf{x})\tilde{\kappa}(\mathbf{y}) := \sum_{i=1}^{n-1} \frac{\delta(\mathbf{x})\kappa_i(\mathbf{y})}{1 + \delta(\mathbf{x})\kappa_i(\mathbf{y})} = \sum_{i=1}^{n-1} \frac{\lambda_i - 1}{\lambda_i} = n - 1 - \sum_{i=1}^{n-1} \frac{1}{\lambda_i}.$$

But, for $\lambda = (\lambda_1, \ldots, \lambda_{n-1}) \in \mathbb{R}^{n-1}$,

$$\frac{\sigma_{n-2}(\lambda)}{\sigma_{n-1}(\lambda)} = \frac{\sum_{1 \le i_1 < \cdots < i_{n-2} \le n-1} \lambda_{i_1} \cdots \lambda_{i_{n-2}}}{\lambda_1 \cdots \lambda_{n-1}} = \sum_{i=1}^{n-1} \frac{1}{\lambda_i}$$

and

$$\frac{\sigma_{n-2}(\lambda)}{\sigma_{n-1}(\lambda)} \ge (n-1)^2 \frac{1}{\sigma_1(\lambda)}$$

with $\sigma_1(\lambda) = \sum \lambda_i$. Therefore

$$n - 1 - \delta(\mathbf{x})\tilde{\kappa}(\mathbf{y}) = \sum_{i=1}^{n-1} \frac{1}{\lambda_i} \ge (n-1)^2 \frac{1}{\sum_{i=1}^{n-1} \lambda_i},$$

and so,

$$\delta(\mathbf{x})\tilde{\kappa}(\mathbf{y}) \le (n-1) \frac{\delta(\mathbf{x}) H(\mathbf{y})}{1 + \delta(\mathbf{x}) H(\mathbf{y})}.$$

□

As an immediate consequence of Lemma 2.4.1, (2.4.5), and (2.5.5) we now know that if $\partial\Omega$ is $C^2$ and weakly mean convex, the distance function $\delta(\mathbf{x})$ is superharmonic in $G(\Omega) = \Omega \setminus \overline{\mathcal{R}(\Omega)}$. In fact, there is an equivalence here which was proved in [107].

**Proposition 2.5.4** *Let $\Omega$ have a $C^2$ boundary. The distance function $\delta(\mathbf{x})$ is superharmonic in $G(\Omega) = \Omega \setminus \overline{\mathcal{R}(\Omega)}$ if and only if $\Omega$ is weakly mean convex. Moreover,*

$$\sup_{\mathbf{x} \in G(\Omega)} [\Delta\delta(\mathbf{x})] = \sup_{\mathbf{y} \in \partial\Omega} (n-1) H(\mathbf{y}), \qquad \mathbf{y} = N(\mathbf{x}). \tag{2.5.6}$$

*Proof* Noting the representation of $\Delta\delta$ given in (2.4.5), we observe that when viewed as a function of $\delta$ only, $\sum \frac{\kappa_i}{1+\delta\kappa_i}$ decreases as $\delta$ increases irrespective of the sign of $\kappa_i$. Therefore, for each $\mathbf{x} \in G(\Omega)$ and $\mathbf{y} = N(\mathbf{x})$

$$\Delta\delta(\mathbf{x}) = \sum_{i=1}^{n-1} \frac{\kappa_i}{1 + \delta\kappa_i} \le \sum_{i=1}^{n-1} \kappa_i = (n-1) H(\mathbf{y}),$$

i.e.,

$$\sup_{\mathbf{x} \in G(\Omega)} [\Delta\delta(\mathbf{x})] \le (n-1) \sup_{\mathbf{y} \in \partial\Omega} [H(\mathbf{y})], \qquad \mathbf{y} = N(\mathbf{x}).$$

Conversely, for $\mathbf{n}(\mathbf{y})$ the unit inward normal at $\mathbf{y} \in \partial\Omega$ and $\mathbf{y} = N(\mathbf{x})$, define

$$\mathbf{x}_t := \mathbf{y} + t\mathbf{n}(\mathbf{y}).$$

Then for all $t > 0$ sufficiently small, $\mathbf{x}_t \in G(\Omega)$, implying that

$$\sup_{\mathbf{x}\in G(\Omega)} [\Delta\delta(\mathbf{x})] \geq \Delta\delta(\mathbf{x}_t) = \sum_{i=1}^{n-1} \frac{\kappa_i}{1 + \delta(\mathbf{x}_t)\kappa_i}.$$

Since $\delta(\mathbf{x}_t) \to 0$ as $t \to 0$, we may conclude that

$$\sup_{\mathbf{x}\in G(\Omega)} [\Delta\delta(\mathbf{x})] \geq \lim_{t\to 0}[\Delta\delta(\mathbf{x}_t)] = \sum_{i=1}^{n-1} \kappa_i = (n-1)H(\mathbf{y}).$$

Thus (2.5.6) holds, which implies the equality in the proposition. The implication of the penultimate sentence in the proposition follows from (2.5.5) □

Finally, we show that for bounded $C^2$ domains $\Omega$, the continuity of the mean curvature $H(\mathbf{y})$ on $\partial\Omega$ is inherited from the continuity of $\Delta\delta$ in $G(\Omega)$.

**Proposition 2.5.5** *If $\partial\Omega \in C^2$ and $\Omega$ is bounded, then the mean curvature $H(\mathbf{y})$ is continuous on $\partial\Omega$.*

*Proof* Let $\mathbf{y},\ \mathbf{y}_0 \in \partial\Omega$ and define

$$\mathbf{x}_0(t) := \mathbf{y}_0 + t\mathbf{n}(\mathbf{y}_0), \qquad \mathbf{x}(t) := \mathbf{y} + t\mathbf{n}(\mathbf{y}),$$

where $\mathbf{n}(\mathbf{y})$ is the unit inward normal at $\mathbf{y} \in \partial\Omega$. For $t$ sufficiently small in order that $t \leq t_0 < \mu$, $\mathbf{x}(t)$ and $\mathbf{x}_0(t)$ are in

$$\Gamma_\mu := \{\mathbf{x} \in \Omega : \delta(\mathbf{x}) < \mu\}$$

and we have that

$$\mathbf{x}(t) - \mathbf{x}_0(t) = \mathbf{y} - \mathbf{y}_0 + t[\mathbf{n}(\mathbf{y}) - \mathbf{n}(\mathbf{y}_0)].$$

Since $\mathbf{n}(\cdot)$ is continuous on $\partial\Omega$ it follows that

$$\mathbf{y} \to \mathbf{y}_0 \quad \Longrightarrow \quad \mathbf{x}(t) \to \mathbf{x}_0(t)$$

uniformly for $t \leq t_0$. By the continuity of $\Delta\delta(\cdot)$ in $G(\Omega)$ established in Lemma 2.4.2 (see also Lemma 14.16 of [68]) it follows that

$$\lim_{\mathbf{y}\to\mathbf{y}_0} \Delta\delta(\mathbf{x}(t)) = \Delta\delta(\mathbf{x}_0(t))$$

uniformly in $t \le t_0$. Hence,

$$\lim_{\mathbf{y}\to\mathbf{y}_0} \{\lim_{t\to 0} \Delta\delta(\mathbf{x}(t))\} = \lim_{t\to 0} \Delta\delta(\mathbf{x}_0(t)),$$

implying that

$$\lim_{\mathbf{y}\to\mathbf{y}_0} H(\mathbf{y}) = H(\mathbf{y}_0).$$

□

## 2.6 Integrability of $\delta^{-m}$

To end this chapter, we establish a connection between a regularity condition on the boundary of a bounded domain $\Omega$ and the rate of decay of $\delta(\mathbf{x})$ as $\mathbf{x}$ tends to the boundary. The regularity condition will be expressed in terms of the inner Minkowski dimension of $\partial\Omega$, while the decay rate of $\delta$ will be measured in terms of the integrability of negative powers of $\delta$. To define the inner Minkowski dimension of $\partial\Omega$, we set

$$\begin{aligned} M^{\lambda}_{\Omega}(\partial\Omega, r) &:= r^{-(n-\lambda)}|(\partial\Omega + B_r(0)) \cap \Omega|, \\ M^{\lambda}_{\Omega}(\partial\Omega) &= \limsup_{r\to 0+} M^{\lambda}_{\Omega}(\partial\Omega, r) \end{aligned}$$

and

$$\dim_{M,\Omega}(\partial\Omega) := \inf\{\lambda : M^{\lambda}_{\Omega}(\partial\Omega) < \infty\}$$

where $\partial\Omega + B_r(0)$ is the set of balls of radius r and centred at a point on the boundary of $\Omega$. The **inner Minkowski dimension** of $\Omega$ is the quantity $\dim_{M,\Omega}(\partial\Omega)$. The corresponding quantities obtained by replacing $|(\partial\Omega + B_r(0)) \cap \Omega|$ by $|(\partial\Omega + B_r(0))|$ are denoted by $M^{\lambda}(\partial\Omega, r), M^{\lambda}(\partial\Omega)$, and $\dim_M(\partial\Omega)$, the last of these being the **Minkowski dimension** of $\partial\Omega$. To establish the connection with the decay of $\delta$ at the boundary, we use the notion of a **Whitney covering** $\mathcal{W}$ of a bounded domain $\Omega$. This is a family of closed cubes $Q$, each with sides parallel to the co-ordinate axes and with side length $\ell_Q = 2^{-k}$ and diameter $d_Q = 2^{-k}\sqrt{n}$ for some $k \in \mathbb{N}$, such that

(1) $\Omega = \bigcup_{Q\in\mathcal{W}} Q$;
(2) the interiors of distinct cubes are disjoint;
(3) for all $Q \in \mathcal{W}$, $1 \le \operatorname{dist}(Q, \partial\Omega)/d_Q \le 4$.

Such a covering is known to exist; see [139], p. 16, Theorem 3. Condition (2) implies that

$$\ell_Q\sqrt{n} \leq \delta(\mathbf{x}) \leq 5\ell_Q\sqrt{n}.$$

Set

$$\mathcal{W}_k := \{Q \in \mathcal{W} : \ell_Q = 2^{-k}\}, \quad k \in \mathbb{N},$$

and let $n(k)$ be the number of cubes in $\mathcal{W}_k$. The way in which $\mathcal{W}$ is constructed then implies that

$$Q \subset \{\mathbf{x} \in \Omega : \sqrt{n}2^{-k} \leq \delta(\mathbf{x}) \leq 5\sqrt{n}2^{-k}\}. \tag{2.6.1}$$

The following result is proved in [32]; see also [49], Lemma 4.3.7.

**Lemma 2.6.1** *Let $\Omega$ be a bounded domain in $\mathbb{R}^n$ and let $0 < \lambda \leq n$. Then $M_\Omega^\lambda(\partial\Omega) < \infty$ if and only if there are positive constants $K$ and $k_0$ such that $n(k) \leq 2^{\lambda k}K$ for all $k \geq k_0,\ k \in \mathbb{N}$.*

*Proof* First suppose that $M_\Omega^\lambda(\partial\Omega) < \infty$. Then there exist $K,\ r_0 > 0$ such that

$$|(\partial\Omega + B_r(0)) \cap \Omega| \leq Kr^{n-\lambda}$$

for all $r \in (0, r_0]$. Take $k \in \mathbb{N},\ k \geq (\log 2)^{-1}\log(12\sqrt{n}/r_0)$ and set $r = 6\sqrt{n}2^{-k}$. Then $2r \leq r_0$. By a standard covering theorem (see, for example, Theorem XI.5.3 in [48]), there are points $x_1, x_2, \cdots, x_m \in \partial\Omega$ and a positive constant $C$, depending only on $n$, such that

$$\partial\Omega \subset \bigcup_{j=1}^m B_r(\mathbf{x}_j), \quad \sum_{j=1}^m \chi_{B_r(\mathbf{x}_j)} \leq C.$$

Every cube $Q \in \mathcal{W}$ is contained in at least one of the balls $B_{2r}(\mathbf{x}_j), j = 1, 2, \cdots, m$. For given $\mathbf{x} \in Q$, choose $\mathbf{y} \in \partial\Omega$ so that $\delta(\mathbf{x}) = |\mathbf{x} - \mathbf{y}| : \mathbf{y} \in B_r(\mathbf{x}_j)$ for some $j \in \{1, 2, \cdots, m\}$. Thus, for every $\mathbf{z} \in Q$ we have

$$|\mathbf{z} - \mathbf{x}_j| \leq |\mathbf{z} - \mathbf{x}| + |\mathbf{x} - \mathbf{y}| + |\mathbf{y} - \mathbf{x}_j| = 12\sqrt{n}2^{-k} = 2r.$$

Let $n_j(k)$ be the number of cubes $Q \in \mathcal{W}_k$ which are contained in $B_{2r}(\mathbf{x}_j)$. Then clearly

$$\begin{aligned} n(k) &\leq \sum_{j=1}^k n_j(k) \leq \sum_{j=1}^k |B_{2r}(\mathbf{x}_j) \cap \Omega)|/|\Omega| \\ &\leq C2^{nk}|(\partial\Omega + B_{2r}(0)) \cap \Omega| \\ &\leq CK(12\sqrt{n})^{n-\lambda}2^{\lambda k}. \end{aligned}$$

Conversely, suppose that $n(k) \le K2^{\lambda k}$ for all $k \ge k_0,\ k \in \mathbb{N}$. We may suppose that $\lambda < n$ since

$$\limsup_{r \to 0} |(\partial\Omega + B_r(0)) \cap \Omega| = |\partial\Omega| < \infty.$$

Fix $r > 0$ with $r \le \sqrt{n}2^{-k_0}$ and choose $k' \ge k_0$ such that

$$\sqrt{n}2^{-k'-1} \le r < \sqrt{n}2^{-k'}.$$

Then by (2.6.1),

$$(\partial\Omega + B_r(0)) \cap \Omega \subset \bigcup_{k \ge k'} \mathcal{W}_k.$$

Hence

$$|(\partial\Omega + B_r(0)) \cap \Omega| \le \sum_{k=k'}^{\infty} K2^{k(\lambda-n)} = \frac{K2^{k'(\lambda-n)}}{1 - 2^{\lambda-n}}.$$

Thus

$$\begin{aligned} r^{-(n-\lambda)}|(\partial\Omega + B_r(0)) \cap \Omega| &\le \frac{K2^{k'(\lambda-n)}}{(1-2^{\lambda-n})(\sqrt{n}2^{-k'-1})^{n-\lambda}} \\ &\le \frac{K2^{n-\lambda}}{(1-2^{\lambda-n})n^{(n-\lambda)/2}} \end{aligned}$$

and the result follows. □

The connection between the Minkowski dimension and the distance function that we were after, can now be given.

**Theorem 2.6.2** *Let $\Omega$ be a bounded domain in $\mathbb{R}^n$. Then the following conditions are equivalent:*

(1) $\dim_{M,\Omega}(\partial\Omega) < n$;
(2) *there exists* $\mu \in (0, n)$ *such that* $\int_\Omega \delta(\mathbf{x})^{-\mu} d\mathbf{x} < \infty$.

*Proof* Suppose that (1) holds. Let $\mathcal{W}$ be a Whitney covering of $\Omega$ and put $\lambda = \dim_{M,\Omega}(\partial\Omega)$. Then if $\mu > 0$,

$$\begin{aligned} \int_\Omega \delta(\mathbf{x})^{-\mu} d\mathbf{x} &= \sum_{Q \in \mathcal{W}} \int_Q \delta(\mathbf{x})^{-\mu} d\mathbf{x} = \sum_{k=1}^{\infty} \sum_{Q \in \mathcal{W}_k} \int_Q \delta(\mathbf{x})^{-\mu} d\mathbf{x} \\ &\le C \sum_{k=1}^{\infty} n(k) 2^{-kn} (2^{-k})^{-\mu} \le C \sum_{k=1}^{\infty} 2^{(\lambda-n+\mu)k}, \end{aligned}$$

for some positive constant $C$. Since $\lambda < n$, the last sum is finite for a suitable $\mu < n - \lambda$.

Conversely, suppose that (2) holds and that $\dim_{M,\Omega}(\partial\Omega) = n$. Then by Lemma 2.6.1, no matter what $K > 0$ and $\lambda \in (0, n)$ are chosen, $n(k) > K2^{\lambda k}$ for all $k$. Thus if we take $\lambda = n - \mu$, there is a sequence of natural numbers $k_j = k_j(\lambda)$ such that $n(k_j) > 2^{\lambda k_j}$. Then

$$\begin{aligned}\int_{\Omega} \delta(\mathbf{x})^{-\mu} d\mathbf{x} &= \sum_{k=1}^{\infty} \sum_{Q \in \mathcal{W}_k} \int_{Q} \delta(\mathbf{x})^{-\mu} d\mathbf{x} \geq C \sum_{k=1}^{\infty} n(k)(2^{-k})^{-\mu} |Q| \\ &\geq C \sum_{j=1}^{\infty} n(k_j)(2^{-k_j})^{-\mu} 2^{-k_j n} > \sum_{j=1}^{\infty} 2^{k_j(\lambda+\mu-n)} = \infty.\end{aligned}$$

This contradiction completes the proof. □

# Chapter 3
# Hardy's Inequality on Domains

## 3.1 Introduction

Let $\Omega$ be a domain (an open, connected set) in $\mathbb{R}^n$ with non-empty boundary, $1 < p < \infty$, and denote by $\delta(\mathbf{x})$ the distance from a point $\mathbf{x} \in \Omega$ to the boundary $\partial\Omega$ of $\Omega$, i.e.,

$$\delta(\mathbf{x}) := \inf\{|\mathbf{x} - \mathbf{y}| : \mathbf{y} \in \mathbb{R}^n \setminus \Omega\}.$$

The basic inequality to be considered in this chapter is

$$\int_\Omega |\nabla f(\mathbf{x})|^p d\mathbf{x} \geq c(n, p, \Omega) \int_\Omega \frac{|f(\mathbf{x})|^p}{\delta(\mathbf{x})^p} d\mathbf{x}, \quad f \in C_0^\infty(\Omega); \tag{3.1.1}$$

equivalently, the inequality is to hold for all $f \in W_0^{1,p}(\Omega)$. We shall say that the inequality is *valid* if there is a positive constant $c(n, p, \Omega)$ which, as indicated, may be dependent on $n, p$ and $\Omega$, but not on $f$.

It was proved by Lewis in [104] that if $n < p < \infty$, the inequality (3.1.1) holds for all proper open subsets of $\mathbb{R}^n$. For $1 < p \leq n$, the situation is more complicated, as assumptions on the boundary of $\Omega$ are necessary. A wide assortment of boundary conditions which are sufficient to ensure a valid inequality may be found in the literature, and many authors have contributed.

## 3.2 Boundary Smoothness

In this section we give brief descriptions of what we regard to be some of the highlights of the results concerning boundary conditions and the validity of the Hardy inequality; a comprehensive up-to-date coverage may be found in [146].



A.A. Balinsky et al., *The Analysis and Geometry of Hardy's Inequality*, Universitext, DOI 10.1007/978-3-319-22870-9_3

**(i)** In [119], Maz'ya showed that (3.1.1) can be characterised in terms of the **p-capacity** $C_p(K, \Omega)$ of compact subsets $K \subset \Omega$ relative to $\Omega$, defined by

$$C_p(K, \Omega) = \inf \left\{ \int_\Omega |\nabla u(\mathbf{x})|^p d\mathbf{x} : u \in C_0^\infty(\Omega),\ u(\mathbf{x}) \geq 1 \text{ for all } \mathbf{x} \in K \right\}. \tag{3.2.1}$$

He proved that (3.1.1) is valid if and only if there exists an absolute constant $C > 0$ such that

$$\int_K \frac{1}{\delta(\mathbf{x})^p} d\mathbf{x} \leq C \cdot C_p(K, \Omega); \tag{3.2.2}$$

a simplified proof is given by Kinnunen and Korte in [86].

**(ii)** The relative $p$-capacity $C_p(K, \Omega)$ also features in a theorem of Lewis in [104] involving the notion of uniform $p$-fatness: a closed set $E \subset \mathbb{R}^n$ is said to be **uniformly** $p$**-fat** if there is a constant $\gamma > 0$ such that for all $\mathbf{x} \in \Omega$ and all $r > 0$,

$$C_p\left(E \cap \overline{B(\mathbf{x}, r)}, B(\mathbf{x}, 2r)\right) \geq \gamma C_p\left(\overline{B(\mathbf{x}, r)}, B(\mathbf{x}, 2r)\right) = \gamma c(n, p) r^{n-p}$$

for some positive constant $c(n, p)$ dependent only on $n$ and $p$. The following examples may help to put this definition in perspective:

(1) If $n < p < \infty$, every non-empty closed set is uniformly $p$-fat.

(2) Every closed set satisfying the interior cone condition is uniformly $p$-fat for every $p \in (1, \infty)$ : *a closed set $E \subset \mathbb{R}^n$ satisfies the interior cone condition if there exists a cone $V$ such that every $\mathbf{x} \in E$ is the vertex of a cone $V_\mathbf{x} \subset E$ which is congruent to $V$, i.e., $V_\mathbf{x} = \mathbf{x} + L_\mathbf{x}(V)$, where $L_\mathbf{x}$ is a rotation operator.*

(3) The complement of a Lipschitz domain is uniformly $p$-fat for every $p \in (1, \infty)$: $\Omega$ is a **Lipschitz domain** if it is a rotation of a set of the form

$$\{\mathbf{x} = (\mathbf{x}', x_n) = (x_1, \cdots, x_{n-1}, x_n) \in \mathbb{R}^n : x_n > \phi(\mathbf{x}')\},$$

where $\phi : \mathbb{R}^{n-1} \to \mathbb{R}$ is Lipschitz continuous.

(4) If there is a constant $\gamma > 0$ such that for all $\mathbf{x} \in E$ and all $r > 0$,

$$|E \cap B(\mathbf{x}, r)| \geq \gamma |B(\mathbf{x}, r)|,$$

then $E$ is uniformly $p$-fat for every $p \in (1, \infty)$.

The aforementioned result of Lewis in [104] is the following:

**Theorem 3.2.1** *If $n < p < \infty$, the inequality (3.1.1) holds for all open sets $\Omega \neq \mathbb{R}^n$. If $1 < p \leq n$ and $\Omega$ is an open set in $\mathbb{R}^n$ which is such that*

*$\mathbb{R}^n \setminus \Omega$ is uniformly p-fat, then the Hardy inequality (3.1.1) is valid. When $p = n$, (3.1.1) holds if and only if $\mathbb{R}^n \setminus \Omega$ is uniformly p-fat.*

In particular then, for $1 < p \leq n$, (3.1.1) is valid for a Lipschitz domain.

**(iii)** The uniform $p$-fatness of $\mathbb{R}^n \setminus \Omega$ was proved in [100, 101] to be equivalent to the **pointwise $q$-Hardy inequality** for some $q \in (1, p)$; this notion was introduced by Hajłasz in [71] and is that there exists a positive constant $c(n, q)$, depending only on $n$ and $q$ such that for all $f \in C_0^\infty(\Omega)$ (extended to all of $\mathbb{R}^n$ by 0),

$$\frac{|f(\mathbf{x})|}{\delta(\mathbf{x})} \leq c(n, q) \left[\mathbb{M}(|\nabla f|^q(\mathbf{x}))\right]^{1/q}, \tag{3.2.3}$$

where $\mathbb{M}f$ is the maximal function defined for $f \in L^1_{loc}(\mathbb{R}^n)$ by

$$\mathbb{M}f(\mathbf{x}) := \sup_{r>0} \frac{1}{|B(\mathbf{x}, r)|} \int_{B(\mathbf{x},r)} |f(\mathbf{y})| d\mathbf{y}.$$

That the pointwise $q$ Hardy inequality for some $q \in (1, p)$ implies the Hardy inequality (3.1.1) is a consequence of the classical result that $\mathbb{M}$ is a bounded map from $L^{p/q}(\mathbb{R}^n)$ into $L^{p/q}(\mathbb{R}^n)$ since $p > q$.

**(iv)** In view of the connection established in Theorem 2.6.2 between integrability properties of $\delta(\cdot)$ and the Minkowski dimension of $\partial\Omega$, the "dimension" of $\partial\Omega$ and the Hardy inequality can be expected to be intimately related. This is known to be the case if the Hausdorff and Aikawa notions of dimension are used. We recall the definitions. For a set $E$ in $\mathbb{R}^n$ and $\lambda \geq 0$, the **$\lambda$-Hausdorff content** of $E$ is defined to be

$$\mathcal{H}^\lambda(E) := \inf\{\sum_{j=1}^\infty r_j^\lambda : E \subseteq \bigcup_{j=1}^\infty B(\mathbf{x}_j, r_j), \mathbf{x}_j \in E, r_j > 0\},$$

where $B(\mathbf{x}_j, r_j)$ is the ball centre $\mathbf{x}_j$ and radius $r_j$. It is readily shown that there is a unique $\lambda_0 \in [0, n]$ such that

$$\mathcal{H}^\lambda(E) = \begin{cases} \infty, & \text{if } \lambda < \lambda_0, \\ 0, & \text{if } \lambda > \lambda_0. \end{cases}$$

The **Hausdorff dimension** of $E$ is then defined by

$$\dim_{\mathcal{H}}(E) := \sup\{\lambda \geq 0 : \mathcal{H}^\lambda(E) = \infty\} = \inf\{\lambda \geq 0 : \mathcal{H}^\lambda = 0\};$$

see [57] for details. The **Aikawa dimension** of a set $E \subseteq \mathbb{R}^n$ is defined as

$$\dim_A(E) := \inf\left\{t > 0 : \int_{B(x,r)} \frac{1}{\delta(\mathbf{y}, E)^{n-t}} d\mathbf{y} < C_t r^t, \ \mathbf{x} \in E, \ r > 0\right\},$$

where $\delta(\mathbf{y}, E)$ is the distance from $\mathbf{y}$ to $E$ and $C_t$ is a constant which may depend on $t$.

In [100] Lehrbäck proved that if $\mathbb{R}^n \setminus \Omega$ is uniformly $p$-fat, then the Hardy inequality (3.1.1) is valid and $\dim_{\mathcal{H}}(\partial\Omega) > n - p$. This is complemented in [101], where Lehrbäck proves that if $\dim_A(\partial\Omega) < n-p$, then (3.1.1) is valid. It is proved in [87, 101] that if (3.1.1) is valid, there exists $\varepsilon = \varepsilon(p, n)$ such that either $\dim_{\mathcal{H}}(\partial\Omega) \geq n - p + \varepsilon$ or $\dim_A(\partial\Omega) \leq n - p - \varepsilon$. In fact this result is local in nature, in the sense that, if $\omega \in \partial\Omega$ and $r > 0$, then either $\dim_{\mathcal{H}}(\partial\Omega \cap B(\omega, r)) > n-p$ or $\dim_A(\partial\Omega \cap B(\omega, r)) < n-p$ whenever (3.1.1) is valid. This means that if $\Omega$ is a punctured disc in $\mathbb{R}^2$ and $p = 2$, there is no valid Hardy inequality.

For any set $E$ it is known that $\dim_{\mathcal{H}}(E) \leq \dim_A(E)$, and indeed, if $E$ is bounded, the Hausdorff, Minkowski and Aikawa dimensions are related by

$$\dim_{\mathcal{H}}(E) \leq \dim_M(E) \leq \dim_A(E);$$

see [57, 102, 113] .

**(v)** A natural question following from Hardy's inequality is if $u \in W^{1,p}(\Omega)$ and $u/\delta \in L^p(\Omega)$ imply that $u \in W_0^{1,p}(\Omega)$. That this is indeed the case is proved in [48], p. 223; furthermore, in [116], Lemma 1.1, it is proved that if $\Omega$ is a bounded domain with a $C^2$ boundary, then the space

$$\tilde{W}^{1,p}(\Omega) := \{u \in W_{loc}^{1,p}(\Omega) : \|\nabla u\|_{L^p(\Omega)} + \|\frac{u}{\delta}\|_{L^p(\Omega)} < \infty\}$$

with norm

$$\|u\|_{\tilde{W}^{1,p}(\Omega)} := \|\nabla u\|_{L^p(\Omega)} + \|\frac{u}{\delta}\|_{L^p(\Omega)}$$

is equivalent to $W_0^{1,p}(\Omega)$ with norm $\|\nabla u\|_{L^p(\Omega)}$. An extension of the quoted result from [48] was obtained by Kinnunen and Martio who prove in [85] that the requirement $u/\delta \in L^p(\Omega)$ can be weakened to $u/\delta \in L^{p,\infty}(\Omega)$, the weak-$L^p$ space defined as the set of functions $f$ satisfying

$$\sup_{\lambda>0} \lambda^p |\{\mathbf{x} \in \Omega : |f(\mathbf{x})| > \lambda\}| < \infty.$$

**(vi)** We now turn from the question of when (3.1.1) is valid to that of finding the best constant when it is. This is naturally associated with the variational problem of determining

$$\mu_p(\Omega) := \inf_{u \in W_0^{1,p}(\Omega)} \frac{\int_\Omega |\nabla u|^p d\mathbf{x}}{\int_\Omega |u/\delta|^p d\mathbf{x}} \tag{3.2.4}$$

and the existence, or otherwise, of a minimizer. Clearly (3.1.1) holds if and only if $\mu_p(\Omega) > 0$; and if $\mu_p(\Omega) > 0$ then the best constant is $\mu_p(\Omega)$. When $n = 1$ it is easily shown that $\mu_p(\Omega) = (1 - 1/p)^p$ (see (3.3.4) below), but when $n > 1$ the situation is more complicated. The problem was resolved for convex domains by Matskewich and Sobolevskii [117] in the case $p = 2$ and by Marcus et al. [116] in general. They proved that, for a convex domain $\Omega$ which is smooth in a neighbourhood of some point, $\mu_p(\Omega) = (1 - 1/p)^p = c_p$, which is equal to the one-dimensional value; see Theorem 3.1 below. Lewis et al. show in [107] that the convexity condition on $\Omega$ to achieve $\mu_p(\Omega) = c_p$ can be relaxed to weak mean convexity; this result is reproduced in Theorem 3.7.14 below.

In [116], it is proved that for all bounded domains in $\mathbb{R}^n$ with boundaries of class $C^2$,

$$\mu_p(\Omega) \leq c_p := \left(\frac{p-1}{p}\right)^p.$$

If there is no minimizer then $\mu_p(\Omega) = c_p$, the existence of a minimizer and $\mu_p(\Omega) = c_p$ being shown to be equivalent in the case $p = 2$.

It follows from the case $n = 1$ that, for the half-space $\mathbb{R}^n_+ := \{\mathbf{x} = (\mathbf{x}', x_n) : \mathbf{x}' \in \mathbb{R}^{n-1}, x_n > 0\}$, we have

$$\mu_p(\mathbb{R}^n_+) = c_p. \tag{3.2.5}$$

For, if $\phi \in C_0^\infty(\mathbb{R}^n_+)$, $|\nabla\phi(\mathbf{x})| \geq |(\partial\phi/\partial x_n)(\mathbf{x})|$ implies that $\mu_p(\mathbb{R}^n_+) \geq \mu_p(\mathbb{R}_+)$, while if $\phi$ is radial, then $|\nabla\phi(\mathbf{x})| = |(\partial\phi/\partial r)(\mathbf{x})|$ and hence $\mu_p(\mathbb{R}^n_+) \leq \mu_p(\mathbb{R}_+)$.

**(vii)** For non-convex domains (and ones not weakly mean convex), the value of $\mu_p(\Omega)$ is not known in general, but for arbitrary planar, simply connected domains $\Omega$, there is the celebrated result of Ancona in [7] that $\mu_2(\Omega) \geq 1/16$; see Sect. 3.4 below. The Ancona result is generalised by Laptev and Sobolev in [98], where the "degree" of convexity of $\Omega$ is quantified in the lower bound obtained for $\mu_2(\Omega)$. This is achieved by establishing a stronger version of Koebe's 1/4 theorem on which Ancona's proof was based.

We note the example of a punctured disc in item **(iv)** above for which there is no valid Hardy inequality. When $\Omega = \mathbb{R}^n \setminus \{0\}$, it is shown in [116], Example 4.1, that it is sufficient to consider the evaluation of $\mu_p(\Omega)$ over radially symmetric functions and then

$$\mu_p(\mathbb{R}^n \setminus \{0\}) = \left|1 - \frac{n}{p}\right|^p. \tag{3.2.6}$$

**(viii)** In [41], Davies introduced a **mean distance function** $\delta_M$, and proved that (3.1.1) holds in the case $p = 2$ for arbitrary domains if $\delta$ is replaced by

$\delta_M$. This was extended to general values of $p$ in $(1, \infty)$ by Tidblom in [144]. The mean distance function will be the subject of the following Sect. 3.2.

**(ix)** An extension of Hardy's inequality of the form

$$\int_\Omega |\nabla u(\mathbf{x})|^2 d\mathbf{x} \geq \frac{1}{4}\int_\Omega \frac{|u(\mathbf{x})|^2}{\delta(\mathbf{x})^2} d\mathbf{x} + C(\Omega)\int_\Omega |u(\mathbf{x})|^2 d\mathbf{x},\ u \in C_0^\infty(\Omega),$$

was established by Brezis and Marcus in [30], for $\Omega$ convex and $C(\Omega) = 1/4\,\mathrm{diam}(\Omega)$. Since then there have been many improvements, notably the sharp result of Avkhadiev and Wirths in [12] in which $C(\Omega) = \lambda_0^2/\delta_0^2$, where $\delta_0$ is the inradius of $\Omega$ and $\lambda_0$ is the first zero in $(0, \infty)$ of the Bessel function equation $J_0(t) - 2J_1(t) = 0$. Section 3.6 will be devoted to such extensions, including analogues in $L^p(\Omega)$.

**(x)** Ward proves in [146] that, for an arbitrary domain $\Omega$, the Schrödinger operator $H = -\Delta + V$ defined on $C_0^\infty(\Omega)$, is essentially self-adjoint if $V$ is locally bounded in $\Omega$ and, close to the boundary of $\Omega$, it satisfies

$$\begin{aligned} V(\mathbf{x}) \geq \frac{1}{\delta(\mathbf{x})^2}\Bigg\{ & 1 - \mu_2(\Omega) - \frac{1}{\log[\delta(\mathbf{x})^{-1}]} \\ & - \frac{1}{\log[\delta(\mathbf{x})^{-1}]\log\log[\delta(\mathbf{x})^{-1}]} - \cdots \\ & - \frac{1}{\log[\delta(\mathbf{x})^{-1}]\log\log[\delta(\mathbf{x})^{-1}]\cdots\log\log\cdots\log[\delta(\mathbf{x})^{-1}]}\Bigg\} \end{aligned} \tag{3.2.7}$$

for some finite number of logarithmic terms. This extends to arbitrary domains, a result of Nenciu and Nenciu in [122] for bounded domains with $C^2$ boundaries. In particular then, $H$ is essentially self-adjoint if

$$V(\mathbf{x}) \geq \frac{1 - \mu_2(\Omega)}{\delta(\mathbf{x})^2}.$$

This recovers the Kalf, Walter, Schmincke, Simon criterion (see [136])

$$V(\mathbf{x}) \geq -\frac{n(n-4)}{4|\mathbf{x}|^2}$$

for the case of $\Omega = \mathbb{R}^2 \setminus \{0\}$, since then $\mu_2(\Omega) = (1 - n/2)^2$ by (1.2.16) and $\delta(\mathbf{x}) = |\mathbf{x}|$.

## 3.3 The Mean Distance Function

### 3.3.1 A Hardy Inequality for General $\Omega$

In [41], Davies introduced the following notion:

**Definition 3.3.1** The **mean distance** function $\delta_M$ is defined by

$$\delta_M(\mathbf{x})^{-2} := n \int_{\mathbb{S}^{n-1}} \rho_\nu(\mathbf{x})^{-2} d\omega(\nu), \tag{3.3.1}$$

where $\rho_\nu(\mathbf{x})$ is the least distance from $\mathbf{x} \in \Omega$ to $\partial\Omega$ in the direction of either $\nu$ or $-\nu$ and $d\omega(\nu)$ is the normalized measure on $\mathbb{S}^{n-1}$, i.e., $\int_{\mathbb{S}^{n-1}} d\omega(\nu) = 1$. Note that the factor $n$ is excluded in [41].

For general $p \in (1, \infty)$, there is the analogue (see [144])

$$\delta_{M,p}(\mathbf{x})^{-p} := \frac{\sqrt{\pi}\,\Gamma\left(\frac{n+p}{2}\right)}{\Gamma\left(\frac{p+1}{2}\right)\Gamma\left(\frac{n}{2}\right)} \int_{\mathbb{S}^{n-1}} \rho_\nu(\mathbf{x})^{-p} d\omega(\nu), \tag{3.3.2}$$

where $\delta_{M,2} = \delta_M$.

The following theorem is proved in [41] in the case $p = 2$, and in [144] for any $p \in (1, \infty)$:

**Theorem 3.3.2** *For all $f \in D_0^{1,p}(\Omega)$, $1 < p < \infty$, and any domain $\Omega$ in $\mathbb{R}^n$,*

$$\int_\Omega |\nabla f(\mathbf{x})|^p d\mathbf{x} \geq \left(\frac{p-1}{p}\right)^p \int_\Omega \frac{|f(\mathbf{x})|^p}{\delta_{M,p}(\mathbf{x})^p} d\mathbf{x}. \tag{3.3.3}$$

*Proof* The root of the result is the one-dimensional inequality

$$\int_a^b |\varphi'(t)|^p dt \geq \left(\frac{p-1}{p}\right)^p \int_a^b \frac{|\varphi(t)|^p}{\rho(t)^p} dt, \quad (\varphi \in C_0^\infty(a,b)), \tag{3.3.4}$$

where $\rho(t) = \min\{|t-a|, |t-b|\}$. To prove this we obtain an inequality in each half of the interval $(a, b)$ separately. Let $\varphi$ be real. With $c := (1/2)(a + b)$,

$$\begin{aligned}\int_a^c \frac{|\varphi(t)|^p}{(t-a)^p} dt &= \int_a^c (t-a)^{-p} \left(\int_a^t [|\varphi(x)|^p]' dx\right) dt \\ &= \int_a^c [|\varphi(x)|^p]' \left(\int_x^c (t-a)^{-p} dt\right) dx\end{aligned}$$

$$= \int_a^c [|\varphi(x)|^p]'(p-1)^{-1}\{(x-a)^{-p+1} - (c-a)^{-p+1}\}dx$$
$$\le \frac{p}{p-1}\int_a^c \frac{|\varphi(x)|^{p-1}|\varphi'(x)|}{(x-a)^{p-1}}dx,$$

since $||\varphi(x)|'| = |\varphi'(x)|$ *a.e.*, by Theorem 1.3.8. Similarly

$$\int_c^b \frac{|\varphi(t)|^p}{(b-t)^p}dt \le \frac{p}{p-1}\int_c^b \frac{|\varphi(x)|^{p-1}|\varphi'(x)|}{(b-x)^{p-1}}dx$$

and the two inequalities combine to give

$$\int_a^b \frac{|\varphi(t)|^p}{\rho(t)^p}dt \le \left(\frac{p}{p-1}\right)\int_a^b \frac{|\varphi(x)|^{p-1}|\varphi'(x)|}{\rho(x)^{p-1}}dx$$
$$\le \left(\frac{p}{p-1}\right)\left(\int_a^b \frac{|\varphi(x)|^p}{\rho(x)^p}dx\right)^{1-1/p}\left(\int_a^b |\varphi'(x)|^p dx\right)^{1/p},$$

whence (3.3.4).

Let $\nu \in \mathbb{S}^{n-1}$, and denote the partial derivative in the direction of $\nu$ by $\partial_\nu$; hence from (3.3.4),

$$\int_{a_\nu}^{b_\nu} |\partial_\nu\varphi(t)|^p dt \ge \left(\frac{p-1}{p}\right)^p \int_{a_\nu}^{b_\nu} \frac{|\varphi(t)|^p}{\rho_\nu(t)^p}dt, \quad (\varphi \in C_0^\infty(a_\nu, b_\nu)), \tag{3.3.5}$$

where $(a_\nu, b_\nu)$ is the interval of intersection of $\Omega$ with the ray in the direction $\nu$. Furthermore, $\partial_\nu\varphi = \nu \cdot \nabla\varphi = |\nabla\varphi|\cos(\nu, \nabla\varphi)$, where for $\omega \in \mathbb{R}^n$, $(\nu, \omega)$ denotes the angle between $\nu$ and $\omega$. On integrating both sides of (3.3.5) with respect to the normalised measure $d\omega(\nu)$ we obtain

$$\int_\Omega \int_{\mathbb{S}^{n-1}} |\cos(\nu, \nabla\varphi(\mathbf{x}))|^p d\omega(\nu)|\nabla\varphi(\mathbf{x})|^p d\mathbf{x}$$
$$\ge \left(\frac{p-1}{p}\right)^p \int_\Omega \int_{\mathbb{S}^{n-1}} \frac{1}{\rho_\nu(\mathbf{x})^p} d\omega(\nu)|\varphi(\mathbf{x})|^p d\mathbf{x}. \tag{3.3.6}$$

For any fixed unit vector $\mathbf{e} \in \mathbb{R}^n$,

$$\int_{\mathbb{S}^{n-1}} |\cos(\nu, \nabla\varphi(\mathbf{x}))|^p d\omega(\nu) = \int_{\mathbb{S}^{n-1}} |\cos(\nu, \mathbf{e})|^p d\omega(\nu),$$

and a calculation gives

$$\int_{\mathbb{S}^{n-1}} |\cos(\nu, \mathbf{e})|^p d\omega(\nu) = \frac{\Gamma\left(\frac{p+1}{2}\right)\Gamma\left(\frac{n}{2}\right)}{\sqrt{\pi}\,\Gamma\left(\frac{n+p}{2}\right)}. \tag{3.3.7}$$

The inequality (3.3.3) follows from (3.3.6) for any real $\varphi \in C_0^\infty(\Omega)$, and hence for all real functions in $D_0^{1,p}(\Omega)$. Suppose $\varphi \in C_0^\infty(\Omega)$ is not necessarily real. Then, since $|\varphi| \in D_0^{1,p}(\Omega)$ and $|\nabla|\varphi|| \leq |\nabla\varphi|$ (see Theorem 1.3.8) (3.3.3) is a consequence of the already established inequality for real functions in $D_0^{1,p}(\Omega)$. The theorem is therefore proved for all $\varphi \in C_0^\infty(\Omega)$ and hence for all $\varphi \in D_0^{1,p}(\Omega)$. □

Thus if the distance function $\delta$ is replaced by the *mean distance* function $\delta_{M,p}$, the resulting Hardy-type inequality is always valid. We shall see in Sect. 3.5 below that this knowledge can be useful in the analysis of the Hardy inequality in many dimensions, by effectively reducing the problem to an easier one in one dimension. Furthermore, the natural quest for conditions which ensure that $\delta_{M,p}$ and $\delta$ are comparable provides a valuable geometric insight into the inequality. It is obviously always true that

$$\delta_{M,p}^p(\mathbf{x}) \geq B(n,p)\delta^p(\mathbf{x}),$$

where

$$B(n,p) = \frac{\Gamma\left(\frac{p+1}{2}\right)\Gamma\left(\frac{n}{2}\right)}{\sqrt{\pi}\,\Gamma\left(\frac{n+p}{2}\right)};$$

thus $B(n,2) = 1/n$. If $\partial\Omega$ is sufficiently regular, an inequality in the reverse direction is available, in which case

$$\delta(\mathbf{x}) \leq B(n,p)^{-1/p}\delta_{M,p}(\mathbf{x}) \leq c\delta(\mathbf{x}) \tag{3.3.8}$$

for some $c > 1$ and the Hardy inequality is valid. An example of this is given in [41], Theorem 18. The boundary $\partial\Omega$ is said to satisfy a $\theta$-*cone* condition if every $\mathbf{x} \in \partial\Omega$ is the vertex of a circular cone $C_{\mathbf{x}}$ of semi-angle $\theta$ which lies entirely in $\mathbb{R}^n \setminus \Omega$. Let $s(\alpha)$ denote the solid angle subtended at the origin by a ball of radius $\alpha < 1$, whose centre is at a distance 1 from the origin. Explicitly

$$s(\alpha) = \frac{1}{2}\int_0^{\arcsin\alpha} \sin^{n-2} t dt \Big/ \int_0^{\pi/2} \sin^{n-2} t dt.$$

**Theorem 3.3.3** *If $\partial\Omega$ satisfies a $\theta$-cone condition, then for all $\mathbf{x} \in \Omega$,*

$$\delta(\mathbf{x}) \leq B(n,p)^{-1/p}\delta_{M,p}(\mathbf{x}) \leq 2[s(\frac{1}{2}\sin\theta)]^{-1/p}\delta(\mathbf{x}).$$

*Proof* Let $\mathbf{x} \in \Omega, \mathbf{y} \in \partial\Omega$ and $\delta(\mathbf{x}) = |\mathbf{x} - \mathbf{y}|$. If $\nu$ is the unit vector directed along the axis of the cone $C_{\mathbf{y}}$ in $\mathbb{R}^n \setminus \Omega$, then the ball with centre $\mathbf{y} + \delta(\mathbf{x})\nu$ and radius $\delta(\mathbf{x})\sin\theta$ lies inside $C_{\mathbf{y}}$ and hence outside $\Omega$ The solid angle $\Lambda$ subtended by this ball at $\mathbf{x}$ is at least $s(\frac{1}{2}\sin\theta)$ and every line from $\mathbf{x}$ within this solid angle meets $\partial\Omega$ at a distance at most $2\delta(\mathbf{x})$ from $\mathbf{x}$. Consequently

$$\frac{B(n,p)}{\delta_{M,p}^p(\mathbf{x})} \geq \frac{1}{|\mathbb{S}^{n-1}|}\int_\Lambda \frac{1}{[2\delta(\mathbf{x})]^p}d\omega(\nu) \geq \frac{s(\frac{1}{2}\sin\theta)}{[2\delta(\mathbf{x})]^p}$$

and the theorem is proved. □

Another result of interest from [43], Exercise 5.7 or [144], p. 2270, is

**Theorem 3.3.4** *If $\Omega$ is convex, then*

$$\delta_{M,p}(\mathbf{x}) \leq \delta(\mathbf{x}), \tag{3.3.9}$$

*and hence*

$$\int_\Omega |\nabla f(\mathbf{x})|^p d\mathbf{x} \geq \left(\frac{p-1}{p}\right)^p \int_\Omega \frac{|f(\mathbf{x})|^p}{\delta(\mathbf{x})^p}d\mathbf{x}. \tag{3.3.10}$$

*Proof* Let $\mathbf{e}$ be a unit vector in $\mathbb{R}^n$ which is such that $\rho_{\mathbf{e}}(\mathbf{x}) = \delta(\mathbf{x})$. Then, if $\Omega$ is convex, it follows that

$$\rho_\nu(\mathbf{x})\cos(\mathbf{e},\nu) \leq \delta(\mathbf{x}).$$

Hence

$$\begin{aligned}\int_{\mathbb{S}^{n-1}} \frac{1}{\rho_\nu(\mathbf{x})^p}d\omega(\nu) &\geq \int_{\mathbb{S}^{n-1}} |\cos(\mathbf{e},\nu)|^p \frac{1}{\delta(\mathbf{x})^p}d\omega(\nu)\\ &= \int_{\mathbb{S}^{n-1}} |\cos(\mathbf{e},\nu)|^p d\omega(\nu)\frac{1}{\delta(\mathbf{x})^p}\\ &= \frac{\Gamma(\frac{p+1}{2})\Gamma(\frac{n}{2})}{\sqrt{\pi}\,\Gamma(\frac{n+p}{2})}\frac{1}{\delta(\mathbf{x})^p},\end{aligned}$$

by (3.3.7), whence (3.3.9). □

In Corollary 3.7.14 we shall prove the result established in [107], Theorem 1.2, that (3.3.10) holds for a domain $\Omega$ which is *weakly mean convex*, a condition which is weaker than convexity; for instance, a ring torus with minor radius $r$ and major radius $R$, with $R \geq r$, is weakly mean convex.

## 3.4 Hardy's Inequality on Convex Domains

### *3.4.1 Optimal Constant*

We shall now prove the result established in [116] that for a convex domain which is smooth in the neighbourhood of one of its boundary points, the optimal constant in (3.2.4) satisfies $\mu_p(\Omega) \leq c_p := \left(\frac{p-1}{p}\right)^p$. When coupled with Theorem 3.3.4 we shall then have $\mu_p(\Omega) = c_p$.

**Theorem 3.4.1** *Let $\Omega \subset \mathbb{R}^n, n \geq 2$, be a convex domain and suppose there is a point $P \in \partial\Omega$ such that $\Omega \in C^2$ in a neighbourhood of $P$. Then*

$$\mu_p(\Omega) := \inf_{u \in W_0^{1,p}(\Omega)} \frac{\int_\Omega |\nabla u|^p d\mathbf{x}}{\int_\Omega |u/\delta|^p d\mathbf{x}} = c_p. \tag{3.4.1}$$

*Proof* In view of Theorem 3.3.4, it is sufficient to prove that $\mu_p(\Omega) \leq c_p$. Let $\Pi$ be a tangent plane at $P \in \partial\Omega$. We may assume, without loss of generality, that $P = 0$, $\Pi = \{\mathbf{x} : x_n = 0\}$, and that there is a line segment $\{(0, x_n) : x_n \in (0, b)\} \subset \Omega$ for some $b > 0$. Let $H$ be the half-space $H := \{x_n > 0\}$ and $\varepsilon \in (0, 1)$, where we have written any point $\mathbf{x} \in \mathbb{R}^n$ in the form $\mathbf{x} = (\mathbf{x}', x_n)$, $\mathbf{x}' \in \mathbb{R}^{n-1}$. From (3.2.5)

$$\mu_p(H) = c_p := \left(\frac{p-1}{p}\right)^p.$$

Hence, with

$$R_H(u) := \frac{\int_H |\nabla u|^p d\mathbf{x}}{\int_H |u/\delta|^p d\mathbf{x}}, \tag{3.4.2}$$

we have that there exists $\phi \in C_0^\infty(H)$ such that $|R_H(\phi) - c_p| < \varepsilon$. Moreover, there exists $A > 0$ such that

$$\operatorname{supp} \phi \subset K := \{\mathbf{x} \in \mathbb{R}^n : x_n > 0, |\mathbf{x}'| < Ax_n\}$$

and a neighbourhood $U$ of 0 such that for all $\mathbf{x} \in U \cap \Omega$,

$$|\text{dist}\,(\mathbf{x}, \Pi) - \delta(\mathbf{x})| \leq o(1)|\mathbf{x}|, \tag{3.4.3}$$

where $o(1) \to 0$ as $\mathbf{x} \to P$. Since $R_H$ and $K$ are invariant with respect to transformations of the form $\mathbf{x} \to a\mathbf{x}$, $a > 0$, we may assume that

$$\operatorname{supp}\phi \subset U \cap \Omega \quad \text{and} \quad \delta(\mathbf{x}) < (1+\varepsilon)x_n \quad \text{for all} \quad \mathbf{x} \in \operatorname{supp}\phi.$$

On collecting all this together, we see that

$$R_\Omega(\phi) \le (1+\varepsilon)R_H(\phi) \le (1+\varepsilon)(c_p+\varepsilon).$$

Since $\varepsilon$ is arbitrary, it follows that $\mu_p(\Omega) \le c_p$ as asserted. □

*Remark 3.4.2* The following results are also proved in [116].

(1) If $\Omega$ is bounded and satisfies the conditions of the Theorem 3.4.1 except that of convexity, then $\mu_p(\Omega) \le c_p$, with equality if the variational problem determined by (3.4.1) has no minimiser; if $p = 2$, there is equality if and only if there is no minimiser.

(2) Let $\Omega_1, \Omega_2$ be bounded Lipschitz domains in $\mathbb{R}^n$ with $\overline{\Omega_1} \subset \Omega_2$; put $\Omega_0 = \mathbb{R}^n \setminus \Omega_1$, $\Omega = \Omega_0 \cap \Omega_2$ and $\Omega^k = \Omega_0 \cap (k\Omega_2)$, $k \in \mathbb{N}$. Then with $\mu_{p,i} = \mu_p(\Omega_i)$, $i = 0, 1, 2$,

$$\frac{\mu_{p,0}\mu_{p,2}}{\mu_{p,0} + \mu_{p,2}} \le \mu_p(\Omega).$$

If $0 \in \Omega$ then

$$\mu_p(\Omega) \le \min\left\{c_p, \left|\frac{n-p}{p}\right|^p\right\}$$

and so $\mu_n(\Omega_0) = 0$. Thus there is no valid Hardy inequality

$$\int_{\mathbb{R}^n\setminus\{0\}} |\nabla f(\mathbf{x})|^n d\mathbf{x} \ge c \int_\Omega \frac{|f(\mathbf{x})|^n}{\delta(\mathbf{x})^n} d\mathbf{x}.$$

(3) If $\Omega$ is either $\mathbb{R}^n \setminus B(0,R)$, $B(0,R) \setminus B(0,r)$ where $0 < r < R$, or $B(0,R) \setminus \{0\}$, then

$$\mu_2(\Omega) = \begin{cases} 0 & \text{if } n = 2, \\ 1/4 & \text{if } n \ge 3. \end{cases}$$

### 3.4.2 A Generalisation on $C_0^\infty(G(\Omega))$, $G(\Omega) = \Omega \setminus \overline{\mathcal{R}}(\Omega)$

In [16], the following inequality was established, which is the analogue for a domain $\Omega$ of the affine invariant inequality (1.2.17) in $\mathbb{R}^n$. It requires $\Delta\delta$ to be defined, which in view of Lemma 2.4.2, means that it applies to functions supported outside the cut locus $\Sigma(\Omega) = \overline{\mathcal{R}}(\Omega)$. Additional smoothness assumptions on the boundary of $\Omega$ enable the class of functions to be extended to $C_0^\infty(\Omega)$, as will be demonstrated in Sect. 3.5 below.

**Theorem 3.4.3** *Let $\Omega$ be an open convex subset of $\mathbb{R}^n$ and $G(\Omega) = \Omega \setminus \overline{\mathcal{R}}(\Omega)$, where $\mathcal{R}(\Omega)$ is the ridge. Then, for all $f \in C_0^\infty(G(\Omega))$,*

$$\int_\Omega |\nabla\delta \cdot \nabla f|^p d\mathbf{x} \geq \left(\frac{p-1}{p}\right)^p \int_\Omega \delta^{-p}|f|^p d\mathbf{x}, \tag{3.4.4}$$

*and hence*

$$\int_\Omega |\nabla f|^p d\mathbf{x} \geq \left(\frac{p-1}{p}\right)^p \int_\Omega \delta^{-p}|f|^p d\mathbf{x}. \tag{3.4.5}$$

*Proof* For any differentiable vector field $V : \mathbb{R}^n \to \mathbb{R}^n$, we have on integration by parts and the application of Hölder's inequality,

$$\begin{aligned}\int_\Omega \mathrm{div} V|f|^p d\mathbf{x} &= -p\mathrm{Re}\int_\Omega (V\cdot\nabla f)|f|^{p-2}\bar{f} d\mathbf{x}\\ &\leq p\left(\int_\Omega |V\cdot\nabla f|^p h^p d\mathbf{x}\right)^{1/p}\left(\int_\Omega |f|^p h^{-p'} d\mathbf{x}\right)^{(p-1)/p}\\ &\leq \varepsilon^p\int_\Omega |V\cdot\nabla f|^p h^p d\mathbf{x} + (p-1)\varepsilon^{-p/(p-1)}\int_\Omega |f|^p h^{-p'} d\mathbf{x},\end{aligned} \tag{3.4.6}$$

where $h$ is any positive function, and we have used Young's inequality

$$ab \leq \frac{(\varepsilon a)^p}{p} + \frac{(\varepsilon^{-1}b)^{p'}}{p'}, \quad p' = p/(p-1),$$

with $\varepsilon > 0$ arbitrary.

We choose $V = \nabla\delta^{-2m}$, where $m$ is to be determined; thus on $G(\Omega)$,

$$\mathrm{div} V = m\delta^{-2(m-1)}\Delta\delta^{-2} + 4m(m-1)\delta^{-2(m+1)}|\nabla\delta|^2. \tag{3.4.7}$$

For any $\mathbf{x} \in G(\Omega)$, rotate the co-ordinate system so that $\mathbf{x} = (\xi_1, \xi')$, where $\xi_1 = \delta(\mathbf{x})$, measured along the line $L$ from $\mathbf{x}$ to its nearest point on $\partial\Omega$, and $\xi' = (\xi_2, \cdots, \xi_n)$ lies in the $(n-1)$-dimensional orthogonal complement $L_{(n-1)}$

of $L$ in $\mathbb{R}^n$. Then, in view of the rotational invariance of the Laplacian,

$$\Delta\delta^{-2} = \partial_1^2\xi_1^{-2} + \Delta'\delta^{-2},$$

where $\Delta'$ is the Laplacian in $L_{(n-1)}$. Since $\Omega$ is assumed to be convex, $\Delta'\delta^{-2} \geq 0$, and hence

$$\Delta\delta^{-2} \geq \partial_1^2\xi_1^{-2} = 6\xi_1^{-4} = 6\delta^{-4}.$$

Consequently, in (3.4.7),

$$\text{div}V \geq 2m(2m+1)\delta^{-2(m+1)}.$$

On substituting this in (3.4.6) and setting $h^{p'} = \delta^{2(m+1)}$, we derive

$$\int_{G(\Omega)} |\nabla\delta^{-2} \cdot \nabla f|^p \delta^{4p-2(m+1)} d\mathbf{x} \geq J(\varepsilon) \int_{G(\Omega)} |f|^p \delta^{-2(m+1)} d\mathbf{x},$$

where

$$\begin{aligned} J(\varepsilon) &= \frac{2(2m+1)}{m^{p-1}}\varepsilon^{-p} - \frac{(p-1)}{m^p}\varepsilon^{-p^2/(p-1)} \\ &\leq \left(\frac{2(2m+1)}{p}\right)^p, \end{aligned}$$

the maximum being attained for $\varepsilon^{-1} = \{2m(2m+1)/p\}^{(p-1)/p}$. It follows that

$$\int_\Omega \left|\nabla\delta^{-2} \cdot \nabla f\right|^p \delta^{4p-2(m+1)} d\mathbf{x} \geq \left(\frac{2(2m+1)}{p}\right)^p \int_\Omega |f|^p \delta^{-2(m+1)} d\mathbf{x},$$

and the choice $m = (p/2) - 1$ completes the proof. □

### 3.4.3 Domains with Convex Complements

The technique used in Theorem 3.4.3 can be used effectively for a domain $\Omega$ whose complement $\Omega^c$ is convex. In this case one has the advantage that $G(\Omega) = \Omega$ since the cut locus is empty by Motzkin's Theorem 2.2.9.

**Theorem 3.4.4** *Let $\Omega$ be a domain in $\mathbb{R}^n$ whose complement $\Omega^c$ is convex, and let* $\delta(\mathbf{x}) := \text{dist}(\mathbf{x}, \Omega^c)$. *Then, for all $f \in C_0^\infty(\Omega)$, and $m \geq 1$,*

$$\int_\Omega \delta^{2(m-1)} \left|\nabla\delta^2 \cdot \nabla f\right|^p d\mathbf{x} \geq \left(\frac{2(2m-1)}{p}\right)^p \int_\Omega \delta^{2(m-1)} |f|^p d\mathbf{x} \tag{3.4.8}$$

*Proof* We set $V(\mathbf{x}) = \nabla\delta^{2m}(\mathbf{x})$ in (3.4.6). Then in any compact subset of $\Omega$,

$$\begin{aligned}\operatorname{div} V &= \sum_{i=1}^{n} \partial_i[\partial_i \delta^{2m}] \\ &= m\delta^{2(m-1)}\Delta\delta^2 + 4m(m-1)\delta^{2(m-1)}|\nabla\delta|^2 \\ &= m\delta^{2(m-1)}\Delta\delta^2 + 4m(m-1)\delta^{2(m-1)}\end{aligned}$$

since $|\nabla\delta| = 1$, a.e. On substituting in (3.4.6) and using the Hölder and Young inequalities, we get

$$\begin{aligned}&\textstyle\int_\Omega \{m\Delta\delta^2 + 4m(m-1)\}\delta^{2(m-1)}|f|^p d\mathbf{x} \\ &\le p\int_\Omega |\nabla\delta^{2m}\cdot\nabla f||f|^{p-1}d\mathbf{x} \\ &\le p\left(\int_\Omega |\nabla\delta^{2m}\cdot\nabla f|^p\delta^{-2(p-1)(m-1)}d\mathbf{x}\right)^{1/p}\left(\int_\Omega \delta^{2(m-1)}|f|^p d\mathbf{x}\right)^{1-1/p} \\ &\le m^p\varepsilon^p\int_\Omega |\nabla\delta^2\cdot\nabla f|^p\delta^{2(m-1)}d\mathbf{x} + (p-1)\varepsilon^{-p/(p-1)}\int_\Omega \delta^{2(m-1)}|f|^p d\mathbf{x}.\end{aligned} \tag{3.4.9}$$

We now proceed as in the proof of Theorem 3.4.3 and define the co-ordinates $\mathbf{x} = (\xi_1, \xi'))$ where $\xi_1 = \delta(\mathbf{x})$. Then, with the same notation, we have that

$$\Delta\delta^2(\mathbf{x}) = \partial_1^2\xi_1^2 + \Delta'\delta^2(\mathbf{x}).$$

Since $\Omega^c$ is convex, $\Delta'\delta^2(\mathbf{x}) \ge 0$, and so $\Delta\delta^2(\mathbf{x}) \ge 2$. It therefore follows from (3.4.9) that

$$\int_\Omega |\nabla\delta^2\cdot\nabla f|^p\delta^{2(m-1)}d\mathbf{x} \ge \int_\Omega \delta^{2(m-1)}K(\varepsilon)|f|^p d\mathbf{x}, \tag{3.4.10}$$

where

$$K(\varepsilon) = \left(\frac{2(2m-1)}{m^{p-1}}\right)\varepsilon^{-p} - \left(\frac{p-1}{m^p}\right)\varepsilon^{-p^2/(p-1)}.$$

It is readily shown that $K(\varepsilon)$ attains its maximum value of $[2(2m-1)/p]^p$ at $\varepsilon = [p/2m(2m-1)]^{(p-1)/p}$. The theorem then follows from (3.4.10). □

**Corollary 3.4.5** *Let $\Omega$ be a domain in $\mathbb{R}^n$ whose complement $\Omega^c$ is convex, and let $\delta(\mathbf{x}) := \operatorname{dist}(\mathbf{x}, \Omega^c)$. Then, for all $g \in C_0^\infty(\Omega)$ and $\gamma > -1/p$,*

$$\int_\Omega \delta^{p(\gamma+1)}|\nabla g|^p d\mathbf{x} \ge (\gamma + 1/p)^p\int_\Omega \delta^{p\gamma}|g|^p d\mathbf{x}. \tag{3.4.11}$$

*Proof* On substituting $f = g/\delta^\alpha$ in (3.4.8), we have

$$|\nabla\delta^2 \cdot \nabla f| \leq 2\{\delta^{-\alpha+1}|\nabla g| + \alpha\delta^{-\alpha}|g|\}$$

and

$$\left\|\delta^{[2(m-1)/p-\alpha+1]}\nabla g\right\| \geq \left[\frac{(2m-1)}{p} - \alpha\right]\left\|\delta^{[2(m-1)/p-\alpha]}g\right\|,$$

where $\|\cdot\|$ is the $L^p(\mathbb{R}^n)$ norm. The corollary follows on setting $\gamma = 2(m-1)/p-\alpha$. □

## 3.5 Non-convex Domains

### 3.5.1 A Strong Barrier on Ω

We show in the following lemma that for any domain $\Omega$, an inequality is satisfied (in the sense of quadratic forms) by the Laplacian in terms of a general vector field; this has a geometrical flavour and leads to the notion of strong barrier on $\Omega$ which has an important role in [7]. Without loss of generality, we shall assume throughout that functions are real.

**Lemma 3.5.1** *Let $\Omega$ be a domain in $\mathbb{R}^n, n \geq 1$ and $\mathbf{V} : \Omega \to \mathbb{R}^n$ a real, differentiable vector field. Then,*

$$-\Delta \geq -\text{div}\mathbf{V} - \|\mathbf{V}\|^2 \tag{3.5.1}$$

*in the sense that for all real $f \in C_0^\infty(\Omega)$,*

$$\int_\Omega |\nabla f|^2 d\mathbf{x} \geq \int_\Omega \left(-\text{div}\mathbf{V} - \|\mathbf{V}\|^2\right) f^2 d\mathbf{x}. \tag{3.5.2}$$

*On choosing $\mathbf{V}(\mathbf{x}) = \nabla[\log s(\mathbf{x})]$, where $s$ is a strictly positive $C^2(\Omega)$ function, it follows that*

$$\int_\Omega |\nabla f|^2 d\mathbf{x} \geq \int_\Omega \left(\frac{-\Delta s}{s}\right) f^2 d\mathbf{x}. \tag{3.5.3}$$

*Let $\Omega$ have a non-empty boundary, and $\delta(\mathbf{x}) := \text{dist}(\mathbf{x}, \partial\Omega)$. If there exists a strictly positive superharmonic function $s$ which is such that $-\Delta s/s \geq \varepsilon/\delta^2$ for* *some positive number* $\varepsilon$*, then*

$$\int_{\Omega}\left|\frac{u(\mathbf{x})}{\delta(\mathbf{x})}\right|^2 d\mathbf{x} \le c_\Omega \int_{\Omega} |\nabla u(\mathbf{x})|^2 d\mathbf{x}, \quad u \in C_0^\infty(\Omega), \tag{3.5.4}$$

*where* $c_\Omega = 1/\varepsilon$.

*Proof* Let $\mathbf{V} = (v_1, v_2, \cdots, v_n)$. Then,

$$\begin{aligned} 0 &\le \sum_{i=1}^{n} \int_{\Omega} \left(\frac{\partial f}{\partial x_i} - f v_i\right)^2 d\mathbf{x} \\ &= \sum_{i=1}^{n} \int_{\Omega} \left\{ \left(\frac{\partial f}{\partial x_i}\right)^2 - 2\left(\frac{\partial f}{\partial x_i}\right)(f v_i) + f^2 v_i^2 \right\} d\mathbf{x} \\ &= \sum_{i=1}^{n} \int_{\Omega} \left\{ \left(\frac{\partial f}{\partial x_i}\right)^2 + f^2 \frac{\partial v_i}{\partial x_i} + f^2 v_i^2 \right\} d\mathbf{x} \end{aligned}$$

and (3.5.2) is proved.

The choice $\mathbf{V}(\mathbf{x}) = \nabla[\log s(\mathbf{x})]$ gives $-\mathrm{div}\mathbf{V} - \|\mathbf{V}\|^2 = -\Delta\, s/s$ and the rest of the lemma follows. □

For a strictly positive superharmonic function $s$ to qualify as a **strong barrier** on $\Omega$ in accordance with [7], it is sufficient for it to satisfy $\Delta s + (\varepsilon/\delta^2)s \le 0$ only in the weak (distributional) sense, that is, for all non-negative $\psi \in C_0^\infty(\Omega)$.

$$\int_{\Omega} \left(\Delta s + (\varepsilon/\delta^2)s\right) \psi d\mathbf{x} = \int_{\Omega} \left(\Delta\psi + (\varepsilon/\delta^2)\psi\right) s d\mathbf{x} \le 0, \tag{3.5.5}$$

which is less than what is required for (3.5.4) above. In the next proposition, which is Proposition 1 in [7], the existence of a strong barrier in this weak sense is equivalent to the validity of the Hardy inequality (3.5.4).

**Theorem 3.5.2** *The Hardy inequality (3.5.4) holds with a finite positive constant* $c_\Omega$*, if and only if there exists a strictly positive superharmonic function* $s$ *on* $\Omega$ *and a positive number* $\varepsilon$ *such that*

$$\Delta s + \frac{\varepsilon}{\delta^2} s \le 0 \tag{3.5.6}$$

*in the weak sense of (3.5.5). The largest value of* $\varepsilon$ *in (3.5.5) is* $1/c_\Omega$*, where* $c_\Omega$ *is the best possible constant in (3.5.4).*

*Proof* Suppose that (3.5.4) is satisfied for some finite positive constant $c_\Omega$. Consider the Hilbert space

$$H = \{f \in L^2_{loc}(\Omega) : \nabla f \in L^2(\Omega), \delta^{-1} f \in L^2(\Omega)\},$$

equipped with the norm

$$\|f\|_H = \left(\|\nabla f\|^2 + \|\delta^{-1}f\|^2\right)^{1/2},$$

and let $H_0$ denote the closure of $C_0^\infty(\Omega)$ in $H$. The quadratic form

$$a[f,g] := \int_\Omega \nabla f \cdot \nabla g d\mathbf{x} - \varepsilon \int_\Omega \delta^{-2} fg d\mathbf{x}$$

is bounded on $H \times H$ since there exists a constant $M > 0$ such that

$$|a[f,g]| \leq M\|f\|_H\|g\|_H.$$

Moreover, $a[\cdot,\cdot]$ is coercive on $H$ if $\varepsilon < 1/c_\Omega$ since

$$a[f,f] \geq (1-\varepsilon c_\Omega)\int_\Omega |\nabla f|^2 d\mathbf{x} \geq \left(\frac{1-\varepsilon c_\Omega}{1+\varepsilon c_\Omega}\right)\|f\|_H^2.$$

Therefore by the **Lax-Milgram** theorem (see [48], Chap. IV), given a non-negative $\psi \in C_0^\infty(\Omega)$, there exists a unique $s \in H_0$ such that

$$a[s,\varphi] = \int_\Omega \psi\varphi d\mathbf{x}, \quad \text{for every} \quad \varphi \in C_0^\infty(\Omega). \tag{3.5.7}$$

Hence $\Delta s + (\varepsilon/\delta^2)s = -\psi$ in the weak sense. Moreover, we claim that $s$ is strictly positive. Suppose otherwise, and set $s^-(\mathbf{x}) := \min\{0, s(\mathbf{x})\}$. We have that $s^- \in H_0$ as in Theorem 1.3.8, and also note that (3.5.7) is valid for all $\varphi \in H_0$ by continuity. Hence

$$0 \leq a[s^-,s^-] = a[s,s^-] = \int_\Omega \psi s^- d\mathbf{x} \leq 0.$$

The claim is therefore verified and (3.5.4) has been shown to imply the existence of a strong barrier on $\Omega$.

Conversely, suppose there exists a strong barrier $s$ on $\Omega$, and let $\omega$ be a relatively compact open subset of $\Omega$. Then $-\Delta - \varepsilon/\delta^2$ defines a self-adjoint operator in $H_0(\omega)$ whose spectrum is discrete and bounded below. Denote its first eigenvalue by $\lambda_0$ and the corresponding eigenfunction by $u_0$. Then

$$T := -\Delta - \lambda_0 - \varepsilon/\delta^2 \geq 0.$$

The form domain $\mathcal{Q}$ of $T$, is the domain of its square root $T^{1/2}$, and if $u \in \mathcal{Q}$, then $|u| \in \mathcal{Q}$ and

$$\begin{aligned}\|T^{1/2}|u|\|^2 &= \int_\omega \left(|\nabla|u||^2 - [\lambda_0 + \varepsilon/\delta^2]|u|^2\right) d\mathbf{x}\\ &\leq \int_\omega \left(|\nabla u|^2 - [\lambda_0 + \varepsilon/\delta^2]|u|^2\right) d\mathbf{x}\\ &= \|T^{1/2}u\|^2.\end{aligned}$$

It follows from [42], Theorem 1.3.2 that if $\varphi \in \mathcal{Q}$ and $\varphi \geq 0$,

$$\begin{aligned}&\textstyle\int_\omega \left(\Delta|u_0| + [\lambda_0 + \varepsilon/\delta^2]|u_0|\right)\varphi d\mathbf{x}\\ &\quad= -\left(T^{1/2}|u_0|, T^{1/2}\varphi\right)\\ &\quad\geq \left(\tfrac{u_0}{|u_0|}Tu_0, \varphi\right) = 0.\end{aligned}$$

Suppose that $\lambda_0$ is negative. Then

$$\begin{aligned}0 &\leq \int_\omega \left\{\Delta|u_0| + [\lambda_0 + (\varepsilon/\delta^2)]|u_0|\right\} s d\mathbf{x}\\ &= \int_\omega \left\{\Delta s + [\lambda_0 + (\varepsilon/\delta^2)]s\right\} |u_0| d\mathbf{x} \leq \lambda_0 \int_\omega s|u_0| d\mathbf{x} < 0\end{aligned}$$

which is a contradiction. It follows that, for all $C_0^\infty(\Omega)$,

$$-\int_\Omega (\Delta\psi)\psi d\mathbf{x} - \varepsilon \int_\Omega \delta^{-2}\psi^2 d\mathbf{x} \geq 0,$$

which is Hardy's inequality (3.5.4) with $c_\Omega = 1/\varepsilon$, since $\varepsilon$ is independent of $\omega$. □

It is proved in [7], Lemma 3 and Theorem 1, that if $n \geq 3$ and there is a constant $c > 0$ such that, for all $\mathbf{x} \in \partial\Omega$ and $r > 0$,

$$c\,|\Omega^c \cap B(\mathbf{x}, r)| \geq cr^{n-2}, \tag{3.5.8}$$

then a strong barrier exists on $\Omega$. When $n \geq 3$, the condition (3.5.8) is equivalent to $\Omega$ being **uniformly $\Delta$-regular**; this property is defined for all $n \geq 2$ and is that there is a constant $\varepsilon_1 \in (0, 1)$ such that, for all $\mathbf{x} \in \partial\Omega$ and $r > 0$, the harmonic measure $\omega_{\mathbf{x},r}$ of $\partial B(\mathbf{x}, r) \cap \Omega$ in $\Omega \cap B(\mathbf{x}, r)$ satisfies $\omega_{\mathbf{x},r} \leq 1 - \varepsilon_1$ on $\partial B(\mathbf{x}, r/2)$.

If $n = 2$, the following three properties are proved to be equivalent in [7], Theorem 2:

(i) $\Omega$ is uniformly $\Delta$-regular,
(ii) there is a strong barrier for $\Omega$,
(iii) Hardy's inequality (3.5.4) holds.

We refer to [7] for the terminology and details.

### 3.5.2 Planar Simply Connected Domains

The best possible constant for weakly mean convex domains was determined in [107], Theorem 1.2; see Corollary 3.6.14 below. However, in general, for non-convex domains (and ones which are not weakly mean convex), the best possible constant $c_\Omega$ in (3.5.4) is not known. Some specific examples of such domains were considered in [44] (see also Tidblom [145]). For example, for $\Omega = \mathbb{R}^2 \setminus \mathbb{R}^+$, with $\mathbb{R}^+ = [0, \infty)$, it was found that $c_\Omega = 4.86902\ldots$. From the many important contributions to the study of Hardy inequalities in [7], the one of special concern to us in this section is Ancona's application of Koebe's one-quarter theorem to prove that for a simply connected domain in the plane, the constant $c_\Omega$ in (3.5.4)is no greater than 16. The proof of this has three ingredients:

*The Riemann mapping theorem.* If $U$ is a non-empty simply connected open subset of the complex plane $\mathbb{C}$ which is not all of $\mathbb{C}$, then there exists a conformal (bijective analytic) map $f$ from $U$ onto the open unit disk $D = \{\mathbf{z} \in \mathbb{C} : |\mathbf{z}| < 1\}$; this is known as a *Riemann map*. A corollary of the theorem is that any two simply connected open subsets of the Riemann sphere which both lack at least two points of the sphere can be conformally mapped into each other.

*Koebe's* $1/4$ *theorem.* Let $D := \{\mathbf{z} \in \mathbb{C} : |\mathbf{z}| < 1\}$, the open unit disc in $\mathbb{C}$, and let $f : D \to \mathbb{C}$ be a Riemann map. Then the image $f(D)$ in $\mathbb{C}$ contains the disc centre $f(0)$ and radius $|f'(0)|/4$.

*Conformal invariance* of the Dirichlet integral $\int_\Omega |\nabla u(\mathbf{x})|^2 d\mathbf{x}$. To see this, let $f : \Omega \to \Omega'$ be conformal, and set $\mathbf{y} = f(\mathbf{x})$, $\mathbf{y} = (y_1, y_2)$ and $\mathbf{x} = (x_1, x_2)$; $\mathbb{R}^2$ and $\mathbb{C}$ are identified by $\mathbf{z} = x_1 + ix_2$. Then, with $f'$ denoting the complex derivative,

$$d\mathbf{y} = \left| \det \left( \frac{\partial(y_1, y_2)}{\partial(x_1, x_2)} \right) \right| d\mathbf{x} = |f'(\mathbf{x})|^2 d\mathbf{x},$$

and

$$\nabla_{\mathbf{x}} u = \nabla_{\mathbf{y}} u \left[ \frac{\partial(y_1, y_2)}{\partial(x_1, x_2)} \right]^t,$$

implying that

$$|\nabla_{\mathbf{x}} u|^2 = |\nabla_{\mathbf{y}} u|^2 |f'(\mathbf{x})|^2.$$

Consequently,

$$\int_{\Omega} |\nabla_{\mathbf{x}}(u \circ f)(\mathbf{x})|^2 \, d\mathbf{x} = \int_{\Omega'} \left|\nabla_{\mathbf{y}} u(\mathbf{y})\right|^2 d\mathbf{y},$$

which confirms the asserted invariance.

To proceed, we need the following consequence of the Koebe theorem.

**Lemma 3.5.3** *Let $\Omega \subset \mathbb{C}$, $\Omega \neq \mathbb{C}$ be a simply connected domain, and let $\mathbb{C}_+ := \{\mathbf{z} = x + iy \in \mathbb{C} : y > 0\}$. Then for any conformal mapping $f : \mathbb{C}_+ \to \Omega$, Koebe's theorem implies that*

$$\delta\left(f(\mathbf{z})\right) \geq \frac{x}{2} |f'(\mathbf{z})|, \quad \mathbf{z} = x + iy. \tag{3.5.9}$$

*Proof* We follow the proof in [98], Theorem 3.2. For arbitrary $\mathbf{z} \in \mathbb{C}_+$, define

$$g(\mathbf{w}) = g_{\mathbf{z}}(\mathbf{w}) = f\left(h(\mathbf{w})\right), \quad h(\mathbf{w}) = \frac{\overline{\mathbf{z}}\mathbf{w} + \mathbf{z}}{1 - \mathbf{w}},$$

where $\mathbf{w} \in D$. For each fixed $\mathbf{z} \in \mathbb{C}_+$ the function $h$ maps $D$ onto $\mathbb{C}_+$, and $h(0) = \mathbf{z}$, $g(0) = f(\mathbf{z})$. Furthermore

$$g'(\mathbf{w}) = \frac{\mathbf{z} + \overline{\mathbf{z}}}{(1 - \mathbf{w})^2} f'\left(h(\mathbf{w})\right),$$

so that $g'(0) = 2\pi f'(\mathbf{z})$. Koebe's theorem now implies that

$$\delta\left(f(\mathbf{z})\right) = \delta\left(g(0)\right) \geq \frac{1}{4}|g'(0)| = \frac{1}{2} x |f'(\mathbf{z})|$$

as required. □

We are now able to prove Ancona's result:

**Theorem 3.5.4** *Let $\Omega \subset \mathbb{R}^2$, $\Omega \neq \mathbb{R}^2$ be a simply connected domain. Then the Hardy inequality (3.5.4) holds with $c_\Omega \leq 16$.*

*Proof* The Hardy inequality on $(0, \infty)$ clearly implies that

$$\int_{\mathbb{C}_+} \frac{|u|^2}{x^2} dx dy \leq 4 \int_{\mathbb{C}_+} |\nabla u|^2 d\mathbf{x}, \quad u \in C_0^\infty(\mathbb{C}_+), \tag{3.5.10}$$

where we have identified $\mathbb{C}_+$ and the upper-half plane with $\mathbf{z} = x+iy$. Let $f : \mathbb{C}_+ \to \Omega$ be the conformal map in Lemma 3.5.3. Then, since canonical maps preserve the Dirichlet integral, we have from (3.5.10),

$$\begin{aligned}\int_\Omega |\nabla u(\mathbf{x})|^2 d\mathbf{x} &= \int_{\mathbb{C}_+} |\nabla (u \circ f)|^2 d\mathbf{z} \\ &\geq \frac{1}{4}\int_{\mathbb{C}_+} \frac{|(u \circ f)|^2}{x^2} d\mathbf{z} \\ &= \int_{\mathbb{C}_+} \frac{|(u \circ f)|^2}{4x^2|f'(\mathbf{z})|^2} |f'(\mathbf{z})|^2 d\mathbf{z} \\ &\geq \frac{1}{16}\int_{\mathbb{C}_+} \frac{|u[f(\mathbf{z})]|^2}{\delta[f(\mathbf{z})]^2} |f'(\mathbf{z})|^2 d\mathbf{z}\end{aligned}$$

on using (3.5.9). As $f'(\mathbf{z})$ is the complex derivative, the substitution $\mathbf{w} = f(\mathbf{z})$, with $\mathbf{w} = u + iv$, gives

$$d\mathbf{w} = \left| \det \left( \frac{\partial(u, v)}{\partial(x, y)} \right) \right| d\mathbf{z} = |f'(\mathbf{z})|^2 d\mathbf{z},$$

and the theorem follows. □

In view of Theorem 3.5.4 and the fact that for convex planar domains $\Omega$ the inequality (3.5.4) holds with $c_\Omega = 4$, it is natural to ask if we can get a value of $c_\Omega$ lying between 4 and 16, for simply connected planar domains with some degree of non-convexity which can be quantified. This question was posed by Laptev and Sobolev in [98]. They introduced two possible "measures" of non-convexity and obtained extensions of Koebe's theorem which led to a positive answer to their question.

We shall briefly describe one of the results in [98], and encourage the reader to consult the paper for further details. For any simply connected domain $\Omega \subset \mathbb{C}$, $\Omega \neq \mathbb{C}$, let us denote by $A(\Omega)$ the class of all conformal maps $f$ from $\Omega$ onto the open unit disk $D$. We denote by $K_\theta$, $\theta \in [0, \pi]$, the open sector

$$K_\theta = \{\mathbf{z} \in \mathbb{C} : \ |\arg \mathbf{z}| < \theta\}.$$

So, $K_\theta$ is symmetric with respect to the real axis and with the angle $2\theta$ at the vertex. We also assume that for any non-zero complex number $\mathbf{z} \in \mathbb{C}$, the argument $\arg \mathbf{z} \in (-\pi, \pi]$. The domains $\Omega$ to be considered are assumed to satisfy the following condition:

**Condition 3.5.1** *There exists a number $\theta \in [0, \pi]$ such that for each $w \in \Omega^c$ one can find a $\phi = \phi_w \in (-\pi, \pi]$ such that*

$$\Omega \in K_\theta(w, \phi_w),$$

*where $K_\theta(w, \phi_w)$ is the transformation of $K_\theta$ through rotation by angle $\phi_w \in (-\pi, \pi]$ and translation by $w$, i.e. $K_\theta(w, \phi_w) = e^{i\phi_w} K_\theta + w$.*

This condition means that the domain $\Omega$ satisfies an exterior cone condition with infinite cone. Since the cone is infinite, Condition 3.5.1 is equivalent to itself if stated for the boundary points $w \in \partial\Omega$ only. If the Condition 3.5.1 is satisfied for some $\theta$, then automatically $\theta \geq \pi/2$ and the equality $\theta = \pi/2$ holds for convex domains.

For domains satisfying Condition 3.5.1, we set

$$r_\theta(\Omega) = \frac{\pi}{4\theta}. \tag{3.5.11}$$

In [98] it was shown that for domains with the Condition 3.5.1 and any $f \in A(\Omega)$, Koebe's theorem holds with $r = r_\theta(\Omega)$. This implies, by the same argument as in Ancona's paper [7], the following stronger version of the Hardy inequality (3.5.4).

**Theorem 3.5.5** *Suppose that the domain $\Omega \subset \mathbb{R}^2$, $\Omega \neq \mathbb{R}^2$ satisfies Condition 3.5.1 with some $\theta \in [\pi/2, \pi]$. Then, for any $\psi \in C_0^\infty(\Omega)$, the Hardy inequality (3.5.4) holds with*

$$c_\Omega = 1/r_\theta^2(\Omega).$$

*Remark 3.5.6* The constant $c_\Omega$ in Theorem 3.5.5 runs from 16 to 4 when $\theta$ varies from $\pi$ to $\pi/2$. For the domain $\Omega = K_\theta$, the theorem does not give the best known result, for in [44], it is shown that $c_\Omega$ remains equal to 4 for the range $\theta \in [0, \theta_0]$, where $\theta_0 \approx 2.428$, which is considerably greater than $\pi/2$.

## 3.6 Extensions of Hardy's Inequality

### *3.6.1 Inequalities of Brezis and Marcus Type in $L^2(\Omega)$*

A considerable amount of interest was generated by the paper of Brezis and Marcus [30] in which the inequality in Theorem 3.3.4 for a convex domain with $p = 2$ is improved by the addition of a positive term to the right-hand side. To be explicit, it is proved in [30] that, for every smooth bounded domain $\Omega \subset \mathbb{R}^n$, there exists $\lambda \in \mathbb{R}$ such that

$$\int_\Omega |\nabla u|^2 d\mathbf{x} - \frac{1}{4}\int_\Omega |u/\delta|^2 d\mathbf{x} \geq \lambda \int_\Omega |u|^2 d\mathbf{x}, \quad u \in H_0^1(\Omega).$$

The largest such constant $\lambda$ is precisely

$$\lambda^*(\Omega) := \inf_{u \in H_0^1(\Omega)} \frac{\int_\Omega |\nabla u|^2 d\mathbf{x} - \frac{1}{4}\int_\Omega |u/\delta|^2 d\mathbf{x}}{\int_\Omega |u|^2 d\mathbf{x}}$$

and this infimum is shown not to be achieved. There are smooth bounded domains for which $\lambda^*(\Omega) < 0$, but for convex domains with a $C^2$-boundary, it is proved that

$$\lambda^*(\Omega) \geq \frac{1}{4\,\mathrm{diam}^2(\Omega)}. \tag{3.6.1}$$

Thus, for $D(\Omega) := \mathrm{diam}(\Omega)$ we have that

$$\int_\Omega |\nabla u|^2 d\mathbf{x} - \frac{1}{4}\int_\Omega |u/\delta|^2 d\mathbf{x} \geq \frac{1}{4D(\Omega)^2}\int_\Omega |u|^2 d\mathbf{x}, \qquad u \in H_0^1(\Omega).$$

In [30] the question was posed as to whether the diameter in (3.6.1) could be replaced by the volume of $\Omega$, i.e., whether

$$\lambda^*(\Omega) \geq \alpha(\mathrm{vol}(\Omega))^{-2/n}, \tag{3.6.2}$$

for some universal constant $\alpha > 0$. This was answered (in the affirmative) in [78] using the mean distance function (3.3.1). We follow the approach in [56] which used much of the analysis from [78]. However, in [56] different one-dimensional inequalities (given in Lemmas 3.6.1 and 3.6.2 below) produce an improved constant $\alpha$.

**Lemma 3.6.1** *Let $u \in C_0^1(0, 2b)$, $\rho(t) := \min\{t, 2b - t\}$, and let $f \in C^1(0, b]$ be monotonic on $[0, b]$. Then for $p > 1$,*

$$\int_0^{2b} |f'(\rho(t))||u(t)|^p dt \leq p^p \int_0^{2b} \frac{|f(\rho(t)) - f(b)|^p}{|f'(\rho(t))|^{p-1}} |u'(t)|^p dt. \tag{3.6.3}$$

*Proof* First let $u := v\chi_{(0,b]}$, the restriction to $(0, b]$ of some $v \in C_0^1(0, 2b)$. For any constant $c$

$$\begin{aligned} -\textstyle\int_0^b [f(t) - c]'|u(t)|^p dt &= -[f(t) - c]|u(t)|^p\Big|_0^b \\ &\quad + \textstyle\int_0^b [f(t) - c]\frac{p}{2}[|u(t)|^2]^{\frac{p}{2}-1}[|u(t)|^2]' dt. \end{aligned}$$

By choosing $c = f(b)$, we have that

$$-\textstyle\int_0^b f'(t)|u(t)|^p dt = p\int_0^b [f(t) - f(b)]|u(t)|^{p-2}\mathrm{Re}[\overline{u(t)}u'(t)]dt. \tag{3.6.4}$$

Similarly, for $u = v\chi_{[b,2b)}$, $v \in C_0^1(0, 2b)$, we have

$$\begin{aligned} &-\textstyle\int_b^{2b} f'(2b - s)|u(s)|^p ds \\ &\quad = p\textstyle\int_b^{2b} [f(2b - s) - f(b)]|u(s)|^{p-2}\mathrm{Re}[\overline{u(s)}u'(s)]ds. \end{aligned}$$

Therefore, since $f$ is monotonic, for any $u \in C_0^1(0, 2b)$

$$\begin{aligned}
&\int_0^{2b} |f'(\rho(t))||u(t)|^p dt \\
&= p\int_0^{2b} |f(\rho(t)) - f(b)||u(t)|^{p-2}\mathrm{Re}[\overline{u(t)}u'(t)]dt \\
&\le p\int_0^b |f'(\rho(t))|^{\frac{p-1}{p}} |u(t)|^{p-1} \tfrac{|f(\rho(t))-f(b)|}{|f'(\rho(t))|^{(p-1)/p}} |u'(t)|dt \\
&\le p\Big[\int_0^b |f'(\rho(t))||u(t)|^p dt\Big]^{\frac{p-1}{p}} \Big[\int_0^b \tfrac{|f(\rho(t))-f(b)|^p}{|f'(\rho(t))|^{p-1}} |u'(t)|^p dt\Big]^{\frac{1}{p}}
\end{aligned}$$

on applying Hölder's inequality. Inequality (3.6.3) now follows. □

**Lemma 3.6.2** *Define* $\mu(t) := 2b - \rho(t)$. *For all* $u \in C_0^1(0, 2b)$

$$\int_0^{2b} |u'(t)|^2 dt \ge \frac{1}{4}\int_0^{2b} \rho(t)^{-2}\Big[1 + \Big(\frac{2\rho(t)}{\mu(t)}\Big)\Big]^2 |u(t)|^2 dt. \tag{3.6.5}$$

*Proof* On setting $f(t) = 1/t$ and $p = 2$ in (3.6.3), we get

$$\int_0^{2b} \rho(t)^{-2}|u(t)|^2 dt \le 4\int_0^{2b} \Big|1 - \frac{\rho(t)}{b}\Big|^2 |u'(t)|^2 dt \tag{3.6.6}$$

for $u \in C_0^1(0, 2b)$. We claim that the substitution $v(t) = [1 - (\frac{\rho(t)}{b})]u(t)$ in (3.6.6) gives

$$\int_0^{2b} |v'(t)|^2 dt \ge \frac{1}{4}\int_0^{2b} \rho(t)^{-2}\Big[1 - \Big(\frac{\rho(t)}{b}\Big)\Big]^{-2} |v(t)|^2 dt. \tag{3.6.7}$$

For

$$\begin{aligned}
|v'(t)|^2 = b^{-2}|u(t)|^2 - b^{-1}\rho'(t)\big[1 - \big(\tfrac{\rho(t)}{b}\big)\big][|u|^2]' \\
+\big[1 - \big(\tfrac{\rho(t)}{b}\big)\big]^2 |u'(t)|^2
\end{aligned}$$

which implies that

$$\int_0^{2b} |v'(t)|^2 dt = \int_0^{2b} \big[1 - \big(\tfrac{\rho(t)}{b}\big)\big]^2 |u'(t)|^2 dt, \tag{3.6.8}$$

since

$$\frac{1}{b}\int_0^{2b} |u(t)|^2 dt = \int_0^b \Big[1 - \frac{t}{b}\Big][|u|^2]' dt - \int_b^{2b} \Big[1 - \frac{2b-t}{b}\Big][|u|^2]' dt.$$

Therefore, (3.6.7) follows from (3.6.6). Since

$$(1-\frac{\rho}{b})^{-2} = [1+\frac{2\rho}{\mu-\rho}]^2 \geq [1+\frac{2\rho}{\mu}]^2,$$

the proof is complete. □

The next theorem corresponds to Theorem 1 in [56] in which weights $\delta(\mathbf{x})^\sigma$ were also included.

**Theorem 3.6.3** *For any $u \in C_0^1(\Omega)$,*

$$\int_\Omega |\nabla u(\mathbf{x})|^2 d\mathbf{x} \geq \frac{1}{4}\int_\Omega \frac{|u(\mathbf{x})|^2}{\delta_M(\mathbf{x})^2} d\mathbf{x} + \frac{3}{2}K(n)\int_\Omega \frac{|u(\mathbf{x})|^2}{|\Omega_\mathbf{x}|^{\frac{2}{n}}} d\mathbf{x}, \tag{3.6.9}$$

*where $K(n) := n\left[\frac{s_{n-1}}{n}\right]^{2/n}$, $s_{n-1} := |\mathbb{S}^{n-1}|$, and*

$$\Omega_\mathbf{x} := \{\mathbf{y} \in \Omega : \mathbf{x} + t(\mathbf{y}-\mathbf{x}) \in \Omega,\ \forall t \in [0,1]\};$$

*i.e., $\Omega_\mathbf{x}$ is the set of all $\mathbf{y} \in \Omega$ that can be "seen" from $\mathbf{x} \in \Omega$.*

*If $\Omega$ is convex, $\Omega_\mathbf{x} = \Omega$ and, for any $u \in C_0^1(\Omega)$,*

$$\int_\Omega |\nabla u(\mathbf{x})|^2 d\mathbf{x} \geq \frac{1}{4}\int_\Omega \frac{|u(\mathbf{x})|^2}{\delta(\mathbf{x})^2} d\mathbf{x} + \frac{3K(n)}{2|\Omega|^{\frac{2}{n}}}\int_\Omega |u(\mathbf{x})|^2 d\mathbf{x}. \tag{3.6.10}$$

*Proof* For each $\mathbf{x} \in \Omega$ and $\nu \in \mathbb{S}^{n-1}$ define

$$\begin{aligned}
\tau_\nu(\mathbf{x}) &:= \min\{s > 0 : \mathbf{x} + s\nu \notin \Omega\},\\
\rho_\nu(\mathbf{x}) &:= \min\{\tau_\nu(\mathbf{x}), \tau_{-\nu}(\mathbf{x})\},\\
\mu_\nu(\mathbf{x}) &:= \max\{\tau_\nu(\mathbf{x}), \tau_{-\nu}(\mathbf{x})\},\\
D_\nu(\mathbf{x}) &:= \tau_\nu(\mathbf{x}) + \tau_{-\nu}(\mathbf{x}).
\end{aligned}$$

We recall that the mean distance function $\delta_M$ is given by

$$\delta_M^{-2}(\mathbf{x}) = n\int_{\mathbb{S}^{n-1}} \rho_\nu^{-2}(\mathbf{x})d\omega(\nu), \tag{3.6.11}$$

and that for a convex $\Omega$, $\delta_M(\mathbf{x}) \leq \delta(\mathbf{x})$ by (3.3.9).

Let $\partial_\nu u$, $\nu \in \mathbb{S}^{n-1}$, denote the derivative of $u$ in the direction of $\nu$, i.e., $\partial_\nu u = \nu \cdot (\nabla u)$. It follows from Lemma 3.6.2 that

$$\int_\Omega |\partial_\nu u|^2 d\mathbf{x} \geq \frac{1}{4}\int_\Omega \rho_\nu(\mathbf{x})^{-2}\Big(1+[\frac{2\rho_\nu(\mathbf{x})}{\mu_\nu(\mathbf{x})}]\Big)^2 |u(\mathbf{x})|^2 d\mathbf{x}. \tag{3.6.12}$$

The last integrand satisfies

$$\begin{aligned}\rho_\nu(\mathbf{x})^{-2}\big(1+\big[\tfrac{2\rho_\nu(\mathbf{x})}{\mu_\nu(\mathbf{x})}\big]\big)^2 &= \rho_\nu(\mathbf{x})^{-2}+\tfrac{4}{\rho_\nu(\mathbf{x})\mu_\nu(\mathbf{x})}+\tfrac{4}{\mu_\nu(\mathbf{x})^2}\\ &\geq \rho_\nu(\mathbf{x})^{-2}+\tfrac{4}{\tau_\nu(\mathbf{x})\tau_{-\nu}(\mathbf{x})}+\tfrac{4}{\tau_\nu(\mathbf{x})^2+\tau_{-\nu}(\mathbf{x})^2}.\end{aligned} \tag{3.6.13}$$

Since

$$\int_{\mathbb{S}^{n-1}} \tau_\nu^2(\mathbf{x})d\omega(\nu)=\int_{\mathbb{S}^{n-1}} \tau_{-\nu}^2(\mathbf{x})d\omega(\nu),$$

we have from the Cauchy-Schwarz inequality,

$$\begin{aligned}\textstyle\int_{\mathbb{S}^{n-1}} \tau_\nu(\mathbf{x})\tau_{-\nu}(\mathbf{x})d\omega(\nu) &\leq \textstyle\int_{\mathbb{S}^{n-1}} \tau_\nu(\mathbf{x})^2 d\omega(\nu)\\ &\leq \big[\textstyle\int_{\mathbb{S}^{n-1}} \tau_\nu(\mathbf{x})^n d\omega(\nu)\big]^{2/n}\\ &= \Big[\tfrac{n}{s_{n-1}}|\Omega_{\mathbf{x}}|\Big]^{2/n}.\end{aligned}$$

Moreover,

$$1\leq \int_{\mathbb{S}^{n-1}}[\tau_\nu(\mathbf{x})\tau_{-\nu}(\mathbf{x})]d\omega(\nu)\int_{\mathbb{S}^{n-1}}[\tau_\nu(\mathbf{x})\tau_{-\nu}(\mathbf{x})]^{-1}d\omega(\nu),$$

and from this we derive

$$\begin{aligned}\textstyle\int_{\mathbb{S}^{n-1}}[\tau_\nu(\mathbf{x})\tau_{-\nu}(\mathbf{x})]^{-1}d\omega(\nu) &\geq \big[\textstyle\int_{\mathbb{S}^{n-1}} \tau_\nu(\mathbf{x})\tau_{-\nu}(\mathbf{x})d\omega(\nu)\big]^{-1}\\ &\geq \Big[\tfrac{n}{s_{n-1}}|\Omega_{\mathbf{x}}|\Big]^{-2/n}.\end{aligned}$$

For the third term in inequality (3.6.13)

$$\textstyle\int_{\mathbb{S}^{n-1}}(\tau_\nu(\mathbf{x})^2+\tau_{-\nu}(\mathbf{x})^2)d\omega(\nu)=2\int_{\mathbb{S}^{n-1}}\tau_\nu(\mathbf{x})^2 d\omega(\nu)$$

and this gives

$$\textstyle\int_{\mathbb{S}^{n-1}}(\tau_\nu(\mathbf{x})^2+\tau_{-\nu}(\mathbf{x})^2)^{-1}d\omega(\nu)\geq \frac{1}{2}\big[\frac{n}{s_{n-1}}|\Omega_{\mathbf{x}}|\big]^{-\frac{2}{n}}.$$

All in all, we have that

$$\begin{aligned}\textstyle\int_{\mathbb{S}^{n-1}}\rho_\nu(\mathbf{x})^{-2}\big[1+\big(\tfrac{2\rho_\nu(\mathbf{x})}{\mu_\nu(\mathbf{x})}\big)\big]^2 d\omega(\nu) &\geq \tfrac{1}{n}\delta_M^{-2}(\mathbf{x})+6\big[\tfrac{n}{s_{n-1}}|\Omega_{\mathbf{x}}|\big]^{-\frac{2}{n}}\\ &= \tfrac{1}{n}\left\{\delta_M(\mathbf{x})^{-2}+6K(n)|\Omega_{\mathbf{x}}|^{-2/n}\right\}.\end{aligned} \tag{3.6.14}$$

Upon combining this fact with (3.6.12) we have

$$\begin{aligned}
&\tfrac{1}{4}\int_\Omega \{\delta_M(\mathbf{x})^{-2} + 6K(n)|\Omega_{\mathbf{x}}|^{-2/n}\}|u(\mathbf{x})|^2 d\mathbf{x} \\
&\le n\int_\Omega \int_{\mathbb{S}^{n-1}} \left|\partial_\nu u(\mathbf{x})\right|^2 d\omega(\nu) d\mathbf{x} \\
&= n\int_\Omega \int_{\mathbb{S}^{n-1}} |\cos(\nu, \nabla u(\mathbf{x}))|^2 d\omega(\nu) \left|\nabla u(\mathbf{x})\right|^2 d\mathbf{x}.
\end{aligned} \tag{3.6.15}$$

On noting that

$$\int_{\mathbb{S}^{n-1}} |\cos(\nu, \nabla u(\mathbf{x}))|^2 d\omega(\nu) = \int_{\mathbb{S}^{n-1}} |\cos(\nu, \mathbf{e})|^2 d\omega(\nu)$$

for any fixed unit vector $\mathbf{e}$ in $\mathbb{R}^n$ and

$$\int_{\mathbb{S}^{n-1}} |\cos(\nu, \mathbf{e})|^2 d\omega(\nu) = \frac{1}{n},$$

the inequality (3.6.9) is seen to follow.

The inequality for $\Omega$ convex follows from the fact that $\Omega_{\mathbf{x}} = \Omega$ for any $\mathbf{x} \in \Omega$ and (3.3.9). □

Filippas et al. [60] estimate $\lambda^*(\Omega)$ in (3.6.1) in terms of the "interior diameter" $D_{int}(\Omega) := 2\sup_{\mathbf{x}\in\Omega} \delta(\mathbf{x})$. Clearly $D_{int}(\Omega) \le D(\Omega)$ and a significant fact is that $\Omega$ need not be bounded, nor have a finite volume, in order for $D_{int}(\Omega)$ to be finite. They prove for $\Omega$ convex that

$$\lambda^*(\Omega) \ge \frac{3}{D_{int}(\Omega)^2}. \tag{3.6.16}$$

In Sect. 3.4 below we shall present a sharp result obtained by Avkhadiev and Wirths [12] by a method reminiscent of those above, in that it is based on one-dimensional inequalities, but not using the mean distance function.

Following Theorem 3.1 in [60], we give here an $L^2$-Hardy inequality in the form introduced in [17]; since $|\nabla\delta| = 1$ the standard form of the inequality is immediate. We make use of the fact, proved in Theorem 2.3.2, that in a convex domain, $-\delta(\mathbf{x})$ is a convex function of $\mathbf{x}$ and $-\Delta\delta(\mathbf{x})$ is a nonnegative Radon measure. The proof remains valid if the requirement that $\Omega$ be convex is replaced by the assumption that $-\Delta\delta \ge 0$ in the distributional sense. For $n = 2$ this is equivalent to convexity, but is a weaker condition than convexity for $n \ge 3$; note the result from [107] in Proposition 2.5.4, that if $\Omega$ has $C^2$ boundary, $-\Delta\delta(\mathbf{x}) \ge 0$ is equivalent to the mean curvature of the boundary of $\Omega$ being non-positive.

**Theorem 3.6.4** *Let $\Omega \subset \mathbb{R}^n$ be a convex domain. Then for any $\alpha > -2$ and all $u \in H_0^1(\Omega)$,*

$$\int_\Omega |\nabla\delta \cdot \nabla u|^2 d\mathbf{x} - \frac{1}{4}\int_\Omega \frac{|u|^2}{\delta^2} d\mathbf{x} \ge \frac{C_\alpha}{D_{int}(\Omega)^{\alpha+2}} \int_\Omega \delta^\alpha |u|^2 dx \tag{3.6.17}$$

*with*

$$C_\alpha = \begin{cases} 2^\alpha(\alpha+2)^2, & \alpha \in (-2,-1), \\ 2^\alpha(2\alpha+3), & \alpha \in [-1,\infty). \end{cases}$$

*Proof* For $u \in C_0^\infty(\Omega)$ and $v(\mathbf{x}) := \delta(\mathbf{x})^{-\frac{1}{2}}u(\mathbf{x})$, it follows that

$$\int_\Omega \delta^\alpha |u|^2 d\mathbf{x} = \int_\Omega \delta^{\alpha+1}|v|^2 d\mathbf{x}, \tag{3.6.18}$$

and since $|\nabla\delta| = 1$ a.e.,

$$\int_\Omega |\nabla\delta\cdot\nabla u|^2 d\mathbf{x} - \frac{1}{4}\int_\Omega \frac{|u|^2}{\delta^2}d\mathbf{x} = \int_\Omega \delta|\nabla\delta\cdot\nabla v|^2 d\mathbf{x} + \frac{1}{2}\int_\Omega(-\Delta\delta)|v|^2 d\mathbf{x}. \tag{3.6.19}$$

Then on integration by parts,

$$\begin{aligned} \int_\Omega \delta^{\alpha+1}|v|^2 d\mathbf{x} &= \tfrac{1}{\alpha+2}\int_\Omega (\nabla\delta^{\alpha+2}\cdot\nabla\delta)|v|^2 d\mathbf{x} \\ &= -\tfrac{1}{\alpha+2}\int_\Omega \delta^{\alpha+2}\mathrm{div}\,(|v|^2\nabla\delta)d\mathbf{x} \\ &= -\tfrac{2}{\alpha+2}\mathrm{Re}\int_\Omega \delta^{\alpha+2}\overline{v}(\nabla\delta\cdot\nabla v)d\mathbf{x} + \tfrac{1}{\alpha+2}\int_\Omega \delta^{\alpha+2}(-\Delta\delta)|v|^2 d\mathbf{x}. \end{aligned}$$

Using the last identity we have for $R_{int} := \frac{1}{2}D_{int}(\Omega)$

$$\begin{aligned} (\alpha+2)\int_\Omega \delta^{\alpha+1}|v|^2 d\mathbf{x} &\le 2\left(\int_\Omega \delta^{\alpha+1}|v|^2 d\mathbf{x}\right)^{\frac{1}{2}}\left(\int_\Omega \delta^{\alpha+3}|\nabla\delta\cdot\nabla v|^2 d\mathbf{x}\right)^{\frac{1}{2}} \\ &\quad + R_{int}^{\alpha+2}\int_\Omega(-\Delta\delta)|v|^2 d\mathbf{x} \\ &\le \varepsilon\int_\Omega \delta^{\alpha+1}|v|^2 d\mathbf{x} + \tfrac{1}{\varepsilon}\int_\Omega \delta^{\alpha+3}|\nabla\delta\cdot\nabla v|^2 d\mathbf{x} \\ &\quad + R_{int}^{\alpha+2}\int_\Omega(-\Delta\delta)|v|^2 d\mathbf{x} \\ &\le \varepsilon\int_\Omega \delta^{\alpha+1}|v|^2 d\mathbf{x} \\ &\quad + R_{int}^{\alpha+2}\left(\tfrac{1}{\varepsilon}\int_\Omega \delta|\nabla\delta\cdot\nabla v|^2 d\mathbf{x} + \int_\Omega(-\Delta\delta)|v|^2 d\mathbf{x}\right) \end{aligned}$$

since $-\Delta\delta \geq 0$. Consequently,

$$(\alpha+2-\varepsilon)\int_\Omega \delta^{\alpha+1}|v|^2 d\mathbf{x} \leq R_{int}^{\alpha+2}\left(\frac{1}{\varepsilon}\int_\Omega \delta|\nabla\delta\cdot\nabla v|^2 d\mathbf{x} + \int_\Omega(-\Delta\delta)|v|^2 d\mathbf{x}\right). \tag{3.6.20}$$

Choose $\varepsilon = \min\{\frac{1}{2}, \frac{\alpha+2}{2}\}$. Then, by (3.6.18) and (3.6.19),

$$\begin{aligned}\varepsilon(\alpha+2-\varepsilon)\int_\Omega \delta^\alpha|u|^2 d\mathbf{x} &\leq R_{int}^{\alpha+2}\left(\int_\Omega \delta|\nabla\delta\cdot\nabla v|^2 d\mathbf{x} + \tfrac{1}{2}\int_\Omega(-\Delta\delta)|v|^2 d\mathbf{x}\right)\\ &= \left(\tfrac{D_{int}(\Omega)}{2}\right)^{\alpha+2}\left(\int_\Omega |\nabla\delta\cdot\nabla u|^2 d\mathbf{x} - \tfrac{1}{4}\int_\Omega \tfrac{|u|^2}{\delta^2}d\mathbf{x}\right)\end{aligned}$$

and (3.6.17) follows by making the appropriate choice of $\varepsilon$. □

Another class of inequalities which extend the Hardy inequality is that of the so-called Hardy-Sobolev-Maz'ya inequalities. An example, which we state in the case $p = 2$ only, is one established by S. Filippas, V. Maz'ya, and A. Tertikas (for $2 \leq p < n$) in [61], that if $\Omega$ has a finite interior diameter, $D_{int}(\Omega)$, and $-\Delta\delta \geq 0$ then

$$\int_\Omega |\nabla u(\mathbf{x})|^2 d\mathbf{x} \geq \frac{1}{4}\int_\Omega \frac{|u(\mathbf{x})|^2}{\delta(\mathbf{x})^2}d\mathbf{x} + C_\Omega\left(\int_\Omega |u(\mathbf{x})|^{2n/(n-2)}d\mathbf{x}\right)^{(n-2)/n}, \tag{3.6.21}$$

for $n \geq 3$ and all $u \in C_0^\infty(\Omega)$. Inequalities of this type will be the subject of Chap. 4.

### 3.6.2 Analogous Results in $L^p(\Omega)$

In the proof of Lemma 3.6.2 a key substitution $v(t) = [1 - (\frac{\rho(t)}{b})]u(t)$ was made that led to the Hardy inequality in Theorem 3.6.3 when $p = 2$. In the absence of such a substitution we treat the case for other values of $p > 1$ using the methods of Tidblom, cf. Theorems 1.1, 2.1, in [144].

**Lemma 3.6.5** *Let $u \in C_0^1(0, 2b)$, $p \in (1, \infty)$. Then*

$$\int_0^{2b} |u'(t)|^p dt \geq \left[\frac{p-1}{p}\right]^p \int_0^{2b}\{\rho(t)^{-p} + (p-1)b^{-p}\}|u(t)|^p dt. \tag{3.6.22}$$

*Proof* By (3.6.4), for a monotonic function $f$ and a positive function $g$,

$$\begin{aligned}\int_0^b |f'(t)||u(t)|^p dt &\leq \int_0^b p|f(t)-f(b)||u(t)|^{p-1}|u'(t)|dt\\ &\leq p\left[\int_0^b g(t)|u'(t)|^p dt\right]^{1/p}\left[\int_0^b \left(\frac{|f(t)-f(b)|^p}{g(t)}\right)^{1/(p-1)}|u(t)|^p dt\right]^{1-1/p}.\end{aligned}$$

Consequently,

$$p^p \int_0^b g(t)|u'(t)|^p dt \geq \frac{\left(\int_0^b |f'(t)||u(t)|^p dt\right)^p}{\left(\int_0^b \left(\frac{|f(t)-f(b)|^p}{g(t)}\right)^{1/(p-1)} |u(t)|^p dt\right)^{p-1}}.$$

From Young's inequality for $q = p/(p-1)$, we have that

$$AB^{\frac{p}{q}} \leq \frac{A^p}{p} + \frac{B^p}{q} \implies A^p/B^{p-1} \geq pA - (p-1)B,$$

which implies, for $A = \int_0^b |f'(t)||u(t)|^p dt$ and $B = \int_0^b \left(\frac{|f(t)-f(b)|^p}{g(t)}\right)^{1/p-1} |u(t)|^p dt$, that

$$\begin{aligned} &p^p \int_0^b g(t)|u'(t)|^p dt \\ &\geq \int_0^b \left\{p|f'(t)| - (p-1)\left(\frac{|f(t)-f(b)|^p}{g(t)}\right)^{1/(p-1)}\right\} |u(t)|^p dt. \end{aligned} \tag{3.6.23}$$

We now choose $f(t) = t^{-p+1}$ and $g(t) = (p-1)^{-(p-1)}$. Then

$$\left(\frac{|f(t)-f(b)|^p}{g(t)}\right)^{1/(p-1)} = (p-1)t^{-p}\left[\left(1-\left(\tfrac{t}{b}\right)^{p-1}\right)^p\right]^{\frac{1}{p-1}}.$$

Consequently, for $t \in (0, b)$,

$$\begin{aligned} &p|f'(t)| - (p-1)\left(\frac{|f(t)-f(b)|^p}{g(t)}\right)^{1/(p-1)} \\ &= (p-1)\left\{p\, t^{-p} - (p-1)t^{-p}\left[\left(1-\left(\tfrac{t}{b}\right)^{p-1}\right)^p\right]^{\frac{1}{p-1}}\right\} \\ &= (p-1)t^{-p}\left\{1 + (p-1)\left(1 - \left[1-\left(\tfrac{t}{b}\right)^{p-1}\right]^{\frac{p}{p-1}}\right)\right\} \\ &\geq (p-1)t^{-p}\left\{1 + (p-1)\left(\tfrac{t}{b}\right)^{p-1}\right\} \\ &\geq (p-1)\left\{t^{-p} + (p-1)\left(\tfrac{1}{b^p}\right)\right\}, \end{aligned} \tag{3.6.24}$$

since

$$\left[1-\left(\frac{t}{b}\right)^{p-1}\right]^{\frac{p}{p-1}} \leq 1 - \left(\frac{t}{b}\right)^{p-1}.$$

As in the proof of Lemma 3.6.1, we let $u := v\chi_{(0,b]}$ for $v \in C_0^1(0, 2b)$, and use (3.6.24) in (3.6.23) to conclude that

$$\int_0^b |u'(t)|^p dt \geq \left[\tfrac{p-1}{p}\right]^p \int_0^b \{\rho(t)^{-p} + (p-1)b^{-p}\} |u(t)|^p dt.$$

A similar analysis on $[b, 2b)$ completes the proof. □

Using Lemma 3.6.5 we are able to prove Theorem 2.1 in [144]and, in the case of $p = 2$, Theorems 3.1 and 3.2, of [78]. Note that inequality (3.6.25) for the case $p = 2$ is weaker than that given by (3.6.9). We shall use the terminology from the proof of Theorem 3.6.3.

**Theorem 3.6.6** *Let* $p \in (1, \infty)$, *and*

$$K(n,p) := (p-1)\left[\frac{s_{n-1}}{n}\right]^{p/n} / B(n,p),$$

*where*

$$B(n,p) := \frac{\Gamma(\frac{p+1}{2}) \cdot \Gamma(\frac{n}{2})}{\sqrt{\pi}\, \Gamma(\frac{n+p}{2})};$$

*thus* $B(n,2) = 1/n$ *and* $K(n,2) = K(n)$. *Then for all* $u \in C_0^1(\Omega)$,

$$\int_\Omega |\nabla u(\mathbf{x})|^p d\mathbf{x} \geq \left(\frac{p-1}{p}\right)^p \left\{ \int_\Omega \frac{|u(\mathbf{x})|^p}{\delta_{M,p}^p(\mathbf{x})} d\mathbf{x} + K(n,p) \int_\Omega \frac{|u(\mathbf{x})|^p}{|\Omega_\mathbf{x}|^{\frac{p}{n}}} d\mathbf{x} \right\}. \tag{3.6.25}$$

*If* $\Omega$ *is convex,*

$$\int_\Omega |\nabla u(\mathbf{x})|^p d\mathbf{x} \geq \left(\frac{p-1}{p}\right)^p \left\{ \int_\Omega \frac{|u(\mathbf{x})|^p}{\delta(\mathbf{x})^p} d\mathbf{x} + \frac{K(n,p)}{|\Omega|^{p/n}} \int_\Omega |u(\mathbf{x})|^p d\mathbf{x} \right\}. \tag{3.6.26}$$

*Proof* From Lemma 3.6.5 we have that for any $\nu \in \mathbb{S}^{n-1}$ and $u \in C_0^1(\Omega)$,

$$\int_\Omega |\partial_\nu u(\mathbf{x})|^p d\mathbf{x} \geq \left[\frac{p-1}{p}\right]^p \int_\Omega \left\{\rho_\nu(\mathbf{x})^{-p} + \frac{(p-1)2^p}{D_\nu(\mathbf{x})^p}\right\} |u(\mathbf{x})|^p d\mathbf{x}. \tag{3.6.27}$$

The next step is to integrate in (3.6.27) over $\mathbb{S}^{n-1}$ with respect to $d\omega(\nu)$ and substitute the identity

$$\begin{aligned} \int_{\mathbb{S}^{n-1}} |\partial_\nu u(\mathbf{x})|^p d\omega(\nu) &= |\nabla u(\mathbf{x})|^p \int_{\mathbb{S}^{n-1}} |\cos(\nu, \nabla u(\mathbf{x}))|^p d\omega(\nu) \\ &= |\nabla u(\mathbf{x})|^p \int_{\mathbb{S}^{n-1}} |\cos(\nu, \mathbf{e})|^p d\omega(\nu) \\ &= B(n,p)|\nabla u(\mathbf{x})|^p, \end{aligned} \tag{3.6.28}$$

where $\mathbf{e}$ is any unit vector in $\mathbb{R}^n$. In order to evaluate the integral of $(2/D_\nu(\mathbf{x}))^p$ with respect to $d\omega(\nu)$, we proceed as in [144]. Since $f(t) = t^p$ is convex for $p > 1$, we have by Jensen's inequality,

$$\begin{aligned} 1 &\le \Big(\int_{\mathbb{S}^{n-1}} \Big(\frac{2}{D_\nu(\mathbf{x})}\Big) d\omega(\nu)\Big)^p \Big(\int_{\mathbb{S}^{n-1}} \Big(\frac{D_\nu(\mathbf{x})}{2}\Big) d\omega(\nu)\Big)^p \\ &\le \Big(\int_{\mathbb{S}^{n-1}} \Big(\frac{2}{D_\nu(\mathbf{x})}\Big)^p d\omega(\nu)\Big)\Big(\int_{\mathbb{S}^{n-1}} \Big(\frac{D_\nu(\mathbf{x})}{2}\Big) d\omega(\nu)\Big)^p, \end{aligned}$$

and hence

$$\int_{\mathbb{S}^{n-1}} \left(\frac{2}{D_\nu(\mathbf{x})}\right)^p d\omega(\nu) \ge \left(\int_{\mathbb{S}^{n-1}} \frac{D_\nu(\mathbf{x})}{2} d\omega(\nu)\right)^{-p}. \tag{3.6.29}$$

But,

$$\begin{aligned} \int_{\mathbb{S}^{n-1}} \frac{D_\nu(\mathbf{x})}{2} d\omega(\nu) &= \frac{1}{2} \int_{\mathbb{S}^{n-1}} (\tau_\nu(\mathbf{x}) + \tau_{-\nu}(\mathbf{x})) d\mathbf{x} \\ &= \int_{\mathbb{S}^{n-1}} \tau_\nu(\mathbf{x}) d\mathbf{x} \\ &\le \left[\int_{\mathbb{S}^{n-1}} \tau_\nu^n(\mathbf{x}) d\mathbf{x}\right]^{\frac{1}{n}} \\ &= \left(\frac{n|\Omega_\mathbf{x}|}{s_{n-1}}\right)^{\frac{1}{n}}. \end{aligned} \tag{3.6.30}$$

Therefore, the conclusion follows on using (3.6.30) in (3.6.29) and (3.6.28) in (3.6.27). □

### 3.6.3 Sharp Results of Avkhadiev and Wirths

We now present the sharp inequality of Avkhadiev and Wirths [12] for convex domains $\Omega$ with finite inradius $\delta_0 := \sup_{\mathbf{x}\in\Omega} \delta(\mathbf{x})$. Let $J_0$ and $J_1$ denote the Bessel functions of order 0 and 1, respectively, with $\lambda_0$ representing the first zero in $(0, \infty)$ of the function

$$g(x) := J_0(x) - 2xJ_1(x) \equiv J_0(x) + 2xJ_0'(x).$$

The next proposition is proved in [12].

**Proposition 3.6.7** *Let $f$ be a real absolutely continuous function in* $[0, 1]$ *such that* $f(0) = 0$ *and* $f' \in L^2[0, 1]$. *If* $f(x) \not\equiv 0$ *then*

$$\int_0^1 f'(x)^2 dx > \frac{1}{4} \int_0^1 \frac{f(x)^2}{x^2} dx + \lambda_0^2 \int_0^1 f(x)^2 dx. \tag{3.6.31}$$

*Proof* Let

$$f_0(x) := \sqrt{x}J_0(\lambda_0 x). \tag{3.6.32}$$

Then,

$$f_0'(x) = \frac{g(\lambda_0 x)}{2\sqrt{x}}.$$

Since $J_n$ is a solution of Bessel's differential equation

$$x^2y'' + xy' + (x^2 - n^2)y = 0,$$

then a calculation shows that $f_0(x)$ is a solution of the differential equation

$$y'' + \left(\frac{1}{4x^2} + \lambda_0^2\right)y = 0.$$

Also, $f_0'(1) = 0$, and since $\lambda_0$ is the first zero of $g$ in $(0, \infty)$, $f_0(x) > 0$ and $f_0'(x) > 0$ for $x \in (0, 1)$.

Since $f_0' \notin L^2[0, 1]$, then $f_0 \not\equiv f$ and

$$\begin{aligned}
0 &< \textstyle\int_0^1 \left(f'(x) - \frac{f_0'(x)}{f_0(x)}f(x)\right)^2 dx \\
&= \textstyle\int_0^1 f'(x)^2dx + \int_0^1 \left(\frac{f_0'(x)^2}{f_0(x)^2} + \left(\frac{f_0'(x)}{f_0(x)}\right)'\right)f(x)^2dx + \lim_{x\to 0+}\frac{f_0'(x)}{f_0(x)}f^2(x) \\
&= \textstyle\int_0^1 f'(x)^2dx - \int_0^1 \left(\frac{1}{4x^2} + \lambda_0^2\right)f(x)^2dx + \lim_{x\to 0+}\frac{f_0'(x)}{f_0(x)}f^2(x).
\end{aligned}$$

In order to show that the limit in the last expression is zero, note that

$$\frac{f_0'(x)}{f_0(x)} = \frac{g(\lambda_0 x)}{2xJ_0(\lambda_0 x)}, \qquad x \in [0, 1],$$

and $f(x)^2/x \to 0$ as $x \to 0^+$ since

$$f(x)^2 \le \left(\int_0^x |f'(t)|dt\right)^2 \le x\int_0^x |f'(t)|^2dt, \tag{3.6.33}$$

which completes the proof. □

The sharpness of $\lambda_0$ in (3.6.31) will now be examined.

**Proposition 3.6.8** *For each $\varepsilon_0 > 0$, there exists a real function $f \in C_0^1(0,2)$ such that $f'(1) = 0$ and*

$$\int_0^1 f'(x)^2 dx < \frac{1}{4}\int_0^1 \frac{f(x)^2}{x^2}dx + (\lambda_0^2 + \varepsilon_0)\int_0^1 f(x)^2 dx. \tag{3.6.34}$$

*Proof* For $f_0$ defined in (3.6.32) and $\varepsilon > 0$, let

$$f_\varepsilon(x) := \begin{cases} x^{\frac{\varepsilon}{2}} f_0(x), & x \in [0,1], \\ f_\varepsilon(2-x), & x \in (1,2]. \end{cases}$$

We shall prove that, for sufficiently small $\varepsilon$, $f_\varepsilon$ satisfies (3.6.34); this will suffice by standard density arguments, since $f_\varepsilon \in H_0^1(0,2)$. Now for $x \in [0,1]$,

$$f_\varepsilon'(x)^2 = \frac{\varepsilon^2}{4}x^{\varepsilon-2}f_0(x)^2 + \frac{\varepsilon}{2}x^{\varepsilon-1}(f_0^2(x))' + x^\varepsilon f_0'(x)^2,$$

implying that

$$\int_0^1 f_\varepsilon'(x)^2 dx = \tfrac{2\varepsilon-\varepsilon^2}{4}\int_0^1 x^{\varepsilon-2}f_0(x)^2 dx + \tfrac{\varepsilon}{2}J_0(\lambda_0)^2 + \int_0^1 x^\varepsilon f_0'(x)^2 dx$$

by the use of (3.6.33). On the other hand, it follows from the identity

$$-f_0'' = (\frac{1}{4x^2} + \lambda_0^2)f_0$$

that

$$\begin{aligned} \int_0^1 (\tfrac{1}{4x^2} + \lambda_0^2)f_\varepsilon(x)^2 dx &= -\int_0^1 x^\varepsilon f_0'' f_0(x)dx \\ &= \int_0^1 x^\varepsilon [f_0'(x)]^2 dx + \tfrac{\varepsilon}{2}J_0(\lambda_0)^2 \\ &\quad -\tfrac{\varepsilon(\varepsilon-1)}{2}\int_0^1 x^{\varepsilon-2}f_0(x)^2 dx. \end{aligned}$$

Consequently, for all $\varepsilon$ sufficiently small,

$$\begin{aligned} &\int_0^1 f_\varepsilon'(x)^2 dx - \int_0^1 (\tfrac{1}{4x^2} + \lambda_0^2)f_\varepsilon(x)^2 dx - \varepsilon_0 \int_0^1 f_\varepsilon(x)^2 dx \\ &= \tfrac{\varepsilon^2}{4}\int_0^1 x^{\varepsilon-2}f_0(x)^2 dx - \varepsilon_0\int_0^1 x^\varepsilon f_0(x)^2 dx < 0 \end{aligned},$$

completing the proof. □

**Proposition 3.6.9** *Let $f$ be a real function in $H_0^1(a, b)$ and $\alpha := (b-a)/2 \in (0, \infty)$. If $f'(x) \not\equiv 0$, then*

$$\int_a^b f'(x)^2 dx > \frac{1}{4}\int_a^b \frac{f(x)^2}{(\min\{x-a, b-x\})^2} dx + \frac{\lambda_0^2}{\alpha^2}\int_a^b f(x)^2 dx, \tag{3.6.35}$$

*where the constants $\frac{1}{4}$ and $\lambda_0^2/\alpha^2$ are sharp.*

*Proof* For $d = (a+b)/2$, (3.6.35) is implied by the inequalities

$$\int_a^d f'(x)^2 dx > \frac{1}{4}\int_a^d \frac{f(x)^2}{(x-a)^2} dx + \frac{\lambda_0^2}{\alpha^2}\int_a^d f(x)^2 dx$$

and

$$\int_d^b f'(x)^2 dx > \frac{1}{4}\int_d^b \frac{f(x)^2}{(b-x)^2} dx + \frac{\lambda_0^2}{\alpha^2}\int_d^b f(x)^2 dx,$$

each of which, by the respective change of variable $x - a = t$ and $b - x = t$, being equivalent to the inequality

$$\int_0^\alpha f'(x)^2 dx > \frac{1}{4}\int_0^\alpha \frac{f(t)^2}{t^2} dx + \frac{\lambda_0^2}{\alpha^2}\int_0^\alpha f(t)^2 dt \tag{3.6.36}$$

for $f \in H_0^1(0, 2\alpha)$, $f \not\equiv 0$. However, the change of variable $t = \alpha x$ in (3.6.36) reduces it to (3.6.31). Proposition 3.6.9 therefore follows from Proposition 3.6.7 and Proposition 3.6.8. □

The one-dimensional results above will now be used to produce results for higher dimensional cases. An essential ingredient in this analysis is an approximation result of Hadwiger [70] for convex domains. In particular, for a convex domain $\Omega \subset \mathbb{R}^n$ and any compact set $K \subset \Omega$, there exists a convex $n$-dimensional polytope $Q$ such that $K \subset \operatorname{int} Q \subset \Omega$; thus, for any $f \in C_0^\infty(\Omega)$, there is a convex $n$-dimensional polytope $Q$ such that

$$\operatorname{supp} f \subset \operatorname{int} Q \subset \Omega. \tag{3.6.37}$$

In this manner, the proof of certain $n$-dimensional Hardy inequalities is reduced to the application of one-dimensional inequalities. We shall give the $L^2(\Omega)$ case of [10], Theorem 11; Avkhadiev considers the general $L^p(\Omega), p > 1$, case and also includes weights of the form $\delta(\mathbf{x})^s$. Results with weights of this type can be used to study the spectral properties of certain elliptic differential operators which have a degeneracy at the boundary of $\Omega$ measured in terms of $\delta(\mathbf{x})$, e.g., see [106], Sect. 3.

**Theorem 3.6.10** *Let $\Omega \subset \mathbb{R}^n$ be an open, convex set with finite inradius $\delta_0 := \sup_{\mathbf{x}\in\Omega} \delta(\mathbf{x})$. If, for $\alpha \in (0, \delta_0]$ and nonnegative constants $b$, $c$,*

$$\int_0^\alpha f'(t)^2 dt \geq b^2 \int_0^\alpha \frac{f(t)^2}{t^2} dt + \frac{c^2}{\delta_0^2} \int_0^\alpha f(t)^2 dt, \qquad f \in H_0^1(0, 2\alpha), \tag{3.6.38}$$

*then*

$$\int_\Omega |\nabla f(\mathbf{x}))|^2 d\mathbf{x} \geq b^2 \int_\Omega \frac{|f(\mathbf{x})|^2}{\delta(\mathbf{x})^2} dt + \frac{c^2}{\delta_0^2} \int_\Omega |f(\mathbf{x})|^2 d\mathbf{x}, \qquad f \in H_0^1(\Omega). \tag{3.6.39}$$

*Proof* Choose $f \in C_0^\infty(\Omega)$ and let $Q$ be an $n$-dimensional polytope satisfying (3.6.37). It will suffice to show that (3.6.39) holds on $Q$. Let $S_1, S_2, \dots S_m$ denote the $(n-1)$-dimensional faces of $Q$. The polytope $Q$ can be decomposed as follows:

$$Q = \cup_{j=1}^m Q_j, \qquad \text{int } Q_j \cap Q_k = \emptyset \text{ for } j \neq k, \tag{3.6.40}$$

in which each $Q_j$ is convex and compact. To see this, let $\mathbf{n}_j(\mathbf{x}')$ be the inward unit normal to $S_j$ at $\mathbf{x}' \in S_j$ and define

$$\alpha_j(\mathbf{x}') := \max\{t \in (0, \infty) : B(\mathbf{x}' + t\mathbf{n}_j(\mathbf{x}'), t) \subset Q\}$$

in which $B(\mathbf{x}, t)$ is the ball with centre $\mathbf{x}$ and radius $t$. It follows that

$$Q_j := S_j \cup \{\mathbf{x} = \mathbf{x}' + t\alpha_j(\mathbf{x}') : t \in (0, \alpha_j(\mathbf{x}')], \mathbf{x}' \in S_j\}$$

is a closed, $n$-dimensional, convex set, and $\{Q_j\}_{j=1}^m$ satisfies (3.6.40). By the convexity of each $Q_j$ we have that $\cup_{j=1}^m \partial Q_j$ has measure zero, implying the important feature that for any $g \in L^1(Q)$

$$\int_Q g(\mathbf{x}) d\mathbf{x} = \sum_{j=1}^m \int_{Q_j} g(\mathbf{x}) d\mathbf{x} = \sum_{j=1}^m \int_{S_j} \int_0^{\alpha_j(\mathbf{x}')} g(\mathbf{x}' + t\mathbf{n}_j(\mathbf{x}')) dt d\mathbf{x}'$$

by Fubuni's theorem. On applying this to (3.6.38) and recalling Theorem 1.3.8, we have that

$$\begin{aligned}
&\textstyle\int_Q \left(\frac{b^2}{\delta(\mathbf{x})^2} + \frac{c^2}{\delta_0^2}\right) |f(\mathbf{x})|^2 d\mathbf{x} \\
&= \textstyle\sum_{j=1}^m \int_{Q_j} \left(\frac{b^2}{\delta(\mathbf{x})^2} + \frac{c^2}{\delta_0^2}\right) |f(\mathbf{x})|^2 d\mathbf{x} \\
&= \textstyle\sum_{j=1}^m \int_{S_j} \int_0^{\alpha_j(\mathbf{x}')} \left(\frac{b^2}{t^2} + \frac{c^2}{\delta_0^2}\right) |f(\mathbf{x}' + t\mathbf{n}_j(\mathbf{x}'))|^2 dt d\mathbf{x}' \\
&\leq \textstyle\sum_{j=1}^m \int_{S_j} \int_0^{\alpha_j(\mathbf{x}')} |\nabla f(\mathbf{x}' + t\mathbf{n}_j(\mathbf{x}'))|^2 dt d\mathbf{x}' \\
&= \textstyle\int_Q |\nabla f(\mathbf{x})|^2 d\mathbf{x},
\end{aligned}$$

where we have used the facts that $\delta(\mathbf{x}) = t$ and $\delta(\mathbf{x}) \le \alpha_j(\mathbf{x}') \le \delta_0$ for $\mathbf{x} = \mathbf{x}' + t\mathbf{n}_j(\mathbf{x}') \in Q_j$, as well as

$$\frac{\partial f}{\partial t}|_{\mathbf{x}=\mathbf{x}'+t\mathbf{n}_j(\mathbf{x}')} = \nabla f \cdot \mathbf{n}_j(\mathbf{x}').$$

That completes the proof. □

We now examine the sharpness of the inequality for $n > 1$.

**Proposition 3.6.11** *Let* $\Omega_n := (0, 2) \times \mathbb{R}^{n-1}$ *for* $n \ge 2$ *and* $\Omega_1 := (0, 2)$. *For any* $\varepsilon_0 > 0$ *there exists* $f_n \in C_0^1(\Omega_n)$ *such that*

$$\int_{\Omega_n} |\nabla f_n|^2 d\mathbf{x} < \frac{1}{4}\int_{\Omega_n} \frac{|f_n(\mathbf{x})|^2}{\delta(\mathbf{x})^2} d\mathbf{x} + (\lambda_0^2 + \varepsilon_0)\int_{\Omega_n} |f_n(\mathbf{x})|^2 d\mathbf{x}. \tag{3.6.41}$$

*Proof* In order to prove the proposition by mathematical induction on the dimension $n$ we note that the case for $n = 1$ is given by Proposition 3.6.8. Suppose that (3.6.41) holds for some dimension $n \ge 1$ and some function $f_n \in C_0^1(\Omega_n)$. Then we must show that it holds for dimension $n + 1$ and some $f_{n+1} \in C_0^1(\Omega_{n+1})$. To this end we define $f_{n+1} := f_{n+1,\varepsilon} \in C_0^1(\Omega_{n+1})$, $\varepsilon > 0$, by

$$f_{n+1,\varepsilon}(\mathbf{x}) = f_n(\mathbf{x}')g_\varepsilon(x_{n+1}), \qquad \mathbf{x}' \in \Omega_n, \quad x_{n+1} \in \mathbb{R}$$

where

$$g_\varepsilon(t) := \begin{cases} 1, & t \in [0, 1/\varepsilon], \\ (1 - (t - 1/\varepsilon)^2)^2, & t \in (1/\varepsilon, 1 + 1/\varepsilon), \\ 0, & t \in [1 + 1/\varepsilon, \infty) \\ g_\varepsilon(-t), & t < 0. \end{cases}$$

Define

$$A_n := \int_{\Omega_n} |\nabla f_n|^2 d\mathbf{x} - \frac{1}{4}\int_{\Omega_n} \frac{|f_n(\mathbf{x})|^2}{\delta(\mathbf{x})^2} d\mathbf{x} - (\lambda_0^2 + \varepsilon_0)\int_{\Omega_n} |f_n(\mathbf{x})|^2 d\mathbf{x}$$

and $g(t) := (1 - t^2)^2$. Note that

$$\delta(\mathbf{x}) = \text{dist}(\mathbf{x}', \partial\Omega_n), \qquad \mathbf{x} = (\mathbf{x}', x_{n+1}) \in \Omega_{n+1}.$$

Calculations show that

$$A_{n+1} = \frac{2}{\varepsilon}A_n + B_n - C_n$$

in which

$$B_n = 2\int_0^1 \int_{\Omega_n} [|\nabla f_n(\mathbf{x}')|^2 g(t)^2 + f_n(\mathbf{x}')^2 g'(t)^2] d\mathbf{x}' dt$$

and

$$C_n = 2\int_0^1 g(t)^2 dt \left[\frac{1}{4}\int_{\Omega_n} \frac{f_n(\mathbf{x}')^2}{\delta^2} d\mathbf{x}' + (\lambda_0^2 + \varepsilon_0)\int_{\Omega_n} f_n(\mathbf{x}')^2 d\mathbf{x}'\right].$$

Since $B_n$ and $C_n$ are not dependent upon $\varepsilon$ and $A_n < 0$ (by the induction assumption), it is clear that $A_{n+1} < 0$ for all $\varepsilon$ sufficiently near zero and positive. That completes the proof of the induction step and the proposition is proved. □

In summary, we have proved the following result which is Theorem 1 in [12]:

**Theorem 3.6.12** *Let $\Omega \subset \mathbb{R}^n$ be convex. If the inradius $\delta_0 < \infty$, then*

$$\int_\Omega |\nabla f(\mathbf{x}))|^2 d\mathbf{x} \geq \frac{1}{4}\int_\Omega \frac{|f(\mathbf{x})|^2}{\delta(\mathbf{x})^2} d\mathbf{x} + \frac{\lambda_0^2}{\delta_0^2}\int_\Omega |f(\mathbf{x})|^2 d\mathbf{x}, \qquad f \in H_0^1(\Omega) \qquad (3.6.42)$$

*where $\lambda_0 = 0.940\ldots$ is the first zero in $(0,\infty)$ of*

$$J_0(t) - 2tJ_1(t).$$

*Inequality (3.6.42) is sharp for $n \geq 1$.*

## 3.7 Hardy Inequalities and Curvature

### *3.7.1 General Inequalities*

Most of the inequalities discussed to this point have required that the domain $\Omega$ be convex. In order to broaden our applications beyond that requirement we will use the connection between the curvature of the boundary $\partial\Omega$ and the distance function $\delta(\mathbf{x})$ as discussed in Chap. 2.

We first establish the following general inequality which will serve as a guide to needed improvements; it also extends Theorem 3.4.3.

**Proposition 3.7.1** *Let $\Omega \subset \mathbb{R}^n, n \geq 2$, be a domain having a ridge $\mathcal{R}(\Omega)$ and a sufficiently smooth boundary for Green's formula to hold. Let $\delta(\mathbf{x}) = \mathrm{dist}(\mathbf{x}, \mathbb{R}^n \setminus \Omega)$. Then for all $f \in C_0^\infty(\Omega \setminus \overline{\mathcal{R}(\Omega)})$ and $p \in (1,\infty)$,*

$$\int_\Omega |\nabla\delta \cdot \nabla f|^p d\mathbf{x} \geq \left(\frac{p-1}{p}\right)^p \int_\Omega \left\{1 - \frac{p\delta\Delta\delta}{p-1}\right\} \frac{|f|^p}{\delta^p} d\mathbf{x}. \qquad (3.7.1)$$

*Proof* For any vector field $V$ we have the identity

$$\int_\Omega (\mathrm{div} V)|f|^p d\mathbf{x} = -p\left[\mathrm{Re}\int_\Omega (V\cdot\nabla f)|f|^{p-2}\bar{f}d\mathbf{x}\right] \tag{3.7.2}$$

for all $f \in C_0^\infty(\Omega \setminus \overline{\mathcal{R}(\Omega)})$. Choose

$$V = -p\nabla\delta/\delta^{p-1}.$$

Then, for any $\varepsilon > 0$,

$$\begin{aligned}\int_\Omega \mathrm{div}V|f|^p d\mathbf{x} &\le p^2\left(\int_\Omega |\nabla\delta\cdot\nabla f|^p d\mathbf{x}\right)^{1/p}\left(\int_\Omega \frac{|f|^p}{\delta^p}d\mathbf{x}\right)^{1-1/p}\\ &\le p\varepsilon^p\int_\Omega |\nabla\delta\cdot\nabla f|^p d\mathbf{x} + p(p-1)\varepsilon^{-p/(p-1)}\int_\Omega \frac{|f|^p}{\delta^p}d\mathbf{x}\end{aligned}$$

which gives, since $\mathrm{div}V = (p-1)p\delta^{-p} - p\delta^{1-p}\Delta\delta$ for $\mathbf{x}\in\Omega\setminus\overline{\mathcal{R}(\Omega)}$,

$$\int_\Omega |\nabla\delta\cdot\nabla f|^p d\mathbf{x} \ge \varepsilon^{-p}\int_\Omega \left[(p-1)-(p-1)\varepsilon^{-p/(p-1)} - \delta\Delta\delta\right]\frac{|f|^p}{\delta^p}d\mathbf{x}.$$

The proof of (3.7.1) is completed on choosing $\varepsilon = [p/(p-1)]^{\frac{(p-1)}{p}}$. □

The requirement that (3.7.1) only applies to functions $f$ that are supported away from the ridge $\mathcal{R}(\Omega)$ can be obviated. Working with examples like those below we are led to some additional requirements ((i) and (ii) below) described in Theorem 2 of [20] and the next Proposition. Subsequently, it is shown that those requirements are met under quite general assumptions.

**Proposition 3.7.2** *Let $\Omega\subset\mathbb{R}^n$, $n\ge 2$, satisfy the hypothesis of Proposition 3.7.1 and furthermore, assume that*

*(i) $\Sigma = \overline{\mathcal{R}(\Omega)}$ has Lebesgue measure zero and is the intersection of a decreasing family of open neighborhoods $\{S_\varepsilon : \varepsilon > 0\}$ with smooth boundaries and a unit inward normal $\eta_\varepsilon(\mathbf{x})$, $\mathbf{x}\in\partial S_\varepsilon$;*

*(ii) for all sufficiently small $\varepsilon$*

$$(\nabla\delta\cdot\eta_\varepsilon)(\mathbf{x}) \ge 0, \qquad \mathbf{x}\in\partial S_\varepsilon; \tag{3.7.3}$$

*and*

*(iii)*

$$\frac{p-1}{p} \ge [\delta\Delta\delta](\mathbf{x}), \qquad \mathbf{x}\in G(\Omega) = \Omega\setminus\overline{\mathcal{R}(\Omega)}. \tag{3.7.4}$$

*Then, for all* $f \in C_0^\infty(\Omega)$ *and* $p \in (1, \infty)$,

$$\int_\Omega |\nabla\delta \cdot \nabla f|^p d\mathbf{x} \geq \left(\frac{p-1}{p}\right)^p \int_\Omega \left\{1 - \frac{p\delta\Delta\delta}{p-1}\right\} \frac{|f|^p}{\delta^p} d\mathbf{x}. \tag{3.7.5}$$

*Proof* We proceed as in the proof of Proposition 3.7.1, but now with $f \in C_0^\infty(\Omega)$, and account for the contribution of the boundary of $S_\varepsilon$. For any vector field $V$, we have the identity

$$\int_{\Omega\setminus S_\varepsilon} (\mathrm{div}V)|f|^p d\mathbf{x} = \int_{\partial S_\varepsilon} (V \cdot \eta_\varepsilon)|f|^p d\mathbf{x} - p\left[\mathrm{Re}\int_{\Omega\setminus S_\varepsilon} (V \cdot \nabla f)|f|^{p-2}\bar{f}\, d\mathbf{x}\right] \tag{3.7.6}$$

for all $f \in C_0^\infty(\Omega)$. Choose

$$V = -p\nabla\delta/\delta^{p-1},$$

which implies that $V \cdot \eta_\varepsilon \leq 0$ according to (3.7.3). Then, for any $a > 0$,

$$\begin{aligned}\int_{\Omega\setminus S_\varepsilon} (\mathrm{div}V)|f|^p d\mathbf{x} &\leq p^2 \left(\int_{\Omega\setminus S_\varepsilon} |\nabla\delta \cdot \nabla f|^p d\mathbf{x}\right)^{1/p} \left(\int_{\Omega\setminus S_\varepsilon} \frac{|f|^p}{\delta^p} d\mathbf{x}\right)^{1-1/p} \\ &\leq pa^p \int_{\Omega\setminus S_\varepsilon} |\nabla\delta \cdot \nabla f|^p d\mathbf{x} + p(p-1)a^{-p/(p-1)} \int_{\Omega\setminus S_\varepsilon} \frac{|f|^p}{\delta^p} d\mathbf{x}\end{aligned}$$

which gives, since $\mathrm{div}V = (p-1)p\delta^{-p} - p\delta^{1-p}\Delta\delta$ for $\mathbf{x} \in G(\Omega)$,

$$\int_{\Omega\setminus S_\varepsilon} |\nabla\delta \cdot \nabla f|^p d\mathbf{x} \geq a^{-p} \int_{\Omega\setminus S_\varepsilon} \left[(p-1) - (p-1)a^{-p/(p-1)} - \delta\Delta\delta\right] \frac{|f|^p}{\delta^p} d\mathbf{x}.$$

On choosing $a = [p/(p-1)]^{\frac{(p-1)}{p}}$ we obtain

$$\begin{aligned}\textstyle\int_\Omega |\nabla\delta \cdot \nabla f|^p d\mathbf{x} &\geq \textstyle\int_{\Omega\setminus S_\varepsilon} |\nabla\delta \cdot \nabla f|^p d\mathbf{x} \\ &\geq \left(\tfrac{p-1}{p}\right)^p \textstyle\int_\Omega \left[1 - \tfrac{p}{p-1}\delta\Delta\delta\right] \tfrac{|f|^p}{\delta^p} \chi_{\Omega\setminus S_\varepsilon} d\mathbf{x}.\end{aligned}$$

The proof concludes on using (3.7.4) and the monotone convergence theorem. □

In Remark 2.2.13, we noted the results from [81, 108] that if $\partial\Omega \in C^{2,1}$, the ridge $\mathcal{R}(\Omega)$ is closed and the cut locus $\Sigma(\Omega) = \overline{\mathcal{R}(\Omega)} = \mathcal{R}(\Omega)$ is null (i.e., it has zero $n$-dimensional Lebesgue measure). We have not been able to prove if this continues to hold if $\partial\Omega \in C^2$, nor found any reference to it in the literature, even in $\mathbb{R}^2$; cf., Remark 3.8 in [115]. In what follows, we shall assume that the domains considered with $C^2$ boundaries have, where necessary, null cut locus, rather than suppose at the outset that they have $C^{2,1}$ boundaries.

If $\Omega$ is weakly mean convex, we may apply (3.7.4) in (2.5.5) to obtain

**Corollary 3.7.3** *Let $\Omega \subset \mathbb{R}^n, n \geq 2$, be a weakly mean convex domain. If (i) and (ii) of Proposition 3.7.2 hold then, for all $f \in C_0^\infty(\Omega)$ and $p \in (1, \infty)$,*

$$\begin{aligned} \int_\Omega |\nabla\delta \cdot \nabla f|^p d\mathbf{x} &\geq \left(\tfrac{p-1}{p}\right)^p \int_\Omega \left\{1 - \tfrac{p}{p-1}\delta(\mathbf{x})\tilde{\kappa}(\mathbf{y})\right\} \tfrac{|f|^p}{\delta^p} d\mathbf{x} \\ &\geq \left(\tfrac{p-1}{p}\right)^p \int_\Omega \left\{1 - \tfrac{p}{p-1}\delta(\mathbf{x})H(\mathbf{y})\right\} \tfrac{|f|^p}{\delta^p} d\mathbf{x}, \end{aligned} \tag{3.7.7}$$

*where $\mathbf{y} \in N(\mathbf{x})$ and $\tilde{\kappa} = \sum_{i=1}^{n-1} \kappa_i/(1 + \delta\kappa_i)$. For a weakly mean convex domain, $\tilde{\kappa} \leq (n-1)H(\mathbf{y}) \leq 0$.*

*Proof* The first inequality follows from (3.7.5) and (2.5.5). The second inequality follows from the observation, already made in the proof of Proposition 2.5.4, that $\tilde{\kappa}_i$ is a decreasing function of $\delta$ irrespective of the signs of the $\kappa_i$ and is not greater than $H(\mathbf{y})$. The remainder of the proof was established in Proposition 2.5.3. □

### 3.7.2 Examples

As first shown in [20], Proposition 3.7.2 can be applied directly to some nonconvex domains such as the torus to obtain Hardy inequalities for those classical nonconvex domains. We present a few of those examples using Corollary 3.7.3, with the more convenient application of mean curvature as in [107].

We begin with an easy example of a cylinder which is a convex domain with infinite volume.

*Example 3.7.4*

Let $\Omega$ be the infinite cylinder $\Omega = B(0, r) \times \mathbb{R}$, where $B(0, r)$ is the ball, radius $r$ and centre the origin in $\mathbb{R}^2$. Clearly $\Omega$ is convex and $\mathcal{R}(\Omega)$ is the $x_3$-axis. The distance function is $\delta(\mathbf{x}) = r - \sqrt{x_1^2 + x_2^2}$,

$$\nabla\delta(\mathbf{x}) = -(r - \delta(\mathbf{x}))^{-1}(x_1, x_2, 0), \quad \eta_\varepsilon = -\varepsilon^{-1}(x_1, x_2, 0) \quad \text{on } \partial S_\varepsilon,$$

and $\Delta\delta = -(r-\delta)^{-1}$, where $S_\varepsilon = \{\mathbf{x} = (x_1^2, x_2^2, x_3^2) \in \Omega : x_1^2 + x_2^2 < \varepsilon\}$. Therefore (3.7.3) and (3.7.4) are seen to be satisfied for all $p > 1$. Inequality (3.7.5) follows: for all $f \in C_0^\infty(\Omega)$ and $p \in (1, \infty)$,

$$\int_\Omega |\nabla\delta \cdot \nabla f|^p d\mathbf{x} \geq \left(\frac{p-1}{p}\right)^p \int_\Omega \left\{1 + \frac{p\delta}{(p-1)(r-\delta)}\right\} \frac{|f|^p}{\delta^p} d\mathbf{x}. \tag{3.7.8}$$

Any one of the many Hardy inequalities for convex domains could have been applied to the infinite cylinder described above. That is not the case for the nonconvex torus in the next example.

*Example 3.7.5*

Let $\Omega$ be the ring torus with minor radius $r$ and major radius $R \geq 2r$ in Example 2.5.2. The mean curvature satisfies

$$H(\mathbf{y}) = \frac{1}{2}(\kappa_1 + \kappa_2) \leq -\frac{R - 2r}{2r(R - r)}, \tag{3.7.9}$$

so that $\Omega$ is weakly mean convex if $R \geq 2r$ and mean convex if $R > 2r$. Rather than using the estimate (2.5.5) as in Corollary 3.7.3 we may use (2.4.5) directly in Proposition 3.7.2 since we know the principal curvatures.

The ridge of the torus is

$$\mathcal{R}(\Omega) = \{\mathbf{x} : \rho(\mathbf{x}) = 0\},$$

where $\rho(\mathbf{x})$ is the distance from the point $\mathbf{x}$ in $\Omega$ to the centre of the cross-section and $\delta(\mathbf{x}) = r - \rho(\mathbf{x})$. Clearly, the ridge is closed with measure zero. Moreover, in the notation of Proposition 3.7.2,

$$S_\varepsilon = \{\mathbf{x} : \rho(\mathbf{x}) < \varepsilon\},$$

and points on the surface of $S_\varepsilon$ are on the level surface $\rho(\mathbf{x}) = \varepsilon$, so that the unit inward normal to $\partial S_\varepsilon$ is $\eta_\varepsilon = -\nabla\rho(\mathbf{x})/|\nabla\rho(\mathbf{x})| = \nabla\delta(\mathbf{x})$. Therefore $\nabla\delta \cdot \nabla\eta_\varepsilon > 0$.

We have proved the following corollary to Proposition 3.7.2.

**Corollary 3.7.6** *Let $\Omega \subset \mathbb{R}^3$ be the interior of a ring torus with minor radius $r$ and major radius $R \geq 2r$. Then the ridge $\mathcal{R}(\Omega)$ is closed and of measure zero. Moreover $\Delta\delta < 0$ in $\Omega \setminus \mathcal{R}(\Omega)$ and*

$$\begin{aligned}\int_\Omega |\nabla\delta \cdot \nabla f|^p d\mathbf{x} \geq& \left(\frac{p-1}{p}\right)^p \int_\Omega \frac{|f|^p}{\delta^p} d\mathbf{x} \\ &+ \left(\frac{p-1}{p}\right)^{p-1} \int_\Omega \left(\frac{1}{(r-\delta)} - \frac{1}{\sqrt{x_1^2 + x_2^2}}\right) \frac{|f|^p}{\delta^{p-1}} d\mathbf{x}\end{aligned} \tag{3.7.10}$$

*for all $f \in C_0^\infty(\Omega)$, where $\mathbf{x} \in \Omega$ has coordinates $(x_1, x_2, x_3)$.*

As a more accessible alternative to (3.7.10), we can use the second inequality in (3.7.7) with (3.7.9) to conclude that for a ring torus $\Omega \subset \mathbb{R}^3$ with minor ring $r$

and major ring $R \geq 2r$, we have for all $f \in C_0^\infty(\Omega)$,

$$\int_\Omega |\nabla\delta \cdot \nabla f|^p d\mathbf{x} \geq \left(\frac{p-1}{p}\right)^p \int_\Omega \frac{|f|^p}{\delta^p} d\mathbf{x} + \left(\frac{p-1}{p}\right)^{p-1} \frac{R-2r}{r(R-r)} \int_\Omega \frac{|f|^p}{\delta^{p-1}} d\mathbf{x}. \tag{3.7.11}$$

Our next example is a domain that is not mean convex. In fact, the mean curvature is everywhere non-negative.

*Example 3.7.7*

We apply Proposition 3.7.1 to the 1-sheeted hyperboloid

$$\Omega = \{(x_1, x_2, x_3) \in \mathbb{R}^3 : x_1^2 + x_2^2 < 1 + x_3^2\}. \tag{3.7.12}$$

This is non-convex and unbounded with infinite volume and infinite interior diameter $D_{int}(\Omega)$. To calculate the principal curvatures, we choose the following parametric co-ordinates for $\mathbf{y} \in \partial\Omega$:

$$\begin{aligned} y_1(s,t) &= \sqrt{s^2+1}\cos t, \\ y_2(s,t) &= \sqrt{s^2+1}\sin t, \\ y_3(s,t) &= s, \end{aligned}$$

for $t \in [0, 2\pi)$ and $s \in (-\infty, \infty)$. A calculation then gives (see [90], p. 132)

$$\kappa_1 = -\frac{1}{[2s^2+1]^{3/2}}, \qquad \kappa_2 = \frac{1}{\sqrt{2s^2+1}},$$

so that the mean curvature

$$H(\mathbf{y}) = \frac{s^2}{[2s+1]^{\frac{3}{2}}} \geq 0.$$

If $\mathbf{y} = N(\mathbf{x}),\ \mathbf{x} \in \Omega \setminus \mathcal{R}(\Omega)$, then by Lemma 2.4.2,

$$\Delta\delta(\mathbf{x}) = \tilde{\kappa} := \sum_{i=1}^{2} \frac{\kappa_i}{1+\delta\kappa_i} = -\frac{1}{w^3-\delta} + \frac{1}{w+\delta}, \tag{3.7.13}$$

where $w = \sqrt{2s^2+1}$ is the distance of $\mathbf{y}$ from the origin, and the ridge is $\mathcal{R}(\Omega) = \{(x_1, x_2, x_3) : x_1 = x_2 = 0,\ x_3 \in (-\infty, \infty)\}$.

To find $\mathbf{y} = N(\mathbf{x})$, we first determine the vector normal to $\partial\Omega$ at $\mathbf{y}$, namely

$$\begin{aligned}\mathbf{y}_s \times \mathbf{y}_t &= \begin{vmatrix} i & j & k \\ \frac{s}{\sqrt{s^2+1}}\cos t & \frac{s}{\sqrt{s^2+1}}\sin t & 1 \\ -\sqrt{s^2+1}\sin t & \sqrt{s^2+1}\cos t & 0 \end{vmatrix} \\ &= [-\sqrt{s^2+1}\cos t]i + [-\sqrt{s^2+1}\sin t]j + sk.\end{aligned}$$

The inward unit normal vector at $\mathbf{y}$ is therefore

$$\mathbf{n} = \{[-\sqrt{s^2+1}\cos t]i + [-\sqrt{s^2+1}\sin t]j + sk\}/\sqrt{2s^2+1}.$$

The distance from $\mathbf{y}$ to the ridge point $p(\mathbf{x})$ of $\mathbf{x}$ is given by $\sqrt{s^2+1}/\cos\theta$, where $\cos\theta = (\mathbf{z}\cdot\mathbf{n})/|\mathbf{z}|$, and

$$\mathbf{z} = [-\sqrt{s^2+1}\cos t]i + [-\sqrt{s^2+1}\sin t]j.$$

Hence

$$\sqrt{s^2+1}/\cos\theta = \sqrt{2s^2+1} = w.$$

Consequently, the near point of $\mathbf{x}$ is the point on the boundary of $\Omega$ which is equidistant from the ridge point $p(\mathbf{x})$ of $\mathbf{x}$ and the origin, which shows that $\delta(\mathbf{x}) \in (0, w)$ for $\mathbf{x} \in G(\Omega)$. Therefore $\Delta\delta(\mathbf{x})$ changes sign in $\Omega$.

We therefore have from Proposition 3.7.1

**Corollary 3.7.8** *Let $\Omega \subset \mathbb{R}^3$ be the 1-sheeted hyperboloid (3.7.12). Then, the mean curvature of $\partial\Omega$ is non-negative, and, for all $f \in C_0^\infty(\Omega \setminus \mathcal{R}(\Omega))$,*

$$\int_\Omega |\nabla\delta\cdot\nabla f|^p d\mathbf{x} \geq \left(\frac{p-1}{p}\right)^p \int_\Omega \frac{|f|^p}{\delta^p} d\mathbf{x} - \left(\frac{p-1}{p}\right)^{p-1} \int_\Omega \tilde{\kappa}\frac{|f|^p}{\delta^{p-1}} d\mathbf{x}, \tag{3.7.14}$$

*where $\tilde{\kappa}$ is given in (3.7.13), with $w = |\mathbf{y}| = \delta(p(\mathbf{x}))$, $\mathbf{y} = N(\mathbf{x})$ and $p(\mathbf{x})$ the ridge point of $\mathbf{x}$.*

### 3.7.3 Proposition 3.7.2 and Domains with $C^2$ Boundaries

The torus and one-sheeted hyperbola are examples of nontrivial, non-convex domains for which (i) and (ii) of Proposition 3.7.2 can be verified easily as shown in [20]. In this section we present methods of Lewis et al. [107] which *a priori* allow for elimination of conditions (i) and (ii) to prove the Hardy inequality given in Proposition 3.7.2. We assume throughout this subsection that $\Omega$ has $C^2$

boundary and null cut locus $\Sigma(\Omega)$. This implies that the function $\bar{s}$ defined in (2.2.8) and (2.2.9) is positive on $\partial\Omega$.

As well as improving Proposition 3.7.2, we also wish to prove that (2.5.5) holds on all of $\Omega$ in the distributional sense, i.e., for all $\varphi \in C_0^\infty(\Omega), \varphi \geq 0$,

$$\int_\Omega \nabla\delta(\mathbf{x}) \cdot \nabla\varphi(\mathbf{x})d\mathbf{x} \geq \int_\Omega \frac{-(n-1)H(\mathbf{y})}{1+\delta(\mathbf{x})H(\mathbf{y})}\varphi(\mathbf{x})d\mathbf{x}. \tag{3.7.15}$$

Since $\varphi$ in (3.7.15) has compact support in a ball $B(0, R)$, for all $R$ sufficiently large, we may assume that there is a bounded domain $\Omega_R$ with $\text{supp}(\varphi) \subset \Omega_R \subset \Omega$, having a $C^2$ boundary and such that the distance function $\delta_R(\mathbf{x})$ in $\Omega_R$ coincides with $\delta(\mathbf{x})$ for all $\mathbf{x} \in \text{supp}(\varphi)$. That is, we may assume, for the sake of this proof, that $\Omega$ is bounded, and hence that the function $\bar{s}$ defined in (2.2.8) is strictly positive.

**Lemma 3.7.9** *Let $\partial\Omega \in C^2$ and for every $k \in C^2(\partial\Omega)$ satisfying*

$$0 < k(\mathbf{y}) < \bar{s}(\mathbf{y}), \quad \mathbf{y} \in \partial\Omega,$$

*let*

$$S := \{\mathbf{y} + k(\mathbf{y})\mathbf{n}(\mathbf{y}) \ : \ \mathbf{y} \in \partial\Omega\}.$$

*Then $S$ is a $C^1$ hypersurface with*

$$\nabla\delta(\mathbf{x}) \cdot \mathbf{n}^S(\mathbf{x}) > 0, \qquad \mathbf{x} \in S, \tag{3.7.16}$$

*where $\mathbf{n}^S(x)$ denotes the unit outward normal of the boundary of*

$$\{\mathbf{y} + tk(\mathbf{y})\mathbf{n}(\mathbf{y}) \ : \ \mathbf{y} \in \partial\Omega,\ 0 < t < 1\}.$$

*Proof* For all $\mathbf{y} \in \partial\Omega$, we have

$$B(m(\mathbf{y}), \bar{s}(\mathbf{y})) \subset \Omega, \quad \text{and} \quad \mathbf{y} \in \partial B(m(\mathbf{y}), \bar{s}(\mathbf{y})), \tag{3.7.17}$$

where $m$ is defined in (2.2.7). For a fixed point $\mathbf{y} \in \partial\Omega$, we may assume, without loss of generality, that $\bar{s}(\mathbf{y}) = 1$. After a translation and rotation, we may assume that $\mathbf{y} = \mathbf{0} \in \mathbb{R}^n$ and the boundary in some neighborhood of $\mathbf{0}$ is given by

$$x_n = g(\mathbf{x}'), \quad \mathbf{x}' = (x_1, \cdots, x_{n-1}),$$

where $g$ is a $C^2$ function in some neighborhood of $\mathbf{0}' \in \mathbb{R}^{n-1}$ satisfying

$$g(\mathbf{0}') = 0, \quad \text{and} \quad \nabla g(\mathbf{0}') = 0,$$

with the Hessian matrix $(\nabla^2 g(\mathbf{0}'))$ being diagonal. The unit inward normal to $\partial\Omega$ at $(\mathbf{x}', g(\mathbf{x}'))$ near $\mathbf{0}$ is given by the graph of

$$\mathbf{n}(\mathbf{x}') := \frac{(-\nabla g(\mathbf{x}'), 1)}{\sqrt{1 + |\nabla g(\mathbf{x}')|^2}}.$$

(cf. [68], Appendix 14.6.) The set $S$ is given locally by

$$X(\mathbf{x}') := (\mathbf{x}', g(\mathbf{x}')) + \tilde{k}(\mathbf{x}')\mathbf{n}(\mathbf{x}'),$$

where $\tilde{k}(\mathbf{x}') = h(\mathbf{x}', g(\mathbf{x}'))$ is a $C^2$ function near $\mathbf{0}'$. We know that $\tilde{k}(\mathbf{0}') < \bar{s}(\mathbf{y}) = 1$.

Clearly $X \in C^1$. Now, we need to show that $S$ has a tangent plane at $X(\mathbf{0}')$. To that end, let

$$\mathbf{e}_1 = (1, 0, \cdots, 0), \ldots, \mathbf{e}_n = (0, \cdots, 0, 1).$$

We have, for $1 \leq \alpha \leq n-1$,

$$\begin{aligned} \tfrac{\partial X}{\partial x_\alpha}(\mathbf{0}') &= \mathbf{e}_\alpha + \tilde{k}_{x_\alpha}(\mathbf{0}')\mathbf{e}_n + \tilde{k}(\mathbf{0}')\tfrac{\partial \mathbf{n}}{\partial x_\alpha}(\mathbf{0}') \\ &= [1 - \tilde{k}(\mathbf{0}')g_{x_\alpha x_\alpha}(\mathbf{0}')]\mathbf{e}_\alpha + \tilde{k}_{x_\alpha}(\mathbf{0}')\mathbf{e}_n. \end{aligned}$$

By (3.7.17) and the fact that $\bar{s}(\mathbf{y}) = 1$, the unit ball centred at $\mathbf{e}_n$ lies in $\{\mathbf{x} : x_n \geq g(\mathbf{x}')\}$ near $\mathbf{0}$. It follows that $g_{x_\alpha x_\alpha}(\mathbf{0}') \leq 1$. Thus

$$1 - \tilde{k}(\mathbf{0}')g_{x_\alpha x_\alpha}(\mathbf{0}') > 0. \tag{3.7.18}$$

Consequently, $S$ has a tangent plane at $X(\mathbf{0}')$. Since $\bar{s}(\mathbf{y}) = 1$, we have

$$\delta(t\mathbf{e}_n) = t, \qquad \text{for } t \in (0, 1),$$

and therefore

$$\nabla\delta(t\mathbf{e}_n) = \mathbf{e}_n, \ 0 < t < 1.$$

Since $\mathbf{n}^S(k(0)\mathbf{e}_n)$ is the outward normal to the set, and $\gamma(t) := tk(\mathbf{0})\mathbf{e}_n$ belongs to the set for $0 < t < 1$, we have

$$\mathbf{n}^S(k(\mathbf{0})\mathbf{e}_n) \cdot \nabla\delta(k(\mathbf{0})\mathbf{e}_n) = \mathbf{n}^S(k(\mathbf{0})\mathbf{e}_n) \cdot \mathbf{e}_n = \frac{1}{k(\mathbf{0})}\mathbf{n}^S(k(\mathbf{0})\mathbf{e}_n) \cdot \gamma'(1) \geq 0.$$

Moreover, in view of (3.7.18),

$$span\,\{\frac{\partial X}{\partial x_\alpha}(\mathbf{0}')\} = span\,\{\mathbf{e}_\alpha + a_\alpha \mathbf{e}_n\}, \quad \text{for some constants } a_\alpha,$$

which does not contain $\mathbf{e}_n$. The inequality (3.7.16) follows. □

We are now able to extend Proposition 2.5.3 from an inequality on $G(\Omega) = \Omega \setminus \Sigma(\Omega)$ to an inequality on the entire domain $\Omega$ if $\partial\Omega \in C^2$ and $|\Sigma(\Omega)| = 0$.

**Theorem 3.7.10** *Let $\Omega \subset \mathbb{R}^n$, $n \geq 2$, have a $C^2$ boundary and null cut locus $\Sigma(\Omega)$. Then*

$$-\Delta\delta(\mathbf{x}) \geq \frac{-(n-1)H(N(\mathbf{x}))}{1+\delta(\mathbf{x})H(N(\mathbf{x}))}, \qquad \mathbf{x} \in \Omega, \tag{3.7.19}$$

*in the distributional sense, i.e., for any $\varphi \in C_0^\infty(\Omega)$, $\varphi \geq 0$, we have*

$$\int_\Omega \nabla\delta(\mathbf{x}) \cdot \nabla\varphi(\mathbf{x})d\mathbf{x} \geq \int_\Omega \frac{-(n-1)(H \circ N)(\mathbf{x})}{1+\delta(\mathbf{x})(H \circ N)(\mathbf{x})}\varphi(\mathbf{x})d\mathbf{x}. \tag{3.7.20}$$

Since $|\Sigma(\Omega)| = 0$, $(H \circ N)(x)$ is a well defined $L^\infty$ function in $\Omega$.

*Proof* We may continue to assume that $\bar{s}$ is strictly positive on $\partial\Omega$. For $\varepsilon > 0$ small, we construct $\bar{s}_\varepsilon \in C^2(\partial\Omega)$ satisfying

$$|\bar{s}_\varepsilon(\mathbf{y}) - \bar{s}(\mathbf{y})| \leq \varepsilon\bar{s}(\mathbf{y}), \qquad \mathbf{y} \in \partial\Omega,$$

the construction being guaranteed by the Stone-Weierstrass theorem. Now, let

$$\Sigma_\varepsilon := \{\mathbf{y} + (1-\varepsilon)\bar{s}_\varepsilon(\mathbf{y})\mathbf{n}(\mathbf{y}) \;:\; \mathbf{y} \in \partial\Omega\}$$

and

$$\Omega_\varepsilon := \{\mathbf{y} + t(1-\varepsilon)\bar{s}_\varepsilon(\mathbf{y})\mathbf{n}(\mathbf{y}) \;:\; \mathbf{y} \in \partial\Omega,\ 0 < t < 1\}. \tag{3.7.21}$$

Clearly, $\partial\Omega_\varepsilon = \Sigma_\varepsilon \cup \partial\Omega$. By Lemma 3.7.9, $\Sigma_\varepsilon$ is a $C^1$ hypersurface satisfying

$$\nabla\delta \cdot \mathbf{n}_\varepsilon \geq 0 \qquad \text{on } \Sigma_\varepsilon, \tag{3.7.22}$$

where $\mathbf{n}_\varepsilon$ is the unit outward normal of $\partial\Omega_\varepsilon$, and we see that

$$\cup_{\varepsilon>0}\Omega_\varepsilon = G(\Omega).$$

Since $\delta \in C^2$ on $\bar{\Omega}_\varepsilon \subset G(\Omega) \cup \partial\Omega$, we may apply Green's formula to obtain

$$\begin{aligned} \int_{\Omega_\varepsilon} \nabla\delta(\mathbf{x}) \cdot \nabla\varphi(\mathbf{x}) d\mathbf{x} &= -\int_{\Omega_\varepsilon} \varphi(\mathbf{x}) \Delta\delta(\mathbf{x}) d\mathbf{x} + \int_{\partial\Omega_\varepsilon} \varphi(\mathbf{s}) \frac{\partial\delta}{\partial\mathbf{n}_\varepsilon}(\mathbf{s}) ds \\ &\geq -\int_{\Omega_\varepsilon} \varphi(\mathbf{x}) \Delta\delta(\mathbf{x}) d\mathbf{x} \\ &\geq -\int_{\Omega_\varepsilon} \frac{(n-1)(H \circ N)(\mathbf{x})}{1 + \delta(H \circ N)(\mathbf{x})} \varphi(\mathbf{x}) d\mathbf{x}, \end{aligned} \tag{3.7.23}$$

where the last two inequalities follow from (3.7.22) and (2.5.5) respectively. Letting $\varepsilon \to 0$ in (3.7.23) completes the proof. □

An immediate consequence of Theorem 3.7.10 is the following extension of Theorem 2.3.2 in which $\Omega$ was assumed to be convex.

**Corollary 3.7.11** *Let $\Omega \subset \mathbb{R}^n$, $n \geq 2$, be weakly mean convex and have a null cut locus. Then $-\Delta\delta \geq 0$ in the distributional sense.*

The next theorem uses the methods of [107] described above to improve Proposition 3.7.2.

**Theorem 3.7.12** *Let $\Omega \subset \mathbb{R}^n, n \geq 2$, be a domain with a $C^2$ boundary and null cut locus. Assume that for $p \in (1, \infty)$*

$$\frac{p-1}{p} \geq [\delta\Delta\delta](\mathbf{x}), \qquad \mathbf{x} \in G. \tag{3.7.24}$$

*Then for all $f \in C_0^\infty(\Omega)$*

$$\int_\Omega |\nabla\delta \cdot \nabla f|^p d\mathbf{x} \geq \left(\frac{p-1}{p}\right)^p \int_\Omega \left\{1 - \frac{p\delta\Delta\delta}{p-1}\right\} \frac{|f|^p}{\delta^p} d\mathbf{x}. \tag{3.7.25}$$

*Proof* It will suffice to show that (i) and (ii) in Proposition 3.7.2 hold.

Let

$$S_\varepsilon := \Omega \setminus \overline{\Omega_\varepsilon}$$

for $\Omega_\varepsilon$ given in (3.7.21). Then (i) and (ii) in Proposition 3.7.2 are satisfied by the family $\{S_\varepsilon : \varepsilon > 0\}$. That completes the proof. □

We have the following as an immediate corollary of Theorems 3.7.10 and 3.7.12.

**Corollary 3.7.13** *Let $\Omega \subset \mathbb{R}^n, n \geq 2$, have a $C^2$ boundary and null cut locus. Assume that for $p \in (1, \infty)$ and $\mathbf{y} = N(\mathbf{x})$, the near point of $\mathbf{x}$,*

$$\frac{p-1}{p} \geq \frac{(n-1)\delta(\mathbf{x})H(\mathbf{y})}{1 + \delta(\mathbf{x})H(\mathbf{y})}, \qquad \mathbf{x} \in G. \tag{3.7.26}$$

*Then for all* $f \in C_0^\infty(\Omega)$

$$\int_\Omega |\nabla\delta \cdot \nabla f|^p d\mathbf{x} \geq \left(\frac{p-1}{p}\right)^p \int_\Omega \left\{1 - \frac{p(n-1)\delta(\mathbf{x})H(\mathbf{y})}{(p-1)[1+\delta(\mathbf{x})H(\mathbf{y})]}\right\} \frac{|f|^p}{\delta^p} d\mathbf{x}. \tag{3.7.27}$$

When $\Omega$ is weakly mean convex, the right side of (3.7.26) is non-positive (see (2.5.1)) so that the inequality is trivially satisfied. However, some positivity of the mean curvature is permissible in order for (3.7.26) to hold. In particular, it is not hard to see that (3.7.26) holds when

$$D_{int} H_M \leq \frac{2(p-1)}{p(n-2)+1}, \tag{3.7.28}$$

where $H_M := \sup_{\mathbf{y}\in\partial\Omega} H(\mathbf{y})$ and the interior diameter of $\Omega$ is given by $D_{int} := 2\sup_{\mathbf{x}\in\Omega} \delta(\mathbf{x})$.

Since $|\nabla\delta| = 1, a.e.$, we have from (3.7.27) that, if $\Omega$ is a weakly mean convex domain, then

$$\int_\Omega |\nabla f|^p d\mathbf{x} \geq \left(\frac{p-1}{p}\right)^p \int_\Omega \frac{|f|^p}{\delta^p} d\mathbf{x}.$$

for all $f \in C_0^\infty(\Omega)$. This yields the following improvement (established in [107], Theorem 1.2) of Theorem 3.4.1 if $|\Sigma(\Omega)| = 0$.

**Corollary 3.7.14** *Let* $\Omega$ *be a weakly mean convex domain in* $\mathbb{R}^n, n \geq 2$, *with* $|\Sigma(\Omega)| = 0$, *and* $1 < p < \infty$. *Then,*

$$\mu_p(\Omega) := \inf_{C_0^\infty(\Omega)} \frac{\int_\Omega |\nabla f|^p d\mathbf{x}}{\int_\Omega |f/\delta|^p d\mathbf{x}} = c_p = \left(\frac{p-1}{p}\right)^p. \tag{3.7.29}$$

*Remark 3.7.15* The weakly mean convexity condition in Corollary 3.7.14 is sharp in the sense that the inequality fails if only $H \leq \varepsilon$ is assumed for $\varepsilon > 0$. This is demonstrated in [107] where counterexamples are given based on ideas from [11] and [116]. See Example 3.8.7 below.

The Brézis-Marcus type results derived for convex domains in Sect. 3.6 can also be extended to weakly mean convex domains. The following is Theorem 4.3 in [107].

**Corollary 3.7.16** *Let* $\Omega$ *be a weakly mean convex domain in* $\mathbb{R}^n, n \geq 2$, *with* $|\Sigma(\Omega)| = 0$, *and* $1 < p < \infty$. *Then, for all* $f \in C_0^\infty(\Omega)$,

$$\int_\Omega |\nabla f|^p d\mathbf{x} \geq \left(\frac{p-1}{p}\right)^p \int_\Omega \frac{|f|^p}{\delta^p} d\mathbf{x} + \lambda(n,p,\Omega) \int_\Omega |f|^p d\mathbf{x}, \tag{3.7.30}$$

*where* $\lambda(n,p,\Omega) = \left(\frac{p-1}{p}\right)^{p-1} \inf_{G(\Omega)} \frac{-\Delta\delta}{\delta^{p-1}} \geq p(n-1)\inf_{\partial\Omega} |H|^p$.

*Proof* We have in (3.7.27)

$$\int_\Omega |\nabla f|^p d\mathbf{x} - \left(\frac{p-1}{p}\right)^p \int_\Omega \frac{|f|^p}{\delta^p} d\mathbf{x} \geq \left(\frac{p-1}{p}\right)^{p-1} \int_\Omega \frac{-(n-1)H}{(1+\delta\, H)} \frac{|f|^p}{\delta^{p-1}} d\mathbf{x}.$$

Now let $g(t) := (at^{p-1} - t^p)^{-1}$, with $a > 0$. The minimum of $g(t)$ in $(0, a)$ is attained at $t_0 = a(p-1)/p$ and this gives that

$$g(t) \geq \frac{p^p}{a^p(p-1)^{p-1}}, t \in (0, a).$$

For $H(N(\mathbf{x})) \neq 0$, choose $a = 1/|H|$ and $t = \delta$ (note (2.5.1)) to give, for $\mathbf{x} \in G(\Omega)$,

$$\frac{-H(N(\mathbf{x}))}{[1+\delta(\mathbf{x})H(N(\mathbf{x}))]\delta^{p-1}(\mathbf{x})} \geq \frac{p^p|H(N(\mathbf{x}))|^p}{(p-1)^{p-1}}.$$

This continues to hold if $H(N(\mathbf{x})) = 0$ and so from (3.7.19),

$$\frac{-\delta\Delta\delta}{\delta^p} \geq \left(\frac{p}{p-1}\right)^{p-1} p(n-1)|H|.$$

The theorem follows from (3.7.25), since it is assumed that $\Sigma(\Omega) = \Omega \setminus G(\Omega)$ is of zero measure. □

The proof of Corollary 3.7.16 above can be adapted to give the following inequality for domains which are not weakly mean convex.

**Corollary 3.7.17** *Let $\Omega$ be a domain with $C^2$ boundary in $\mathbb{R}^n, n \geq 2$, $|\Sigma(\Omega)| = 0$, and $H_0 := \sup_{\mathbf{y}\in\partial\Omega} H(\mathbf{y}) > 0$. Then, for all $f \in C_0^\infty(\Omega)$ and $1 < p < \infty$,*

$$\int_\Omega |\nabla f(\mathbf{x})|^p d\mathbf{x} + \left(\frac{p-1}{p}\right)^{p-1} (n-1)H_0 \int_\Omega \frac{|f(\mathbf{x})|^p}{\delta(\mathbf{x})^{p-1}} d\mathbf{x}$$
$$\geq \left(\frac{p-1}{p}\right)^p \int_\Omega \frac{|f(\mathbf{x})|^p}{\delta(\mathbf{x})^p} d\mathbf{x}. \tag{3.7.31}$$

*Proof* The proof follows that of Corollary 3.7.16, on observing that

$$\frac{(n-1)H}{1+\delta H} \leq (n-1)H \leq (n-1)H_0.$$

□

For an application of Corollary 3.7.17 we return to Example 3.7.7 and obtain a Hardy-type inequality with a domain that is not weakly mean convex.

*Example 3.7.18* In Example 3.7.7 we studied the 1-sheeted hyperboloid (3.7.12), and showed that the mean curvature

$$H(\mathbf{y}(s,t)) = \frac{s^2}{(2s^2+1)^{\frac{3}{2}}}, \qquad s \in (-\infty, \infty), \quad t \in [0, 2\pi).$$

Calculations show that this function of $s$ assumes its maximum at $s = \pm 1$ so that $H_0 = 3^{-\frac{3}{2}}$. Consequently, for $\Omega$ given by (3.7.12)

$$\int_\Omega |\nabla f(\mathbf{x})|^p d\mathbf{x} + \frac{2}{3^{\frac{3}{2}}}\left(\frac{p-1}{p}\right)^{p-1} \int_\Omega \frac{|f(\mathbf{x})|^p}{\delta(\mathbf{x})^{p-1}} d\mathbf{x} \geq \left(\frac{p-1}{p}\right)^p \int_\Omega \frac{|f(\mathbf{x})|^p}{\delta(\mathbf{x})^p} d\mathbf{x} \tag{3.7.32}$$

for $1 < p < \infty$ and all $f \in C_0^\infty(\Omega)$.

## 3.8 Doubly Connected Domains

Domains that are not weakly mean convex present special problems. A domain $\Omega \subset \mathbb{R}^2 \equiv \mathbb{C}$ is *doubly connected* if its boundary is a disjoint union of 2 simple curves. If it has a smooth boundary then it can be mapped conformally onto an annulus $\Omega_{\rho,R} = B_R \setminus \overline{B_\rho} = \{z \in \mathbb{C} : \rho < |z| < R\}$, for some $\rho, R$; see [149], Theorem 1.2. In order to proceed, we need the following inequality on an annulus, which is an analogue of Theorem 1 in [11].

**Theorem 3.8.1** *Let $\Omega_1$, $\Omega_2$, be convex domains in $\mathbb{R}^n, n \geq 2$, with $C^2$ boundaries and $\bar{\Omega}_1 \subset \Omega_2$. For $\mathbf{x} \in \Omega := \Omega_2 \setminus \bar{\Omega}_1$ denote the distances of $\mathbf{x}$ to $\partial\Omega_1, \partial\Omega_2$ by $\delta_1, \delta_2$, respectively. Then, for all $f \in C_0^\infty(\Omega \setminus \mathcal{R}(\Omega))$*

$$\begin{aligned}\int_{\Omega_2\setminus\bar{\Omega}_1} |\nabla f(\mathbf{x})|^2 d\mathbf{x} \geq \frac{1}{4}\int_{\Omega_2\setminus\bar{\Omega}_1} \Bigg\{ & \frac{(n-1)(n-3)}{|\mathbf{x}|^2} + \frac{1}{\delta_1^2} + \frac{1}{\delta_2^2} \\ & - \frac{2\Delta\delta_1}{\delta_1} - \frac{2\Delta\delta_2}{\delta_2} - \frac{2\nabla\delta_1\cdot\nabla\delta_2}{\delta_1\delta_2} \\ & + 2(n-1)\frac{\mathbf{x}\cdot\nabla\delta_1}{|\mathbf{x}|^2\delta_1} + 2(n-1)\frac{\mathbf{x}\cdot\nabla\delta_2}{|\mathbf{x}|^2\delta_2}\Bigg\} |f(\mathbf{x})|^2 d\mathbf{x}.\end{aligned} \tag{3.8.1}$$

*Proof* Our starting point is again (3.7.2), which, for a differentiable vector field $V$ and arbitrary $\varepsilon > 0$, yields the inequality

$$\int_{\Omega_2\setminus\bar{\Omega}_1} (\mathrm{div}V)|f(\mathbf{x})|^2 d\mathbf{x} \le \varepsilon^2 \int_{\Omega_2\setminus\bar{\Omega}_1} |\nabla f|^2 d\mathbf{x} + \varepsilon^{-2} \int_{\Omega_2\setminus\bar{\Omega}_1} |V|^2|f|^2 d\mathbf{x}. \tag{3.8.2}$$

Guided by the proof of Corollary 1 in [11], the theorem follows on setting

$$V(\mathbf{x}) = 2(n-1)\frac{\nabla|\mathbf{x}|}{|\mathbf{x}|} - 2\frac{\nabla\delta_1(\mathbf{x})}{\delta_1(\mathbf{x})} - 2\frac{\nabla\delta_2(\mathbf{x})}{\delta_2(\mathbf{x})} \tag{3.8.3}$$

and $\varepsilon = 2$. □

Notice that $\delta_1(\mathbf{x})$ and $\delta_2(\mathbf{x})$ in Theorem 3.8.1 coincide with $\delta(\mathbf{x})$ only for certain values of $\mathbf{x}$.

In the next corollary we account for behavior near the ridge in order to prove an inequality for $f \in C_0^\infty(\Omega)$.

**Corollary 3.8.2** *For all* $f \in C_0^\infty(B_R \setminus B_\rho)$

$$\begin{aligned} \int_{B_R\setminus B_\rho} |\nabla f(\mathbf{x})|^2 d\mathbf{x} &\ge \frac{1}{4}\int_{B_R\setminus B_\rho} \left\{ \frac{(n-1)(n-3)}{|\mathbf{x}|^2} + \left(\frac{1}{\delta_1(\mathbf{x})} + \frac{1}{\delta_2(\mathbf{x})}\right)^2 \right\} |f(\mathbf{x})|^2 d\mathbf{x} \\ &\ge \frac{1}{4}\int_{B_R\setminus B_\rho} \left\{ \frac{(n-1)(n-3)}{|\mathbf{x}|^2} + \frac{1}{\delta^2(\mathbf{x})} \right\} |f(\mathbf{x})|^2 d\mathbf{x} \end{aligned} \tag{3.8.4}$$

*in which* $\delta_1(\mathbf{x}) = \mathrm{dist}(\mathbf{x}, \partial B_\rho)$, $\delta_2(\mathbf{x}) = \mathrm{dist}(\mathbf{x}, \partial B_R)$, *and* $\delta(\mathbf{x}) = \mathrm{dist}(\mathbf{x}, \partial(B_R \setminus B_\rho))$.

*Proof* For $\Omega_1 = B_\rho$ and $\Omega_2 = B_R$, $R > \rho$, the ridge $\mathcal{R}(B_R \setminus B_\rho) = \partial B_c$, $c := (R+\rho)/2$, has measure zero. In this case we use (3.7.6) with $S_\varepsilon = B_{c+\varepsilon} \setminus B_{c-\varepsilon}$ and $p = 2$. Namely

$$\int_{\Omega\setminus S_\varepsilon} (\mathrm{div}V)|f|^2 d\mathbf{x} = \int_{\partial S_\varepsilon} (V\cdot\eta_\varepsilon)|f|^2 d\mathbf{x} - 2\left[\mathrm{Re}\int_{\Omega\setminus S_\varepsilon} (V\cdot\nabla f)\bar{f}\, d\mathbf{x}\right]. \tag{3.8.5}$$

For $V$ defined in (3.8.3), consider the terms appearing in $V\cdot\eta_\varepsilon$ in which $\eta_\varepsilon$ is the inward normal on $\partial S_\varepsilon$,

$$\begin{aligned} \int_{\partial S_\varepsilon} |f|^2 \tfrac{\partial\delta_1}{\partial\eta_\varepsilon} ds &= \int_{\partial B_{c+\varepsilon}} |f|^2 \nabla\delta_1 \cdot \tfrac{-\mathbf{x}}{|\mathbf{x}|} ds + \int_{\partial B_{c-\varepsilon}} |f|^2 \nabla\delta_1 \cdot \tfrac{\mathbf{x}}{|\mathbf{x}|} ds \\ &= -\int_{\partial B_{c+\varepsilon}} |f|^2 ds + \int_{\partial B_{c-\varepsilon}} |f|^2 ds \end{aligned}$$

and

$$\begin{aligned} \int_{\partial S_\varepsilon} |f|^2 \tfrac{\partial\delta_2}{\partial\eta_\varepsilon} ds &= \int_{\partial B_{c+\varepsilon}} |f|^2 \nabla\delta_2 \cdot \tfrac{-\mathbf{x}}{|\mathbf{x}|} ds + \int_{\partial B_{c-\varepsilon}} |f|^2 \nabla\delta_2 \cdot \tfrac{\mathbf{x}}{|\mathbf{x}|} ds \\ &= \int_{\partial B_{c+\varepsilon}} |f|^2 ds - \int_{\partial B_{c-\varepsilon}} |f|^2 ds. \end{aligned}$$

We therefore have that

$$0 = \int_{\partial S_\varepsilon} |f|^2 \frac{\partial \delta_1}{\partial \eta_\varepsilon} ds + \int_{\partial S_\varepsilon} |f|^2 \frac{\partial \delta_2}{\partial \eta_\varepsilon} ds$$

A similar calculation shows that

$$\int_{\partial S_\varepsilon} |f|^2 \frac{\partial |\mathbf{x}|}{\partial \eta_\varepsilon} ds = 0$$

and we conclude from (3.8.3) that

$$\int_{\partial S_\varepsilon} (V \cdot \eta_\varepsilon)|f|^2 d\mathbf{x} = 0.$$

Then, using this fact in (3.8.5) and applying (3.8.2) (with $\varepsilon$ replaced by $\alpha$), we obtain

$$\int_{\Omega \setminus S_\varepsilon} (\mathrm{div} V - \alpha^{-2}|V|^2)|f(\mathbf{x})|^2 d\mathbf{x} \leq \alpha^2 \int_\Omega |\nabla f|^2 d\mathbf{x}. \tag{3.8.6}$$

Observe that in this case $\delta_1 = |\mathbf{x}| - \rho$, $\delta_2 = R - |\mathbf{x}|$,

$$\Delta \delta_1 = \frac{n-1}{|\mathbf{x}|}, \quad \Delta \delta_2 = -\frac{n-1}{|\mathbf{x}|}$$

and $\nabla \delta_1 = -\nabla \delta_2 = \mathbf{x}/|\mathbf{x}|$, implying that

$$(\mathrm{div} V) - \frac{1}{4}|V|^2 = \frac{(n-1)(n-3)}{|\mathbf{x}|^2} + \left(\frac{1}{\delta_1} + \frac{1}{\delta_2}\right)^2.$$

On choosing $\alpha = 2$ and allowing $\varepsilon \to 0$ in (3.8.6), we establish the first inequality (3.8.4). The last inequality follows since

$$\frac{1}{\delta_1} + \frac{1}{\delta_2} = \begin{cases} \frac{1}{\delta} + \frac{1}{\delta_2} & \text{in } B_c \setminus B_\rho, \\ \frac{1}{\delta_1} + \frac{1}{\delta} & \text{in } B_R \setminus B_c. \end{cases}$$

□

**Lemma 3.8.3** *Let $\Omega_1 \subset \Omega_2 \subset \mathbb{C}$ and $B_\rho \subset B_R \subset \mathbb{C}$, $0 < \rho < R$, where $B_r$ is the disc of radius $r$ centred at the origin. Let*

$$F : \Omega_2 \setminus \bar{\Omega}_1 \to B_R \setminus \overline{B_\rho}$$

*be analytic and univalent. Then for $\mathbf{z} = x_1 + ix_2$, $\mathbf{x} = (x_1, x_2) \in \Omega_2 \setminus \bar{\Omega}_1$,*

$$\mathfrak{F}(\mathbf{z}) := -\frac{|F'(\mathbf{z})|^2}{|F(\mathbf{z})|^2} + |F'(\mathbf{z})|^2 \left\{ \frac{1}{|F(\mathbf{z})| - \rho} + \frac{1}{R - |F(\mathbf{z})|} \right\}^2 \tag{3.8.7}$$

*is invariant under scaling, rotation, and inversion. Hence, $\mathfrak{F}$ does not depend on the choice of the mapping $F$, but only on the geometry of $\Omega_2 \setminus \bar{\Omega}_1$.*

*Proof* The fact that $\mathfrak{F}$ is invariant under scaling and rotations is straightforward. To see that it is also invariant under inversions, suppose that $F(\mathbf{z}) = 1/G(\mathbf{z})$. Then, under inversion, $\mathfrak{F}(\mathbf{z})$ becomes

$$\begin{aligned}
&-\frac{|G'(\mathbf{z})|^2}{|G(\mathbf{z})|^2} + \frac{|G'(\mathbf{z})|^2}{|G(\mathbf{z})|^4}\left\{\frac{1}{\frac{1}{|G(\mathbf{z})|}-\rho^{-1}} + \frac{1}{R^{-1}-\frac{1}{|G(\mathbf{z})|}}\right\}^2 \\
&= -\frac{|G'(\mathbf{z})|^2}{|G(\mathbf{z})|^2} + \frac{|G'(\mathbf{z})|^2}{|G(\mathbf{z})|^2}\left\{\frac{\rho}{\rho-|G(\mathbf{z})|} + \frac{R}{|G(\mathbf{z})|-R}\right\}^2 \\
&= -\frac{|G'(\mathbf{z})|^2}{|G(\mathbf{z})|^2} + \frac{|G'(\mathbf{z})|^2}{|G(\mathbf{z})|^2}\left\{\frac{(\rho-R)|G(\mathbf{z})|}{(\rho-|G(\mathbf{z})|)(|G(\mathbf{z})|-R)}\right\}^2 \\
&= -\frac{|G'(\mathbf{z})|^2}{|G(\mathbf{z})|^2} + |G'(\mathbf{z})|^2\left\{\frac{1}{\rho-|G(\mathbf{z})|} + \frac{1}{|G(\mathbf{z})|-R}\right\}^2
\end{aligned}$$

implying that $\mathfrak{F}$ is invariant under inversions. The rest of the lemma follows from [88], p. 133. □

In applying the last Lemma we regard $\Omega_1$, $\Omega_2$ as domains in $\mathbb{R}^2$ with $\mathbf{z} = x + iy$ and $\mathbf{x} = (x, y)$.

**Theorem 3.8.4** *For* $\Omega := \Omega_2 \setminus \bar{\Omega}_1 \subset \mathbb{R}^2$,

$$\int_\Omega |\nabla u(\mathbf{x})|^2 d\mathbf{x} \geq \frac{1}{4}\int_\Omega \mathfrak{F}(\mathbf{x})|u(\mathbf{x})|^2 d\mathbf{x}.$$

*Proof* From (3.8.4), it follows that for all $u \in H_0^1(B_R \setminus \overline{B_\rho})$,

$$\int_{B_R\setminus\overline{B_\rho}} |\nabla u(\mathbf{y})|^2 d\mathbf{y} \geq \frac{1}{4}\int_{B_R\setminus\overline{B_\rho}} \left[\frac{-1}{|\mathbf{y}|^2} + \left(\frac{1}{\delta_\rho(\mathbf{y})} + \frac{1}{\delta_R(\mathbf{y})}\right)^2\right] |u(\mathbf{y})|^2 d\mathbf{y},$$

where $\delta_\rho(\mathbf{y}) := |\mathbf{y}| - \rho$ and $\delta_R(\mathbf{y}) := R - |\mathbf{y}|$. Let $F : \Omega \to \Omega_{\rho,R}$ be analytic and univalent, and set $\mathbf{y} = F(\mathbf{x})$, with $\mathbf{y} = (y_1, y_2)$, $\mathbf{x} = (x_1, x_2)$. Then, as we saw in Sect. 3.4.2, with $F'$ denoting the complex derivative,

$$d\mathbf{y} = \left|\det\left(\frac{\partial(y_1, y_2)}{\partial(x_1, x_2)}\right)\right| d\mathbf{x} = |F'(\mathbf{x})|^2 d\mathbf{x},$$

and

$$\nabla_{\mathbf{x}} u = \nabla_{\mathbf{y}} u \left[\frac{\partial(y_1, y_2)}{\partial(x_1, x_2)}\right]^t,$$

implying that

$$|\nabla_{\mathbf{x}} u|^2 = |\nabla_{\mathbf{y}} u|^2 |F'(\mathbf{x})|^2.$$

The theorem follows from Lemma 3.8.3. □

*Example 3.8.5* Let $\Phi(z) = (z-1)(z+1)$ and

$$\Omega = \{z : \rho^2 < |\Phi(z)| < R^2\}$$

for $0 < \rho < R$. The function $F(z) = \sqrt{\Phi(z)}$ is analytic and univalent in $\Omega$ and

$$F : \Omega \to \Omega_{\rho,R}.$$

A calculation gives

$$\begin{aligned}\mathfrak{F}(z) &= -\frac{|z|^2}{|z^2-1|^2} \\ &+ \frac{|z|^2}{|z^2-1|}\frac{(R-\rho)^2}{(\sqrt{|z|^2-1}-\rho)^2(R-\sqrt{|z|^2-1})^2}.\end{aligned}$$

We end this chapter with an example of a doubly connected domain formed with ellipsoids. First, we need a result of Avkhadiev [10].

**Lemma 3.8.6** *Let $\rho < R$ and $\Omega_2 := B_R \setminus B_\rho \subset \mathbb{R}^2$. The best constant $\lambda(\Omega_2)$ in the Hardy inequality*

$$\int_{\Omega_2} |\nabla u|^2 d\mathbf{x} \geq \lambda(\Omega_2) \int_{\Omega_2} \frac{|u(\mathbf{x})|^2}{\delta(\mathbf{x})^2} d\mathbf{x}, \qquad u \in H_0^1(\Omega_2),$$

*satisfies*

$$\frac{2}{\pi} \ln \frac{R}{\rho} \leq \frac{1}{\lambda(\Omega_2)} \leq \ln \frac{R}{\rho} + k_0$$

*where* $k_0 = \frac{\Gamma(\frac{1}{4})^4}{2\pi^2} = 8.75\ldots$.

*Example 3.8.7* For $n \geq 3$ and any $\varepsilon > 0$ there exist ellipsoids $E_1, E_2$ with $\overline{E_2} \subset E_1 \subset \mathbb{R}^n$, and a function $f \in C_0^1(E_1 \setminus \overline{E_2})$, such that

$$\int_{E_1 \setminus E_2} |\nabla f|^2 d\mathbf{x} \leq \varepsilon \int_{E_1 \setminus E_2} \frac{|f(\mathbf{x})|^2}{\delta(\mathbf{x})^2} d\mathbf{x} \tag{3.8.8}$$

where $\delta(\mathbf{x})$ is the distance from $\mathbf{x} \in E_1 \setminus E_2$ to the boundary of $E_1 \setminus E_2$. Moreover, the mean curvature $H(N(\mathbf{x})) \leq \varepsilon$ for all $\mathbf{x} \in E_1 \setminus \overline{E_2}$.

*Proof* Note that $\lambda(\Omega_2) \to 0$ as $R/\rho \to \infty$; more precisely, for $\varepsilon > 0$ and $R > e^{\frac{\pi}{\varepsilon}} \rho$, we have that $\lambda(\Omega_2) < \varepsilon/2$. Since $\lambda(\Omega_2)$ is the best constant in Example 3.8.6, for each $\varepsilon > 0$, there is a title function $f_\varepsilon \in C_0^1(\Omega_2)$ such that

$$\int_{\Omega_2} |\nabla f_\varepsilon(\mathbf{x}')|^2 d\mathbf{x}' \leq \frac{\varepsilon}{2} \int_{\Omega_2} \frac{|f_\varepsilon(\mathbf{x}')|^2}{\delta(\mathbf{x}')^2} d\mathbf{x}', \qquad \mathbf{x}' := (x_1, x_2). \tag{3.8.9}$$

For each $s > 0$, construct a function $g_s \in C_0^1(\mathbb{R}^1)$ to satisfy

$$g_s(t) = \begin{cases} 1, & |t| \le s, \\ 0, & |t| > 1+s, \\ (1-(|t|-s)^2)^2, & s < |t| \le 1+s. \end{cases}$$

Note that

$$\int_{-\infty}^{\infty} g_s(t)^2 dt = 2s + c_1 \quad \text{and} \quad \int_{-\infty}^{\infty} [g_s'(t)]^2 dt = c_2,$$

in which $c_1$ and $c_2$ do not depend upon $s$. Define $\Omega_n := \Omega_2 \times \mathbb{R}^{n-2}$ and

$$f(\mathbf{x}) := f_\varepsilon(\mathbf{x}')\Pi_{j=3}^n g_s(x_j), \qquad \mathbf{x} = (x_1, \dots, x_n).$$

Then, $f \in C_0^1(\Omega_n)$ and

$$|\nabla f|^2 = |\nabla f_\varepsilon|^2 \Pi_{j=3}^n g_s(x_j)^2 + |f_\varepsilon|^2 \sum_{k=3}^n \left[ g_s'(x_k)^2 \Pi_{j=3, j\neq k}^n g_s(x_j)^2 \right].$$

Therefore,

$$\int_{\mathbb{R}^n} |\nabla f(\mathbf{x})|^2 d\mathbf{x} = (2s + c_1)^{n-2} \int_{\mathbb{R}^2} |\nabla f_\varepsilon|^2 d\mathbf{x}' + C(2s + c_1)^{n-3} \int_{\mathbb{R}^2} |f_\varepsilon|^2 d\mathbf{x}', \tag{3.8.10}$$

where $c_1$ and $C$ do not depend upon $s$.

Observe that the distance $\delta(\mathbf{x})$ from $\mathbf{x} \in \Omega_n$ to the boundary $\partial\Omega_n$, is just the distance from $\mathbf{x}'$ to $\partial\Omega_2$, i.e., $\delta(\mathbf{x}')$. Hence,

$$\begin{aligned} \int_{\Omega_n} \frac{|f(\mathbf{x})|^2}{\delta(\mathbf{x})^2} d\mathbf{x} &= \int_{\Omega_2} \frac{|f_\varepsilon(\mathbf{x}')|^2}{\delta(\mathbf{x}')^2} d\mathbf{x}' \Pi_{j=3}^n \int_{-\infty}^{\infty} |g_s(x_j)|^2 dx_j \\ &= (2s + c_1)^{n-2} \int_{\Omega_2} \frac{|f_\varepsilon(\mathbf{x}')|^2}{\delta(\mathbf{x}')^2} d\mathbf{x}'. \end{aligned}$$

Using the fact that

$$\delta(\mathbf{x}) = \delta(\mathbf{x}') \le R - \rho < \rho(e^{\pi\varepsilon} - 1),$$

Eqs. (3.8.9) and (3.8.10) we have

$$\begin{aligned} \int_{\mathbb{R}^n} |\nabla f(\mathbf{x})|^2 d\mathbf{x} &= (2s + c_1)^{n-2} \int_{\mathbb{R}^2} |\nabla f_\varepsilon|^2 d\mathbf{x}' + C(2s + c_1)^{n-3} \int_{\mathbb{R}^2} |f_\varepsilon|^2 d\mathbf{x}' \\ &\le (2s + c_1)^{n-2} \frac{\varepsilon}{2} \int_{\mathbb{R}^2} \frac{|f_\varepsilon|^2}{\delta(\mathbf{x}')^2} d\mathbf{x}' + C(2s + c_1)^{-1} \int_{\mathbb{R}^n} |f|^2 d\mathbf{x} \\ &\le \left[\frac{\varepsilon}{2} + \frac{C\rho^2(e^{\pi\varepsilon} - 1)^2}{2s + c_1}\right] \int_{\mathbb{R}^n} \frac{|f|^2}{\delta(\mathbf{x})^2} d\mathbf{x} \\ &\le \varepsilon \int_{\mathbb{R}^n} \frac{|f|^2}{\delta(\mathbf{x})^2} d\mathbf{x} \end{aligned}$$

for $s$ large.

Now, choose ellipsoids $E_1$ and $E_2$ defined respectively by

$$\frac{x_1^2 + x_2^2}{R^2} + \frac{x_3^2 + \cdots x_n^2}{b^2} \quad \text{and} \quad \frac{x_1^2 + x_2^2}{a^2} + \frac{x_3^2 + \cdots x_n^2}{b^2},$$

with $a := \frac{1}{2}[\rho + \min_{\mathbf{x}' \in \operatorname{supp}(f_\varepsilon)} |\mathbf{x}'|]$ and $b$ sufficiently large in order that $\operatorname{supp} f \subset E_1 \setminus E_2$. Inequality (3.8.8) follows.

Recall that the principal radius is the reciprocal of the principal curvature, i.e., $r_i = 1/\kappa_i$. Therefore, if we rescale by replacing $\mathbf{x}$ with $\sigma\mathbf{x}$ for some constant $\sigma > 0$, then $\delta(\sigma\mathbf{x}) = \sigma\delta(\mathbf{x})$, indicating that the rescaled principal curvature is $\kappa_i/\sigma$. Such a scaling leaves inequality (3.8.8) invariant, but the new mean curvature scales to $H(\mathbf{x})/\sigma$. Consequently, we can rescale in order that $H(\mathbf{y}) < \varepsilon, \mathbf{y} \in \partial E_2$. Since $E_1$ is convex, then $H(\mathbf{y}) < 0, \mathbf{y} \in \partial E_1$. It follows that $H(N(\mathbf{x})) < \varepsilon$ for all $\mathbf{x} \in E_1 \setminus \overline{E_2}$. □

As mentioned earlier, a consequence of Example 3.8.7 is the fact that the weakly mean convexity requirement, $H(\mathbf{y}) \leq 0$, in Corollary 3.7.14 of Chap. 3 cannot be replaced by the global condition $H(\mathbf{y}) < \varepsilon$ for any arbitrarily small $\varepsilon > 0$. In contrast, Theorem 4 of [10] shows that we are able to obtain a Hardy inequality with a sharp constant if $E_2$ in Example 3.8.7 is replaced by a ball $B_\rho$ that approximates the ellipsoid $E_1$ according to the inequality

$$(n-2)\delta(\mathbf{x}) \leq \rho, \qquad \mathbf{x} \in E_1 \setminus \overline{B_\rho}.$$

If $n = 3$ and $E_1 = B_R$, the inequality reduces to $R \leq 3\rho$. See (3.8.4) for the case in which $E_1 = B_R$ and $E_2 = B_\rho$ with $\rho < R$.

# Chapter 4
# Hardy, Sobolev, Maz'ya (HSM) Inequalities

## 4.1 Introduction

From the Hardy and Sobolev inequalities

$$\|\nabla u\|_{p,\Omega}^p \geq C_H \|u/\delta\|_{p,\Omega}^p, \quad \|\nabla u\|_{p,\Omega}^p \geq C_S \|u\|_{p^*,\Omega}^p, \quad u \in D_0^{1,p}(\Omega),$$

where $\delta(\mathbf{x}) = \mathrm{dist}(\mathbf{x}, \partial\Omega)$, $C_H$, $C_S$ are the optimal constants and $p^* = np/(n-p)$, it follows that for $0 < \alpha \leq C_H$,

$$\begin{aligned} \|\nabla u\|_{p,\Omega}^p - \alpha \|u/\delta\|_{p,\Omega}^p &\geq (1 - \alpha/C_H) \|\nabla u\|_{p,\Omega}^p \\ &\geq (1 - \alpha/C_H)\, C_S \|u\|_{p^*,\Omega}^p. \end{aligned} \tag{4.1.1}$$

In this chapter, we discuss the existence of inequalities involving the left-hand side of (4.1.1) with $\alpha = C_H$, and of the form

$$\|\nabla u\|_{p,\Omega}^p - C_H \|u/\delta\|_{p,\Omega}^p \geq C(n,p,\Omega) \|u\|_{q,\Omega}^p$$

for some $q \in (1, p^*]$. Such inequalities are known as **Hardy, Sobolev, Maz'ya** (which we abbreviate to HSM) inequalities. An early example was provided by Maz'ya in [118], Corollary 3, p. 97, where the following is proved: denoting points in $\mathbb{R}^{n+m}$ by $\mathbf{x} = (\mathbf{y}, \mathbf{z})$, $\mathbf{y} \in \mathbb{R}^n$, $\mathbf{z} \in \mathbb{R}^m$, with $n + m > 2$,

$$\int_{\mathbb{R}^{n+m}} \left( |\nabla u|^2 - \frac{(m-2)^2}{4} \frac{|u|^2}{|\mathbf{y}|^2} \right) d\mathbf{x} \geq K_{n,m} \left( \int_{\mathbb{R}^{n+m}} |u|^{\frac{2(n+m)}{n+m-2}} d\mathbf{x} \right)^{\frac{n+m-2}{n+m}},$$



A.A. Balinsky et al., *The Analysis and Geometry of Hardy's Inequality*,
Universitext, DOI 10.1007/978-3-319-22870-9_4

for all $u \in C_0^\infty(\mathbb{R}^{n+m})$, subject to the condition that $u(\mathbf{y}, 0) = 0$ in the case $m = 1$. With $\mathbb{R}_+^n := \{\mathbf{x} \in \mathbb{R}^n : x_n > 0\} = \mathbb{R}^{n-1} \times \mathbb{R}_+$, it follows that for $n \geq 3$,

$$\int_{\mathbb{R}_+^n} \left( |\nabla u|^2 - \frac{|u|^2}{4x_n^2} \right) d\mathbf{x} \geq K_{n,2} \left( \int_{\mathbb{R}_+^n} |u|^{\frac{2n}{n-2}} d\mathbf{x} \right)^{\frac{n-2}{n}}, \tag{4.1.2}$$

for all $u \in C_0^\infty(\mathbb{R}_+^n)$. These inequalities are refinements of both the Hardy and Sobolev inequalities on $\mathbb{R}^{n+m}$ and $\mathbb{R}_+^n$ respectively.

We shall return to the proof of (4.1.2) later, in Corollary 4.3.2.

## 4.2 An HSM Inequality of Brezis and Vázquez

In [31], Brezis and Vázquez proved the following theorem:

**Theorem 4.2.1** *Let $\Omega$ be a bounded domain in $\mathbb{R}^n$, $n > 2$, and $1 < q < 2^* = 2n/(n-2)$. Then, for every $u \in H_0^1(\Omega)$,*

$$\|u\|_q^2 \leq c|\Omega|^{2\left\{\frac{1}{q} - \frac{1}{2^*}\right\}} \int_\Omega \left( |\nabla u|^2 - \left( \frac{n-2}{2} \right)^2 \frac{|u|^2}{|\mathbf{x}|^2} \right) d\mathbf{x} \tag{4.2.1}$$

*for some positive constant $c$.*

The proof of (4.2.1) in [31] depends on the following result which is of independent interest.

**Theorem 4.2.2** *For any bounded domain $\Omega \subset \mathbb{R}^n$, $n \geq 2$, and every $u \in H_0^1(\Omega)$,*

$$\int_\Omega |\nabla u|^2 d\mathbf{x} - \left( \frac{n-2}{2} \right)^2 \int_\Omega \frac{|u|^2}{|\mathbf{x}|^2} d\mathbf{x} \geq H_2 \left( \frac{\omega_n}{|\Omega|} \right)^{\frac{2}{n}} \int_\Omega |u|^2 d\mathbf{x}. \tag{4.2.2}$$

*The constant $H_2$ is the first eigenvalue of the Laplacian on the unit ball in $\mathbb{R}^2$, hence positive and independent of $n$. The constants $(n-2)^2/4$ and $H_2$ are optimal when $\Omega$ is a ball.*

*Proof* It suffices to prove (4.2.2) on $C_0^1(\Omega)$. For once it is established on $C_0^1(\Omega)$, it follows that

$$\|u\| := \left\{ \int_\Omega \left( |\nabla u|^2 - \left( \frac{n-2}{2} \right)^2 \frac{|u|^2}{|\mathbf{x}|^2} \right) d\mathbf{x} \right\}^{1/2} \tag{4.2.3}$$

is a norm on $C_0^1(\Omega)$, and (4.2.2) continues to hold on the completion, $H(\Omega)$ say, of $C_0^1(\Omega)$ with respect to this norm. Furthermore, in view of (4.2.2), we have the continuous embeddings

$$D_0^{1,2}(\Omega) \hookrightarrow H(\Omega) \hookrightarrow L^2(\Omega)$$

and $D_0^{1,2}(\Omega)$ coincides with $H_0^1(\Omega)$. Hence, in particular, (4.2.2) holds on $H_0^1(\Omega)$.

From the list of properties of symmetric non-increasing rearrangements after Definition 1.3.10, we note that

$$\|\nabla u\|_{p,\Omega} \geq \|\nabla |u|\|_{p,\Omega} \geq \|\nabla |u|^{\star}\|_{p,\Omega}$$

and

$$\int_{\Omega} \frac{|u(\mathbf{x})|^2}{|\mathbf{x}|^2} d\mathbf{x} \leq \int_{\Omega^{\star}} \frac{(|u(\mathbf{x})|^{\star})^2}{|\mathbf{x}|^2} d\mathbf{x}$$

where $\Omega^{\star}$ is the ball $B_R$ centre the origin with volume $|\Omega|$, hence $\omega_n R^n = |\Omega|$. Also the $L^2(\Omega)$ norm of $u$ on $\Omega$ is equal to that of $|u|^{\star}$ on $B_R$. It follows that the symmetric non-increasing rearrangement decreases the left-hand side of (4.2.2) while the right-hand side is unchanged. As the symmetric rearrangement $|u|^{\star}$ of $|u|$ is a non-negative, radial function it will suffice to prove (4.2.2) in the radially symmetric case.

For $n = 2$ the result is the Friedrichs inequality and $H_2$ is the first eigenvalue of the Dirichlet Laplacian on $B_1$. Equality is satisfied by the corresponding eigenfunction, which is the Bessel function $J_0(zr)$, where $z \approx 2.4048$ is the first zero of $J_0$. Then $H_2 = z^2 \approx 5.7832$.

Suppose $n \geq 3$ and that $u$ is a non-negative radial function in the ball $B_1$. Define the function

$$\tilde{u}(\mathbf{x}) = |\mathbf{x}|^{-\frac{n-2}{2}}, \qquad n > 2,$$

which is not in $D^1(B_1)$ and satisfies

$$\Delta \tilde{u}(\mathbf{x}) + \left(\frac{n-2}{2}\right)^2 |\mathbf{x}|^{-2} \tilde{u}(\mathbf{x}) = 0.$$

We use $\tilde{u}$ to make a dimension reduction from $n$ to 2 dimensions by introducing a new variable

$$v(r) := u(r) r^{\frac{n-2}{2}}, \qquad r = |\mathbf{x}|.$$

It is readily shown that

$$\begin{aligned}
&\int_{B_R}\left[|\nabla u(\mathbf{x})|^2-\left(\frac{n-2}{2}\right)^2\frac{|u(\mathbf{x})|^2}{|\mathbf{x}|^2}\right]d\mathbf{x}\\
&=n\omega_n\left[\int_0^R|v'(r)|^2rdr-\frac{n-2}{2}\int_0^R[v^2(r)]'dr\right]\\
&=n\omega_n\int_0^R|v'(r)|^2rdr. \qquad (4.2.4)
\end{aligned}$$

since $u\in C_0^1(B_R)$. The penultimate integral in (4.2.4) can be bounded below using the Friedrichs inequality in two dimensions,

$$\int_0^R|v'(r)|^2rdr\geq\frac{H_2}{R^2}\int_0^R|v(r)|^2rdr. \qquad (4.2.5)$$

The proof is completed on noting that

$$\int_{B_R}|u(\mathbf{x})|^2d\mathbf{x}=n\omega_n\int_0^R|v(r)|^2rdr.$$

□

*Remark 4.2.3* The inequality (4.2.2) implies that

$$\int_\Omega|\nabla u(\mathbf{x})|^2d\mathbf{x}>\left(\frac{n-2}{2}\right)^2\int_\Omega\frac{|u(\mathbf{x})|^2}{|\mathbf{x}|^2}d\mathbf{x},$$

for any non-trivial function $u\in H_0^1(\Omega)$. Also, in (4.2.2), $H_2(\omega_n/|\Omega|)$ is not attained in $H_0^1(\Omega)$, for that would imply equality in (4.2.5) with $R=1$ and $v(r)=cJ_0(zr)$. Hence $u(\mathbf{x})=cJ_0(zr)/r^{(n-2)/2}$, which is not in $H_0^1(B_1)$.

Before embarking on the proof, we first note that in the case $n=2$, we have the Sobolev inequality

$$\int_{B_R}\exp\left(\frac{|f(\mathbf{x})|}{c_1\|\nabla f(\mathbf{x})\|_{2,B_R}}\right)^2d\mathbf{x}\leq c_2|B_R| \qquad (4.2.6)$$

for all $f\in D_0^1(B_R)$, where $c_1,\ c_2$ are positive constants; see [49], Theorem V.3.16. It therefore follows that, for all $p\in[1,\infty)$,

$$\int_{B_R}\left(\frac{|f(\mathbf{x})|}{\|\nabla f\|_{2,B_R}}\right)^{2p}d\mathbf{x}\leq c|B_R|$$

for some positive constant $c$, and hence, on applying the Cauchy-Schwarz inequality,

$$\begin{aligned}\|f\|_{p,B_R} &\le |B_R|^{1/2p}\|f\|_{2p,B_R}\\ &\le cR^{2/p}\|\nabla f\|_{2,B_R};\end{aligned}$$

if $f$ is radial,

$$\int_0^R |f(r)|^p rdr \le cR^2\left(\int_0^R |f'(r)|^2 rdr\right)^{p/2}. \tag{4.2.7}$$

**Proof of Theorem 4.2.1** For the radial functions $u$ and $v$ in the proof of Theorem 4.2.2, on setting $H = [(n-2)/2]^2$, we have from (4.2.4) and (4.2.7),

$$\begin{aligned}\int_{B_R}\left(|\nabla u(\mathbf{x})|^2 - H\frac{u^2(\mathbf{x})}{|\mathbf{x}|^2}\right)d\mathbf{x} &= n\omega_n\int_0^R [v'(r)]^2 rdr\\ &\ge cR^{-4/p}\|v\|_{p,B_R}^2\\ &= cR^{-4/p}\left(\int_0^R u(r)^p r^{(n-2)p/2} rdr\right)^{2/p}.\end{aligned} \tag{4.2.8}$$

Furthermore, for $\alpha > 0$ (to be determined), and $1 < q < p$,

$$\begin{aligned}\left(\frac{1}{n\omega_n}\|u\|_q^q\right)^{2/q} &= \left(\int_0^R |u|^q r^\alpha r^{n-2-\alpha} rdr\right)^{2/q}\\ &\le \left(\int_0^R |u|^p r^{\alpha p/q} rdr\right)^{2/p}\left(\int_0^R r^{(n-2-\alpha)\gamma} rdr\right)^{2/q\gamma}, \end{aligned} \tag{4.2.9}$$

where we have used Hölder's inequality with $1/\gamma + q/p = 1$; $\gamma > 1$ since $q < p$. The strategy of the proof is to choose $\alpha$ such that the last integral in (4.2.9) is finite, and the right-hand side of (4.2.9) is less than a constant multiple of the right-hand side of (4.2.8). Thus we choose $\alpha$ so that $\alpha p/q = p(n-2)/2$, i.e.,

$$\alpha = q(n-2)/2,$$

and we require $n - 2 - (n-2)q/2 > -2/\gamma = -2(1 - q/p)$, i.e.,

$$n > q\left(\frac{n-2}{2} + \frac{2}{p}\right). \tag{4.2.10}$$

We now choose $2/p = [(n-2)/2]\varepsilon$, where $\varepsilon$ is sufficiently small that $q/p = q[(n-2)/4]\varepsilon < 1$; then (4.2.10) becomes

$$n > q\left(\frac{n-2}{2}\right)(1+\varepsilon),$$

which is satisfied for small enough $\varepsilon$ as $q < 2n/(n-2) = 2^*$.

From (4.2.8) and (4.2.9), we now have

$$\begin{aligned}\int_{B_R}\left(|\nabla u(\mathbf{x})|^2 - H\frac{u^2(\mathbf{x})}{|\mathbf{x}|^2}\right)d\mathbf{x} &\geq cR^{\{-\frac{4}{p}-\frac{2}{q}(n-2-\alpha)-\frac{4}{q\gamma}\}}\|u\|^2_{q,B_R}\\ &= cR^{2n(\frac{n-2}{2n}-\frac{1}{q})}\|u\|^2_{q,B_R}.\end{aligned}$$

Therefore

$$\|u\|^2_{q,B_R} \leq cR^{2n(\frac{1}{q}-\frac{1}{2^*})}\int_{B_R}\left(|\nabla u(\mathbf{x})|^2 - H\frac{u^2(\mathbf{x})}{|\mathbf{x}|^2}\right)d\mathbf{x}.$$

The theorem follows on recalling that $|\Omega| = \omega_n R^n$. □

The following result is similar to (1.4) in [63], where it is observed that $\varepsilon > 0$ is necessary.

**Corollary 4.2.4** *For any bounded domain $\Omega \subset \mathbb{R}^n$, $n > 2$, and every $u \in H_0^1(\Omega)$*

$$\int_\Omega\left(|\nabla u|^2 - \left(\frac{n-2}{2}\right)^2\frac{|u|^2}{|\mathbf{x}|^2}\right)d\mathbf{x} \geq C(2,n,\varepsilon)\|u\|^2_{\frac{2n}{n-2+\varepsilon}} \tag{4.2.11}$$

*where $C(2,n,\varepsilon) \to 0$ as $\varepsilon \to 0$.*

*Proof* Let $q = 2n/(n-2+\varepsilon)$ for arbitrary small $\varepsilon > 0$. Then $\alpha = \frac{n(n-2)}{n-2+\varepsilon}$ in (4.2.9), and we choose $p$ to be such that $\gamma = 1+\varepsilon$. The last integral in (4.2.9) becomes

$$\left(\int_0^R r^{(n-2-\alpha)\gamma}rdr\right)^{1/\gamma} = \left(\int_0^R r^{\frac{(n-2)(1+\varepsilon)(-2+\varepsilon)}{n-2+\varepsilon}+1}dr\right)^{1/(1+\varepsilon)}$$

which converges to 0 as $\varepsilon \to 0$. The proof of (4.2.11) follows by using this fact in the proof of Theorem 4.2.1. □

*Remark 4.2.5* Theorem 4.2.1 is reminiscent of Theorem 2.3 in [63] for fractional HSM inequalities. The fractional analogue of the Hardy inequality is

$$\int_{\mathbb{R}^n}\frac{|u(\mathbf{x})|^2}{|\mathbf{x}|^{2s}}d\mathbf{x} \leq C_{s,n}^{-1}\int_{\mathbb{R}^n}|\mathbf{p}|^{2s}|\hat{u}(\mathbf{p})|^2d\mathbf{p}, \quad u \in C_0^\infty(\mathbb{R}^n), \tag{4.2.12}$$

which is valid for $0 < 2s < n$, and where the sharp constant is

$$C_{s,n} := 2^{2s} \frac{\Gamma^2([n+2s]/4)}{\Gamma^2([n-2s]/4)},$$

established in [77, 151]. The aforementioned result proved in [63] is the following:

**Theorem 4.2.6** *Let* $0 < s < \min\{1, n/2\}$ *and* $1 \leq q < 2_s^* := 2n/(n-2s)$. *Then there exists a positive constant* $C_{q,n,s}$ *such that for any domain* $\Omega \subset \mathbb{R}^n$ *of finite measure* $|\Omega|$,

$$\|u\|_q^2 \leq C_{q,n,s} |\Omega|^{2\left(\frac{1}{q} - \frac{1}{2_s^*}\right)} \left( \int_{\mathbb{R}^n} |\mathbf{p}|^{2s} |\hat{u}(\mathbf{p})|^2 d\mathbf{p} - C_{s,n} \int_{\mathbb{R}^n} \frac{|u(\mathbf{x})|^2}{|\mathbf{x}|^{2s}} d\mathbf{x} \right), \tag{4.2.13}$$

*for all* $u \in C_0^\infty(\Omega)$.

It is noted in [31] that $q$ must be strictly less than the critical exponent $2_s^*$.

## 4.3 A General HSM Inequality in $L^p(\Omega)$

In [60], Filippas, Maz'ya and Tertikas proved that for a bounded convex domain $\Omega$ with a $C^2$ boundary, there exists a constant $C = C(\Omega)$ depending on $\Omega$, such that

$$\int_\Omega \left( |\nabla u|^2 - \frac{|u|^2}{4\delta(\mathbf{x})^2} \right) d\mathbf{x} \geq C \left( \int_\Omega |u|^{\frac{2n}{n-2}} d\mathbf{x} \right)^{\frac{n-2}{n}}, \quad u \in C_0^\infty(\Omega), \tag{4.3.1}$$

and posed the problem: can $C$ be chosen to be independent of $\Omega$? This was answered in the affirmative by Frank and Loss in [62] who proved it as a consequence of a more general inequality, which holds for an arbitrary domain $\Omega \subsetneqq \mathbb{R}^n$; this is that for $n \geq 3$, there exists a positive constant $K_n$, independent of $\Omega$, such that

$$\int_\Omega \left( |\nabla u|^2 - \frac{|u|^2}{4\delta_M(\mathbf{x})^2} \right) d\mathbf{x} \geq K_n \left( \int_\Omega |u|^{\frac{2n}{n-2}} d\mathbf{x} \right)^{\frac{n-2}{n}}, \quad u \in C_0^\infty(\Omega), \tag{4.3.2}$$

where $\delta_M = \delta_{M,2}$ is the mean distance function of Definition 3.3.1. Since $\delta_M(\mathbf{x}) \leq \delta(\mathbf{x})$ if $\Omega$ is convex, by Theorem 3.3.4, (4.3.1) follows from (4.3.2) with $C \leq K_n$. As noted in [62], an application of Hölder's inequality to the right-hand side of (4.3.1) yields the inequality

$$\int_\Omega \left( |\nabla u|^2 - \frac{|u|^2}{4\delta(\mathbf{x})^2} \right) d\mathbf{x} \geq K_n |\Omega|^{-2/n} \int_\Omega |u|^2 d\mathbf{x}, \quad u \in C_0^\infty(\Omega), \tag{4.3.3}$$

which was discussed in Sect. 3.3.1.

Another important consequence of (4.3.2) is Maz'ya's inequality (4.1.2) when $\Omega = \mathbb{R}^n_+$. In the case $n = 3$, the optimal constant in this inequality was obtained in [28]: to be specific, they proved that with $\mathbb{R}^3_+ = \mathbb{R}^2 \times \mathbb{R}_+$, for all $f \in C_0^\infty(\mathbb{R}^3_+)$,

$$\int_{\mathbb{R}^3_+} \nabla|f(\mathbf{x})|^2 d\mathbf{x} \geq \int_{\mathbb{R}^3_+} \frac{|f(\mathbf{x})|^2}{4|\mathbf{x}|^2} d\mathbf{x} + S_3 \left( \int_{\mathbb{R}^3_+} |f(\mathbf{x})|^6 d\mathbf{x} \right)^{1/3}, \tag{4.3.4}$$

where $S_3$ is optimal, and given by

$$S_3 = 3(\pi/2)^{4/3}, \tag{4.3.5}$$

which is the sharp constant in the Sobolev inequality (1.3.6) in three dimensions. The inequality is also shown to be strict for non-zero functions $f$.

In [62], an $L^p$ analogue of (4.3.2) is obtained for $p \geq 2$, and it is this result which will be the centre-piece of this section. It is

**Theorem 4.3.1** *(Frank and Loss [62]) Let $p \in [2, n)$. Then there is a constant $K_{n,p}$, depending only upon $n$ and $p$, such that for any domain $\Omega \subsetneq \mathbb{R}^n$ and all $u \in C_0^\infty(\Omega)$*

$$\int_\Omega \left( |\nabla u|^p - \left(\frac{p-1}{p}\right)^p \frac{|u|^p}{\delta_{M,p}(\mathbf{x})^p} \right) d\mathbf{x} \geq K_{n,p} \left( \int_\Omega |u|^{\frac{np}{n-p}} d\mathbf{x} \right)^{\frac{n-p}{n}}. \tag{4.3.6}$$

By Theorem 3.3.4, an immediate consequence is

**Corollary 4.3.2** *Let $\Omega \subsetneq \mathbb{R}^n$, $n \geq 3$, be a convex domain. Then there exists a positive constant $K_{n,p}$, $p \in [2, n)$, depending only upon $n$ and $p$, such that*

$$\int_\Omega \left( |\nabla u|^p - \left(\frac{p-1}{p}\right)^p \frac{|u|^p}{\delta(\mathbf{x})^p} \right) d\mathbf{x} \geq K_{n,p} \left( \int_\Omega |u|^{\frac{np}{n-p}} d\mathbf{x} \right)^{\frac{n-p}{n}}$$

*for all $u \in C_0^\infty(\Omega)$.*

Before embarking on the proof of Theorem 4.3.1, we need some preparatory results.

**Lemma 4.3.3** *For all $a, b \in \mathbb{C}$, and $p \geq 2$,*

$$|a+b|^p \geq |a|^p + p|a|^{p-2}\mathrm{Re}[\bar{a}b] + c_p|b|^p \tag{4.3.7}$$

*for some $c_p \in (0, 1]$.*

*Proof* Let $x = |a+b|^2$, $y = |a|^2$ and $f(t) = t^{p/2}$, $t \in (0, \infty)$. Then since $f''(t) \geq 0$, we have by Taylor's theorem,

$$f(x) \geq f(y) + (x-y)f'(y)$$

and so

$$|a+b|^p \geq |a|^p + \left(|b|^2 + 2\mathrm{Re}[\overline{a}b]\right)\frac{p}{2}|a|^{p-2}$$
$$= |a|^p + p\mathrm{Re}[\overline{a}b]|a|^{p-2} + \frac{p}{2}|b|^2|a|^{p-2}. \qquad (4.3.8)$$

Suppose that $|a| \geq c|b|$ for some positive constant $c < 1$ to be determined later. Then, from (4.3.8)

$$|a+b|^p \geq |a|^p + p\mathrm{Re}[\overline{a}b]|a|^{p-2} + \frac{p}{2}c^{p-2}|b|^p \qquad (4.3.9)$$

and so (4.3.7) is satisfied.

Let $z = |a|^2 + |b|^2$. Then

$$f(x) \geq f(z) + (x-z)f'(z)$$

yields

$$|a+b|^p \geq \left(|a|^2 + |b|^2\right)^{p/2} + 2\mathrm{Re}[\overline{a}b]\frac{p}{2}\left(|a|^2 + |b|^2\right)^{\frac{p-2}{2}}, \qquad (4.3.10)$$

and on the right-hand side

$$\left(|a|^2 + |b|^2\right)^{p/2} \geq |a|^p + |b|^p.$$

Suppose now that $|a| \leq c|b|$. Then in (4.3.10), for some constant $K$ (independent of $|a|$ and $|b|$),

$$\left(|a|^2 + |b|^2\right)^{\frac{p-2}{2}} \leq (|a| + |b|)^{p-2} = |b|^{p-2}\left\{1 + \frac{|a|}{|b|}\right\}^{p-2}$$
$$\leq K(|b|^{p-2}). \qquad (4.3.11)$$

Hence from (4.3.10),

$$|a+b|^p \geq |a|^p + |b|^p + p\mathrm{Re}[\overline{a}b]|a|^{p-2}$$
$$+ p\mathrm{Re}[\overline{a}b]\left\{\left(|a|^2 + |b|^2\right)^{\frac{p-2}{2}} - |a|^{p-2}\right\}$$

and

$$p\left|\mathrm{Re}[\overline{a}b]\left\{\left(|a|^2 + |b|^2\right)^{\frac{p-2}{2}} - |a|^{p-2}\right\}\right|$$
$$\leq p|a||b|\left(K|b|^{p-2} + [c|b|]^{p-2}\right) \leq pc|b|^p(K + c^{p-2}).$$

Therefore, for $c$ sufficiently small, (4.3.7) is again satisfied and the proof is complete. □

**Proposition 4.3.4** *Let $q \geq p \geq 2$. There is a constant $C_{p,q}$ (where $C_{2,q} \leq (q+2)^2$) such that for every $f \in C_0^\infty(-1,1)$ and for every $t \in [-1,1]$,*

$$|f(t)|^{q(p-1)+p} \leq C_{p,q} \int_{-1}^{1} \left(|f'|^p - \left(\tfrac{p-1}{p}\right)^p \tfrac{|f|^p}{(1-|s|)^p}\right) ds \cdot \left(\int_{-1}^{1} |f|^q ds\right)^{p-1}. \tag{4.3.12}$$

*Proof* This is Proposition 2.5 in [62]. Let $f(t) = (1-|t|)^{(p-1)/p} g(t)$. Then on using (4.3.7), we have

$$\begin{aligned}
&\int_{-1}^{1} \left(|f'|^p - \left(\tfrac{p-1}{p}\right)^p \tfrac{|f|^p}{(1-|s|)^p}\right) ds \\
&= \int_{-1}^{1} \left[\left|(1-|s|)^{(p-1)/p} g' - \tfrac{p-1}{p}\operatorname{sgn}(s)(1-|s|)^{-1/p} g\right|^p - \left(\tfrac{p-1}{p}\right)^p \tfrac{|g|^p}{1-|s|}\right] ds \\
&\geq \int_{-1}^{1} \left[-p\left(\tfrac{p-1}{p}\right)^{p-1} \operatorname{sgn}(s)|g|^{p-2}\mathrm{Re}[\bar{g}g'] + c_p(1-|s|)^{p-1}|g'|^p\right] ds \\
&= \int_{-1}^{1} \left\{\tfrac{p}{2}\left(\tfrac{p-1}{p}\right)^{p-1} \operatorname{sgn}(s)\tfrac{2}{p}\left(\left[|g|^2\right]^{p/2}\right)' + c_p(1-|s|)^{p-1}|g'|^p\right\} ds \\
&= 2\left(\tfrac{p-1}{p}\right)^{p-1} |g(0)|^p + c_p \int_{-1}^{1}(1-|s|)^{p-1}|g'|^p ds
\end{aligned}$$

We shall show that, for $d = 2c_p^{-1}(\frac{p-1}{p})^{p-1}$ and some positive constant $C = C(p,q)$,

$$\begin{aligned}
|g(t)|^{q(p-1)+p} \leq{}& C(1-|t|)^{-\frac{(p-1)(p+q(p-1))}{p}} \left(\int_{-1}^{1}(1-|s|)^{p-1}|g'|^p ds + d|g(0)|^p\right) \\
&\times \left(\int_{-1}^{1} |g|^q (1-|s|)^{\frac{q(p-1)}{p}} ds\right)^{p-1}.
\end{aligned} \tag{4.3.13}$$

It will then follow that

$$\begin{aligned}
&2\left(\tfrac{p-1}{p}\right)^{p-1}|g(0)|^p + c_p \int_{-1}^{1}(1-|s|)^{p-1}|g'|^p ds \\
&\geq C^{-1} c_p \left(\int_{-1}^{1} |f|^q ds\right)^{1-p} |f(t)|^{q(p-1)+p}
\end{aligned}$$

which is (4.3.12) with $C_{p,q} = C/c_p$.

By symmetry it suffices to show (4.3.13) only for $t \in [0,1]$. Since $(1-t)^{[(p-1)(q(p-1)+p)]/p^2}$ is decreasing in $[0,1]$ we have that

$$\begin{aligned}
&|g(t)|^{\frac{q(p-1)+p}{p}} - |g(0)|^{\frac{q(p-1)+p}{p}} \\
&\leq \frac{[q(p-1)+p]}{p} \int_0^t |g|^{\frac{q(p-1)}{p}} |g'| ds
\end{aligned}$$

$$\le \frac{[q(p-1)+p]}{p(1-t)^{\frac{(p-1)(q(p-1)+p)}{p^2}}} \int_0^1 |g|^{\frac{q(p-1)}{p}} |g'|(1-s)^{\frac{(p-1)(q(p-1)+p)}{p^2}} ds$$

$$\le \frac{[q(p-1)+p]}{p(1-t)^{\frac{(p-1)(q(p-1)+p)}{p^2}}}$$

$$\times \left(\int_0^1 |g'|^p(1-s)^{p-1} ds\right)^{\frac{1}{p}} \left(\int_0^1 |g|^q (1-s)^{\frac{q(p-1)}{p}} ds\right)^{\frac{p-1}{p}}.$$

Thus it remains to show that

$$\begin{aligned} |g(0)|^{q(p-1)+p} \;&\le\; C\left(\int_{-1}^1 |g'|^p(1-|t|)^{p-1} dt + d|g(0)|^p\right) \\ &\times \left(\int_{-1}^1 |g|^q (1-|t|)^{\frac{q(p-1)}{p}} dt\right)^{p-1}. \end{aligned} \tag{4.3.14}$$

To that end we choose a free parameter $T \in (0,1)$ and a Lipschitz function $\chi$ with $0 \le \chi \le 1$, $\chi(0) = 1$, $\chi(t) = 0$ when $|t| \in [T, 1]$. Let

$$L := \left(\int_{-1}^1 |\chi'(s)|^{\frac{pq}{q-p}} ds\right)^{\frac{q-p}{pq}}.$$

Choose another parameter $A$, to be fixed later and depending upon $T$ and $L$, and distinguish two cases as to whether or not the inequality

$$|g(0)|^q \le A^{\frac{p}{p-1}} \int_{-1}^1 |g(s)|^q (1-|s|)^{\frac{q(p-1)}{p}} ds \tag{4.3.15}$$

holds.

If (4.3.15) holds, then as an immediate consequence we have that

$$\begin{aligned} |g(0)|^{\frac{q(p-1)+p}{p}} &\le A|g(0)| \left(\int_{-1}^1 |g(t)|^q (1-|t|)^{\frac{q(p-1)}{p}} dt\right)^{\frac{p-1}{p}} \\ &\le A d^{-\frac{1}{p}} \left(\int_{-1}^1 |g'|^p (1-|t|)^{p-1} dt + d|g(0)|^p\right)^{\frac{1}{p}} \\ &\times \left(\int_{-1}^1 |g|^q (1-|t|)^{\frac{q(p-1)}{p}} dt\right)^{\frac{p-1}{p}}, \end{aligned}$$

which implies (4.3.14).

Suppose

$$|g(0)|^q > A^{\frac{p}{p-1}} \int_{-1}^{1} |g(s)|^q (1-|s|)^{\frac{q(p-1)}{p}} ds, \tag{4.3.16}$$

and define $g_0 := \chi g$. Since $g_0(T) = g_0(-T) = 0$

$$\begin{aligned}
|g_0(0)|^{\frac{q(p-1)+p}{p}} &\le \frac{q(p-1)+p}{2p} \int_{-T}^{T} |g_0|^{\frac{q(p-1)}{p}} |g_0'| ds \\
&\le \frac{q(p-1)+p}{2p} (1-T)^{-\frac{(p-1)(p+q(p-1))}{p^2}} \\
&\quad \times \int_{-1}^{1} |g_0|^{\frac{q(p-1)}{p}} |g_0'| (1-|s|)^{\frac{(p-1)(p+q(p-1))}{p^2}} ds \\
&\le \frac{q(p-1)+p}{2p(1-T)^{\frac{(p-1)(p+q(p-1))}{p^2}}} \left( \int_{-1}^{1} |g_0'|^p (1-|s|)^{p-1} ds \right)^{\frac{1}{p}} \\
&\quad \times \left( \int_{-1}^{1} |g_0|^q (1-|s|)^{\frac{q(p-1)}{p}} ds \right)^{\frac{p-1}{p}}.
\end{aligned}$$

We recall that $g_0(0) = g(0)$, and the last term satisfies

$$\int_{-1}^{1} |g_0|^q (1-|s|)^{\frac{q(p-1)}{p}} ds \le \int_{-1}^{1} |g|^q (1-|s|)^{\frac{q(p-1)}{p}} ds.$$

For the integral involving $g'$, we use the triangle inequality for $L^p$:

$$\begin{aligned}
&\left( \int_{-1}^{1} |g_0'|^p (1-|s|)^{p-1} ds \right)^{\frac{1}{p}} \\
&\le \left( \int_{-1}^{1} |g'|^p \chi^p (1-|s|)^{p-1} ds \right)^{\frac{1}{p}} + \left( \int_{-1}^{1} |g|^p |\chi'|^p (1-|s|)^{p-1} ds \right)^{\frac{1}{p}} \\
&\le \left( \int_{-1}^{1} |g'|^p (1-|s|)^{p-1} ds \right)^{\frac{1}{p}} + L \left( \int_{-1}^{1} |g|^q (1-|s|)^{\frac{q(p-1)}{p}} ds \right)^{\frac{1}{q}} \\
&\le \left( \int_{-1}^{1} |g'|^p (1-|s|)^{p-1} ds \right)^{\frac{1}{p}} + L A^{-\frac{p}{q(p-1)}} |g(0)| \\
&\le 2^{\frac{p-1}{p}} \left( \int_{-1}^{1} |g'|^p (1-|s|)^{p-1} ds + L^p A^{-\frac{p^2}{q(p-1)}} |g(0)|^p \right)^{\frac{1}{p}};
\end{aligned}$$

in the penultimate inequality we have used (4.3.16). Choosing $A$ large enough in order that $L^p A^{-\frac{p^2}{q(p-1)}} \le d$, we arrive at (4.3.14) again, and hence the proof is complete. □

**Corollary 4.3.5** *Let $q \ge p \ge 2$. For every open set $\Omega \subsetneqq \mathbb{R}$ and all $f \in C_0^\infty(\Omega)$, there exists a positive constant $C_{p,q}$ such that*

$$\sup_{t\in\Omega} |f(t)|^{q(p-1)+p} \le C_{p,q} \int_\Omega \left( |f'|^p - \left(\frac{p-1}{p}\right)^p \frac{|f|^p}{dist(t,\Omega^c)^p} \right) dt \times \left( \int_\Omega |f|^q dt \right)^{p-1}. \tag{4.3.17}$$

*Proof* First note that (4.3.17) follows from (4.3.12) for any interval by a translation and a dilation. The extension to an arbitrary open set $\Omega \subsetneqq \mathbb{R}$ follows from the fact that every proper open subset of $\mathbb{R}$ is the union of countably many, pairwise disjoint, open intervals. □

In order to pass from the one-dimensional inequality of Corollary 4.3.5 to $n$ dimensions, Frank and Loss [62] apply an argument of Gagliardo and Nirenberg (see Sect. 4.5 in [52]), which we now describe. We use the notation

$$\hat{\mathbf{x}}_j := (x_1, \dots, x_{j-1}, x_{j+1}, \dots, x_n) \in \mathbb{R}^{n-1}.$$

**Lemma 4.3.6** *For $n \ge 2$ and $f_1, \dots, f_n \in L^{n-1}(\mathbb{R}^n)$, the function $f$ defined by $f(\mathbf{x}) := f_1(\hat{\mathbf{x}}_1)\cdots f_n(\hat{\mathbf{x}}_n)$, belongs to $L^1(\mathbb{R}^n)$ and*

$$\|f\|_{L^1(\mathbb{R}^n)} \le \Pi_{j=1}^n \|f_j\|_{L^{n-1}(\mathbb{R}^{n-1})}. \tag{4.3.18}$$

*Proof* The proof is trivial for $n = 2$. For $n \ge 3$, we have by Hölder's inequality

$$\int_{\mathbb{R}^n} |f(\mathbf{x})| d\mathbf{x} = \int_{\mathbb{R}^{n-1}} |f_1| \left( \int_{\mathbb{R}} |f_2 \cdots f_n| dx_1 \right) d\hat{\mathbf{x}}_1$$

$$\le \|f_1\|_{L^{n-1}(\mathbb{R}^{n-1})} \left( \int_{\mathbb{R}^{n-1}} \left[ \int_{\mathbb{R}} |f_2 \cdots f_n| dx_1 \right]^{\frac{n-1}{n-2}} d\hat{\mathbf{x}}_1 \right)^{\frac{n-2}{n-1}}.$$

Similar successive applications of Hölder's inequality gives

$$\|f\|_{L^1(\mathbb{R}^n)} \le \|f_1\|_{L^{n-1}(\mathbb{R}^{n-1})} \cdots \|f_k\|_{L^{n-1}(\mathbb{R}^{n-1})}$$
$$\times \left( \int_{\mathbb{R}^{n-k}} \left\{ \int_{\mathbb{R}^k} |f_{k+1} \cdots f_n|^{\frac{n-1}{n-k}} dx_1 \cdots dx_k \right\}^{\frac{n-k}{n-k-1}} dx_{k+1} \cdots dx_n \right)^{\frac{n-k-1}{n-1}}$$

for $k = 1, 2, \cdots, n-2$. Thus with $k = n-2$,

$$\|f\|_{L^1(\mathbb{R}^n)} \le \|f_1\|_{L^{n-1}(\mathbb{R}^{n-1})} \cdots \|f_{n-2}\|_{L^{n-1}(\mathbb{R}^{n-1})} \times \left( \int_{\mathbb{R}^2} \left\{ \int_{\mathbb{R}^{n-2}} |f_{n-1} f_n|^{\frac{n-1}{2}} dx_1 \cdots dx_{n-2} \right\}^2 dx_{n-1} dx_n \right)^{\frac{1}{n-1}} \le \Pi_{j=1}^n \|f_j\|_{L^{n-1}(\mathbb{R}^{n-1})},$$

as asserted. □

**Proof of Theorem 4.3.1** Let $\{\mathbf{e}_1, \ldots, \mathbf{e}_n\}$ be the standard basis in $\mathbb{R}^n$ and define

$$\rho_j(\mathbf{x}) = \rho_{\mathbf{e}_j} := \inf\{|t| : \mathbf{x} + t\mathbf{e}_j \in \Omega^c\}.$$

Let

$$g_j(\hat{\mathbf{x}}_j) := \int_{\mathbb{R}} \left( \left| \frac{\partial u(\mathbf{x})}{\partial x_j} \right|^p - \left( \frac{p-1}{p} \right)^p \frac{|u(\mathbf{x})|^p}{\rho_j(\mathbf{x})^p} \right) dx_j$$

and

$$h_j(\hat{\mathbf{x}}_j) := \int_{\mathbb{R}} |u(\mathbf{x})|^q d\mathbf{x}_j, \qquad q = \frac{np}{n-p},$$

for $j = 1, \ldots, n$. By Corollary 4.3.5, as $q(p-1) + p = p^2(n-1)/(n-p)$,

$$|u(\mathbf{x})| \le (C_{p,q} g_j(\hat{\mathbf{x}}_j) h_j(\hat{\mathbf{x}}_j)^{p-1})^{\frac{n-p}{p^2(n-1)}}.$$

Therefore,

$$|u(\mathbf{x})|^n \le C_{p,q}^{\frac{n(n-p)}{p^2(n-1)}} \Pi_{j=1}^n (g_j(\hat{\mathbf{x}}_j) h_j(\hat{\mathbf{x}}_j)^{p-1})^{\frac{n-p}{p^2(n-1)}};$$

so

$$|u(\mathbf{x})|^q \le C_{p,q}^{\frac{n}{p(n-1)}} \Pi_{j=1}^n (g_j(\hat{\mathbf{x}}_j) h_j(\hat{\mathbf{x}}_j)^{p-1})^{\frac{1}{p(n-1)}},$$

and, by Lemma 4.3.6,

$$\begin{aligned} \int_{\mathbb{R}^n} |u(\mathbf{x})|^q dx &\le C_{p,q}^{\frac{n}{p(n-1)}} \int_{\mathbb{R}^n} \Pi_{j=1}^n (g_j(\hat{\mathbf{x}}_j) h_j(\hat{\mathbf{x}}_j)^{p-1})^{\frac{1}{p(n-1)}} d\mathbf{x} \\ &\le C_{p,q}^{\frac{n}{p(n-1)}} \Pi_{j=1}^n \left[ \int_{\mathbb{R}^{n-1}} (g_j(\hat{\mathbf{x}}_j) h_j(\hat{\mathbf{x}}_j)^{p-1})^{\frac{1}{p}} d\hat{\mathbf{x}}_j \right]^{\frac{1}{n-1}}. \end{aligned} \tag{4.3.19}$$

Also, note that for $j = 1, \ldots, n$

$$\|h_j\|_{L^1(\mathbb{R}^{n-1})} = \int_{\mathbb{R}^n} |u(\mathbf{x})|^q d\mathbf{x}.$$

We now apply Hölder's inequality followed by the arithmetic-geometric inequality to get

$$\begin{aligned}
\Pi_{j=1}^n \int_{\mathbb{R}^{n-1}} g_j(\hat{\mathbf{x}}_j)^{\frac{1}{p}} h_j(\hat{\mathbf{x}}_j)^{\frac{p-1}{p}} d\hat{\mathbf{x}}_j &\le \Pi_{j=1}^n \|g_j\|_{L^1(\mathbb{R}^{n-1})}^{\frac{1}{p}} \|h_j\|_{L^1(\mathbb{R}^{n-1})}^{\frac{p-1}{p}} \\
&= \|u\|_q^{\frac{nq(p-1)}{p}} \Pi_{j=1}^n \|g_j\|_{L^1(\mathbb{R}^{n-1})}^{\frac{1}{p}} \\
&\le \|u\|_q^{\frac{nq(p-1)}{p}} \left[\frac{1}{n} \Sigma_{j=1}^n \|g_j\|_{L^1(\mathbb{R}^{n-1})}^{1/p}\right]^n \\
&\le \|u\|_q^{\frac{nq(p-1)}{p}} \left[\frac{1}{n} \Sigma_{j=1}^n \|g_j\|_{L^1(\mathbb{R}^{n-1})}\right]^{\frac{n}{p}},
\end{aligned}$$

where $\|\cdot\|_q$ denotes the $L^q(\mathbb{R}^n)$ norm. On using the last inequality in (4.3.19), we conclude that

$$\int_{\mathbb{R}^n} |u(\mathbf{x})|^q dx \le C_{p,q}^{\frac{n}{p(n-1)}} \|u\|_q^{\frac{nq(p-1)}{p(n-1)}} \left[\frac{1}{n} \Sigma_{j=1}^n \|g_j\|_{L^1(\mathbb{R}^{n-1})}\right]^{\frac{n}{p(n-1)}},$$

which gives

$$\|u\|_q^{\frac{q(n-p)}{p(n-1)}} \le C_{p,q}^{\frac{n}{p(n-1)}} \left[\frac{1}{n} \Sigma_{j=1}^n \|g_j\|_{L^1(\mathbb{R}^{n-1})}\right]^{\frac{n}{p(n-1)}},$$

and thus, since $q = np/(n-p)$,

$$\|u\|_q^p \le \frac{C_{p,q}}{n} \Sigma_{j=1}^n \|g_j\|_{L^1(\mathbb{R}^{n-1})}.$$

On recalling the definition of $g_j$, this is the inequality

$$\left(\int_{\mathbb{R}^n} |u(\mathbf{x})|^q d\mathbf{x}\right)^{\frac{p}{q}} \le \frac{C_{p,q}}{n} \Sigma_{j=1}^n \int_{\mathbb{R}^n} \left(\left|\frac{\partial u(\mathbf{x})}{\partial x_j}\right|^p - \left(\frac{p-1}{p}\right)^p \frac{|u(\mathbf{x})|^p}{\rho_j(\mathbf{x})^p}\right) d\mathbf{x}.$$

Thus, if we use the notation of Sect. 3.3.1, namely, that, for $\nu \in \mathbb{S}^{n-1}$, $\rho_\nu(\mathbf{x})$ denotes the distance from $\mathbf{x} \in \Omega$ to $\partial\Omega$ in the direction $\nu$ or $-\nu$, $\partial_\nu$ the derivative along $\nu$ and $d\omega(\nu)$ the normalised measure on $\mathbb{S}^{n-1}$, then

$$\begin{aligned}
&\left(\int_{\mathbb{R}^n} |u(\mathbf{x})|^q d\mathbf{x}\right)^{\frac{p}{q}} \\
&\le C_{p,q} \int_{\mathbb{R}^n} \int_{\mathbb{S}^{n-1}} \left(|\partial_\nu u(\mathbf{x})|^p - \left(\frac{p-1}{p}\right)^p \frac{|u(\mathbf{x})|^p}{\rho_\nu(\mathbf{x})^p}\right) d\omega(\nu) d\mathbf{x}.
\end{aligned}$$

Since $\partial_\nu u(\mathbf{x}) = \nabla u(\mathbf{x}) \cdot \nu = |\nabla u(\mathbf{x})| \cos(\nabla u(\mathbf{x}), \nu)$, we have from (3.3.7)

$$\begin{aligned}
&\int_{\mathbb{R}^n} \int_{\mathbb{S}^{n-1}} |\partial_\nu u(\mathbf{x})|^p d\omega(\nu) d\mathbf{x} \\
&= \int_{\mathbb{R}^n} \int_{\mathbb{S}^{n-1}} |\cos(\nabla u(\mathbf{x}), \nu)|^p |\nabla u(\mathbf{x})|^p d\omega(\nu) d\mathbf{x} \\
&= \frac{\Gamma(\frac{p+1}{2})\Gamma(\frac{n}{2})}{\sqrt{\pi}\,\Gamma(\frac{n+p}{2})} \int_{\mathbb{R}^n} |\nabla u(\mathbf{x})|^p d\mathbf{x} \\
&= B_{n,p} \int_{\mathbb{R}^n} |\nabla u(\mathbf{x})|^p d\mathbf{x}.
\end{aligned}$$

Hence

$$\begin{aligned}
&\left( \int_{\mathbb{R}^n} |u(\mathbf{x})|^q d\mathbf{x} \right)^{\frac{p}{q}} \\
&\leq C_{p,q} B_{n,p} \int_{\mathbb{R}^n} \left\{ |\nabla u(\mathbf{x})|^p - \left( \frac{p-1}{p} \right)^p \left[ B_{n,p}^{-1} \int_{\mathbb{S}^{n-1}} \frac{d\omega(\nu)}{\rho_\nu(\mathbf{x})^p} \right] |u(\mathbf{x})|^p \right\} d\mathbf{x} \\
&= C_{p,q} B_{n,p} \int_{\mathbb{R}^n} \left\{ |\nabla u(\mathbf{x})|^p - \left( \frac{p-1}{p} \right)^p \frac{|u(\mathbf{x})|^p}{\delta_{M,p}(\mathbf{x})^p} \right\} d\mathbf{x}
\end{aligned}$$

which gives (4.3.6).

## 4.4 Weakly Mean Convex Domains

We explore further results of Filippas et al. in [61]. Conditions designated by (C) and (R) were of central importance, and we begin by introducing and discussing these.

Let $K$ be a $C^2$ manifold without boundary embedded in $\mathbb{R}^n$, of co-dimension $k$, $1 \leq k < n$. When $k = 1$ assume that $K = \partial\Omega$ and when $1 < k < n$ assume that $K \cap \bar{\Omega} \neq \emptyset$. In general $\delta(\mathbf{x}) := \operatorname{dist}(\mathbf{x}, K)$. For $p > 1$ and $p \neq k$ condition (C) is defined as follows:

$$-\Delta_p \delta^{\frac{p-k}{p-1}} \geq 0 \qquad \text{on } \Omega \setminus K, \qquad\qquad (C)$$

where $\Delta_p u := \operatorname{div}(|\nabla u|^{p-2} \nabla u)$ is the $p$-Laplacian.

We shall only be concerned with the case $k = 1$ in which condition (C) reduces to the requirement that $-\Delta\delta \geq 0$ on $\Omega$. From Proposition 2.5.4, we know that $-\Delta\delta \geq 0$ in $\Omega \setminus \Sigma(\Omega)$ if and only if $\Omega$ is weakly mean convex. Moreover, by Corollary 3.7.11, $-\Delta\delta \geq 0$ in $\Omega$ in the distributional sense if $\Omega$ is weakly mean convex and has null cut locus $\Sigma(\Omega)$.

In condition (R) there exists $\varepsilon^* > 0$ and a positive constant $c_0$ such that for all $\mathbf{x} \in \Omega_\varepsilon := \{\mathbf{x} \in \Omega : \delta(\mathbf{x}) < \varepsilon\}$

$$|\delta(\mathbf{x})\Delta\delta(\mathbf{x})| \leq c_0\, \delta(\mathbf{x}), \quad \text{for each } \varepsilon \in (0, \varepsilon^*]. \qquad (R)$$

Condition (R) allows for some unbounded domains, e.g., $\Omega = \Omega_0 \times \mathbb{R}$ with $\Omega_0 \subset \mathbb{R}^{n-1}$ convex and bounded.

**Lemma 4.4.1** *Let $\Omega$ be a weakly mean convex domain with a null cut locus. Then condition (C) holds for $k = 1$. Condition (R) holds if either $\Omega$ is bounded or $B := \sup_{\mathbf{y}\in\partial\Omega} \sum_{j=1}^{n-1} |\kappa_j(\mathbf{y})| < \infty$.*

*Proof* It follows from Corollary 3.7.11 that (C) holds in the distributional sense for the case $k = 1$.

Suppose $\Omega$ is bounded. We recall from Remark 2.4.6, that there exists $\varepsilon > 0$ such that $\Omega_\varepsilon := \{\mathbf{x} \in \Omega : \delta(\mathbf{x}) < \varepsilon\} \subset G(\Omega)$. By Lemma 2.4.2, $\Delta\delta$ is continuous in $G(\Omega)$, and by Proposition 2.5.4, $\sup_{\mathbf{x}\in G(\Omega)} |\Delta\delta(\mathbf{x})| = (n-1) \sup_{\mathbf{y}\in\partial\Omega} |H(\mathbf{y})|$, where $\Delta\delta(\mathbf{x})$ and $H(\mathbf{y})$ are non-positive. Also, as $\mathbf{x} \to \mathbf{y} = N(\mathbf{x})$ in $G(\Omega)$, $\delta(\mathbf{x}) \to 0$ and so $\Delta\delta(\mathbf{x}) \to (n-1)H(\mathbf{y})$, the convergence being uniform in a closed neighbourhood of $\partial\Omega$. Hence, given any $\eta > 0$, we may choose $\varepsilon^*$ sufficiently small such that $\Omega_{\varepsilon^*} \subset G(\Omega)$ and

$$|\Delta\delta(\mathbf{x})| < (n-1)|H(\mathbf{y})| + \eta \leq H_0 + \eta, \quad \text{for } \mathbf{x} \in \Omega_{\varepsilon^*},$$

where $H_0 := \max_{\mathbf{y}\in\partial\Omega} |H(\mathbf{y})| < \infty$. Now, let $c_0 = (n-1)H_0 + \eta$ and (R) follows.

Suppose now that $B := \sup_{\mathbf{y}\in\partial\Omega} \sum_{j=1}^{n-1} |\kappa_j(\mathbf{y})| < \infty$. Choose $\varepsilon$ in order that $\Omega_\varepsilon \subset G(\Omega)$ and $\varepsilon^* := \varepsilon B < 1$. For each $\mathbf{x} \in \Omega_\varepsilon$ and the corresponding $\mathbf{y} = N(\mathbf{x})$, let $J_+ := \{j : \kappa_j(\mathbf{y}) \geq 0\}$ and $J_- := \{j : \kappa_j(\mathbf{y}) < 0\}$. Then for each $\mathbf{x}$

$$\begin{aligned} -\Delta\delta &= -\sum_{j\in J_+} \tfrac{\kappa_j}{1+\delta\kappa_j} - \sum_{j\in J_-} \tfrac{\kappa_j}{1-\delta|\kappa_j|} \\ &\leq -\sum_{j\in J_+} \tfrac{\kappa_j}{1+\varepsilon^*} - \sum_{j\in J_-} \tfrac{\kappa_j}{1-\varepsilon^*}. \end{aligned}$$

Let $H_\pm := \frac{1}{(n-1)} \sum_{j\in J_\pm} \kappa_j$. Then

$$\begin{aligned} -\Delta\delta &\leq (n-1)\left\{\tfrac{-1}{1+\varepsilon^*}H_+ + \tfrac{-1}{1-\varepsilon^*}H_-\right\} \\ &= (n-1)\left\{\tfrac{-H+\varepsilon^*(H_+-H_-)}{1-(\varepsilon^*)^2}\right\} \\ &\leq \tfrac{1}{1-(\varepsilon^*)^2}\left[\sum_{j=1}^n |\kappa_j| + \varepsilon^* \sum_{j=1}^n |\kappa_j|\right] \\ &\leq \tfrac{B}{1-\varepsilon^*} = \tfrac{B}{1-\varepsilon B}. \end{aligned}$$

Therefore condition (R) holds with $c_0 = \frac{B}{(1-\varepsilon B)}$. □

The next lemma is Lemma 2.2 in [61].

**Lemma 4.4.2** *Let $\Omega$ be a domain satisfying condition (R). Then for any $C \in (0, \frac{1}{2}n\omega_n^{1/n})$ and $a > 0$, there exists $\varepsilon_0 = \varepsilon_0(a/c_0)$ such that for all $\varepsilon \in (0, \varepsilon_0]$,*

$$C\|\delta^a v\|_{\frac{n}{n-1},\Omega_\varepsilon} \le \int_{\Omega_\varepsilon} \delta^a |\nabla v| d\mathbf{x}, \qquad v \in C_0^\infty(\Omega_\varepsilon), \tag{4.4.1}$$

*where $\Omega_\varepsilon := \{\mathbf{x} \in \Omega : \delta(\mathbf{x}) < \varepsilon\}$*

*Proof* By the Gagliardo-Nirenberg-Sobolev inequality (1.3.7),

$$n\omega_n^{1/n}\|f\|_{\frac{n}{n-1},\Omega_\varepsilon} \le \|\nabla f\|_{1,\Omega_\varepsilon}, \qquad f \in W_0^{1,1}(\Omega_\varepsilon), \tag{4.4.2}$$

where $\omega_n$ is the volume of the unit ball in $\mathbb{R}^n$. On substituting $f = \delta^a v$ for $v \in C_0^\infty(\Omega_\varepsilon)$, we have

$$n\omega_n^{1/n}\|\delta^a v\|_{\frac{n}{n-1},\Omega_\varepsilon} \le \int_{\Omega_\varepsilon} \delta^a |\nabla v| d\mathbf{x} + \int_{\Omega_\varepsilon} a\delta^{a-1}|v| d\mathbf{x}. \tag{4.4.3}$$

The last term may be estimated as follows: for $\varepsilon < \varepsilon_0$

$$\begin{aligned} a\int_{\Omega_\varepsilon} \delta^{a-1}|v|d\mathbf{x} &= \int_{\Omega_\varepsilon} (\nabla\delta^a \cdot \nabla\delta)\,|v|d\mathbf{x} \\ &= -\int_{\Omega_\varepsilon} \delta^a(\Delta\delta)|v|d\mathbf{x} - \int_{\Omega_\varepsilon} \delta^a(\nabla\delta \cdot \nabla|v|)d\mathbf{x} \\ &\le c_0 \int_{\Omega_\varepsilon} \delta^a|v|d\mathbf{x} + \int_{\Omega_\varepsilon} \delta^a|\nabla|v||d\mathbf{x} \\ &\le c_0\varepsilon \int_{\Omega_\varepsilon} \delta^{a-1}|v|d\mathbf{x} + \int_{\Omega_\varepsilon} \delta^a|\nabla v|d\mathbf{x}, \end{aligned}$$

where we have used condition (R). Thus,

$$(a - c_0\varepsilon)\int_{\Omega_\varepsilon} \delta^{a-1}|v|d\mathbf{x} \le \int_{\Omega_\varepsilon} \delta^a|\nabla v|d\mathbf{x},$$

and on substituting in (4.4.3),

$$n\omega_n^{1/n}\|\delta^a v\|_{\frac{n}{n-1},\Omega_\varepsilon} \le \left\{1 + \frac{a}{a - c_0\varepsilon}\right\}\int_{\Omega_\varepsilon} \delta^a|\nabla v|d\mathbf{x}.$$

The proof is completed by choosing $\varepsilon$. □

The following theorem is an alternative form of Theorem 2.5 in [61].

**Theorem 4.4.3** *Let $\Omega$ be a weakly mean convex domain in $\mathbb{R}^n$ with a null cut locus, and suppose that it satisfies condition (R) with $D = \sup_{\mathbf{x}\in\Omega}\delta(\mathbf{x}) < \infty$. Then, for any $p \in (1, n)$ there exists a positive constant $C = C(n, p, c_0 D)$ such that*

$$\int_\Omega \delta^{p-1}|\nabla v|^p d\mathbf{x} + \int_\Omega (-\Delta\delta)|v|^p d\mathbf{x} \ge C\|\delta^{\frac{p-1}{p}} v\|_{\frac{np}{n-p},\Omega}$$

*for all $v \in C_0^\infty(\Omega)$.*

*Proof* Let $\alpha \in C^{\infty}[0,\infty)$ be a non-increasing function with the property that $|\alpha'(t)| \le K_0$ and

$$\alpha(t) = \begin{cases} 1, & t \in [0, \frac{1}{2}), \\ 0, & t \ge 1. \end{cases}$$

Then the function $\phi$ defined by $\phi_\varepsilon(\mathbf{x}) := \alpha(\delta(\mathbf{x})/\varepsilon)$ is in $C_0^2(\Omega_\varepsilon)$ with $|\nabla\phi_\varepsilon| \le \frac{K_0}{\varepsilon}$. We set $v = \phi_\varepsilon v + (1-\phi_\varepsilon)v$ for every $v \in C_0^\infty(\Omega)$ and apply Lemma 4.4.2 to obtain, for arbitrary $a > 0$ and $C \in (0, n\omega_n^{1/n}/2)$,

$$C\|\delta^a\phi_\varepsilon v\|_{\frac{n}{n-1},\Omega_\varepsilon} \le \int_{\Omega_\varepsilon} \delta^a|\nabla(\phi_\varepsilon v)|d\mathbf{x}, \qquad v \in C_0^\infty(\Omega). \tag{4.4.4}$$

By the Gagliardo-Nirenberg-Sobolev inequality (1.3.7),

$$n\omega_n^{1/n}\|(1-\phi_\varepsilon)v\|_{\frac{n}{n-1},\Omega} \le \|\nabla[(1-\phi_\varepsilon)v]\|_{1,\Omega}, \qquad v \in C_0^\infty(\Omega).$$

Since, for $\mathbf{x} \in \text{supp}(1-\phi_\varepsilon)v$, $\delta(\mathbf{x}) \in [\frac{\varepsilon}{2}, D]$, where $D := \sup_{\mathbf{x}\in\Omega}\delta(\mathbf{x})$, it follows that

$$\frac{n\omega_n^{1/n}}{D^a}\|\delta^a(1-\phi_\varepsilon)v\|_{\frac{n}{n-1},\Omega} \le (2/\varepsilon)^a \int_\Omega \delta^a|\nabla[(1-\phi_\varepsilon)v]|d\mathbf{x}. \tag{4.4.5}$$

On combining (4.4.4) and (4.4.5), we have for some positive constant $C = C(n,p,a,\varepsilon/D)$,

$$\begin{aligned} C\|\delta^a v\|_{\frac{n}{n-1},\Omega} &\le \int_{\Omega\setminus\Omega_{\varepsilon/2}} \delta^a|\nabla[(1-\phi_\varepsilon)v]|d\mathbf{x} + \int_{\Omega_\varepsilon}\delta^a|\nabla(\phi_\varepsilon v)|d\mathbf{x} \\ &\le \int_{(\Omega\setminus\Omega_{\varepsilon/2})\cup\Omega_\varepsilon} \delta^a|\phi_\varepsilon||\nabla v|d\mathbf{x} \\ &\quad + 2\int_{\Omega_\varepsilon\setminus\Omega_{\varepsilon/2}} \delta^a|\nabla\phi_\varepsilon||v|d\mathbf{x} \\ &\le \int_\Omega \delta^a|\nabla v|d\mathbf{x} + 2K_0\int_{\Omega_\varepsilon\setminus\Omega_{\varepsilon/2}} \delta^{a-1}|v|d\mathbf{x}, \end{aligned}$$

where we have used the fact that $\delta(\mathbf{x}) < \varepsilon$ for $\mathbf{x} \in \Omega_\varepsilon \setminus \Omega_{\varepsilon/2}$. As a consequence, we have the $L^1$ estimate

$$C\|\delta^a v\|_{\frac{n}{n-1},\Omega} \le \int_\Omega \delta^a|\nabla v|d\mathbf{x} + \int_{\Omega_\varepsilon\setminus\Omega_{\varepsilon/2}} \delta^{a-1}|v|d\mathbf{x}. \tag{4.4.6}$$

In order to derive the $L^p$ estimate for $p > 1$, we substitute $v^s$, $s = \frac{p(n-1)}{n-p}$, for $v$ in (4.4.6) to obtain

$$C\left(\int_\Omega \delta^{\frac{an}{n-1}}|v|^{\frac{np}{n-p}}d\mathbf{x}\right)^{\frac{n-1}{n}} \leq s\int_\Omega \delta^{a\frac{n-p}{p(n-1)}}|\nabla v|\delta^{a\frac{n(p-1)}{p(n-1)}}|v|^{\frac{n(p-1)}{n-p}}d\mathbf{x} + \int_{\Omega_\varepsilon\setminus\Omega_{\varepsilon/2}} \delta^{(a-1)c_1}|v|\delta^{(a-1)c_2}|v|^{\frac{n(p-1)}{n-p}}d\mathbf{x},$$

where $c_1 + c_2 = 1$. By the choice

$$(a-1)c_1 := \frac{a(n-p)}{p(n-1)} - 1, \quad (a-1)c_2 = \frac{an(p-1)}{p(n-1)},$$

and the application of Hölder's inequality, we derive

$$\begin{aligned} &C\left(\int_\Omega \delta^{\frac{an}{n-1}}|v|^{\frac{np}{n-p}}d\mathbf{x}\right)^{\frac{n-1}{n}} \\ &\leq \left[s\left(\int_\Omega \delta^{\frac{a(n-p)}{(n-1)}}|\nabla v|^p d\mathbf{x}\right)^{1/p} + \left(\int_{\Omega_\varepsilon\setminus\Omega_{\varepsilon/2}} \delta^{\frac{a(n-p)}{n-1}-p}|v|^p d\mathbf{x}\right)^{1/p}\right] \\ &\quad \times \left(\int_\Omega \delta^{\frac{an}{(n-1)}}|v|^{\frac{np}{n-p}}d\mathbf{x}\right)^{\frac{p-1}{p}}, \end{aligned}$$

which yields,

$$\begin{aligned} &C\left(\int_\Omega \delta^{\frac{an}{n-1}}|v|^{\frac{np}{n-p}}d\mathbf{x}\right)^{\frac{n-p}{np}} \\ &\leq s\left(\int_\Omega \delta^{\frac{a(n-p)}{(n-1)}}|\nabla v|^p d\mathbf{x}\right)^{1/p} + \left(\int_{\Omega_\varepsilon\setminus\Omega_{\varepsilon/2}} \delta^{\frac{a(n-p)}{n-1}-p}|v|^p d\mathbf{x}\right)^{1/p}. \end{aligned} \tag{4.4.7}$$

Now select $a = \frac{(n-1)(p-1)}{n-p} > 0$. The inequality $|x+y|^p \leq 2^{p-1}(|x|^p + |y|^p)$ then gives

$$C(n,p,\tfrac{\varepsilon}{D})\|\delta^{\frac{p-1}{p}}v\|^p_{\frac{np}{n-p},\Omega} \leq \int_\Omega \delta^{p-1}|\nabla v|^p d\mathbf{x} + \int_{\Omega_\varepsilon\setminus\Omega_{\varepsilon/2}} \delta^{-1}|v|^p d\mathbf{x}. \tag{4.4.8}$$

We need an estimate for the last term in (4.4.8). For $\theta > 0$

$$(\frac{\varepsilon}{2})^{p\theta}\int_{\Omega_\varepsilon\setminus\Omega_{\varepsilon/2}} \delta^{-1}|v|^p d\mathbf{x} \leq \int_{\Omega_\varepsilon\setminus\Omega_{\varepsilon/2}} \delta^{-1+p\theta}|v|^p d\mathbf{x} \leq \int_\Omega \delta^{-1+p\theta}|v|^p d\mathbf{x}.$$

The identity

$$\operatorname{div}(\delta^{\theta p}\nabla\delta) = \theta p\delta^{-1+p\theta} + \delta^{\theta p}\Delta\delta$$

and integration by parts leads to

$$\begin{aligned}&\theta p \int_\Omega \delta^{-1+p\theta}|v|^p d\mathbf{x}\\&= \int_\Omega \delta^{\theta p}(-\Delta\delta)|v|^p d\mathbf{x} + \int_\Omega \operatorname{div}(\delta^{\theta p}\nabla\delta)|v|^p d\mathbf{x}\\&= \int_\Omega \delta^{\theta p}(-\Delta\delta)|v|^p d\mathbf{x} - p\int_\Omega \delta^{\theta p}|v|^{p-1}[\nabla\delta\cdot\nabla|v|]d\mathbf{x},\end{aligned}$$

and since $|\nabla|v|| \leq |\nabla v|$, the last term satisfies

$$\begin{aligned}&p\left|\int_\Omega \delta^{\theta p}|v|^{p-1}[\nabla\delta\cdot\nabla|v|]d\mathbf{x}\right|\\&\leq p\varepsilon_2\int_\Omega \delta^{-1+\theta p}|v|^p d\mathbf{x} + pC_{\varepsilon_2}\int_\Omega \delta^{p-1+p\theta}|\nabla v|^p d\mathbf{x},\end{aligned}$$

for an arbitrary $\varepsilon_2 > 0$. Therefore

$$\begin{aligned}&(\theta p - p\varepsilon_2)\int_\Omega \delta^{-1+p\theta}|v|^p d\mathbf{x}\\&\leq \int_\Omega \delta^{\theta p}(-\Delta\delta)|v|^p d\mathbf{x} + pC_{\varepsilon_2}\int_\Omega \delta^{p-1+p\theta}|\nabla v|^p d\mathbf{x},\end{aligned}$$

so that for $\varepsilon_2 < \theta$

$$C(p,\theta)\int_\Omega \delta^{-1+p\theta}|v|^p d\mathbf{x} \leq \int_\Omega \delta^{\theta p}(-\Delta\delta)|v|^p d\mathbf{x} + \int_\Omega \delta^{p-1+p\theta}|\nabla v|^p d\mathbf{x}.$$

From this and $\sup_{\mathbf{x}\in\Omega}\delta(\mathbf{x}) = D$, we infer that

$$\begin{aligned}C(p,\theta)\left(\tfrac{\varepsilon}{D}\right)^{p\theta}\int_{\Omega_\varepsilon\setminus\Omega_{\varepsilon/2}} \delta^{-1}|v|^p d\mathbf{x} &\leq C(p,\theta)D^{-p\theta}\int_\Omega \delta^{-1+p\theta}|v|d\mathbf{x}\\&\leq \int_\Omega(-\Delta\delta)|v|^p d\mathbf{x} + \int_\Omega \delta^{p-1}|\nabla v|^p d\mathbf{x}. \qquad (4.4.9)\end{aligned}$$

We now choose $\theta = 1$ and combine (4.4.9) with (4.4.8) to complete the proof. □

We are now in a position to prove an extension of Corollary 4.3.2 to weakly mean convex domains (cf. Theorem 5.3 of [107]).

**Theorem 4.4.4** *Let $\Omega$ be a weakly mean convex domain in $\mathbb{R}^n$ with a null cut locus, and suppose that it satisfies condition (R) with $D = \sup_{\mathbf{x}\in\Omega}\delta(\mathbf{x}) < \infty$. Then, for any $p \in [2, n)$, there exists a constant $C = C(n, p, c_0 D)$ such that*

$$\int_\Omega |\nabla u|^p d\mathbf{x} - \left(\frac{p-1}{p}\right)^p \int_\Omega \frac{|u|^p}{\delta^p}d\mathbf{x} \geq C\left(\int_\Omega |u|^{\frac{np}{n-p}}d\mathbf{x}\right)^{\frac{n-p}{n}} \qquad (4.4.10)$$

*for all $u \in C_0^\infty(\Omega)$.*

*Proof* Set $u(\mathbf{x}) = \delta(\mathbf{x})^{\frac{p-1}{p}}w(\mathbf{x})$, so that

$$\nabla|u| = \frac{p-1}{p}\delta^{\frac{p-1}{p}-1}|w|\nabla\delta + \delta(\mathbf{x})^{\frac{p-1}{p}}\nabla|w| =: A + B, \qquad (4.4.11)$$

say. For $A = (a_j)_1^n,\ B = (b_j)_1^n \in \mathbb{R}^n$, define the norm and inner product

$$|A| = \left(\sum_{j=1}^{n} |a_j|^2\right)^{1/2}, \quad A \cdot B = \sum_{j=1}^{n} a_j b_j.$$

Then, we claim that, for $p \geq 2$

$$|A+B|^p - |A|^p \geq c_p|B|^p + p|A|^{p-2}A \cdot B \tag{4.4.12}$$

for some $c_p \in (0,1]$. The proof follows similar lines to that for Lemma 4.3.3, after first dividing the inequality by $|A|$ to consider it in the form

$$\left|\frac{A}{|A|} + \frac{B}{|B|}\right|^p - 1 \geq c_p \left(\frac{|B|}{|A|}\right)^p + p\left(\frac{A}{|A|} \cdot \frac{B}{|A|}\right).$$

On applying (4.4.12) to (4.4.11) we obtain

$$\begin{aligned}
&|\nabla u|^p - \left(\tfrac{p-1}{p}\right)^p \delta^{-1}|w|^p \\
&\quad \geq c_p\delta^{p-1}|\nabla|w||^p + p\left(\tfrac{p-1}{p}\right)^{p-2} \delta^{-\frac{p-2}{p}}|w|^{p-2}\left(\tfrac{p-1}{p}\right)\delta^{\frac{p-2}{p}}|w|\nabla|w| \cdot \nabla\delta) \\
&\quad = c_p\delta^{p-1}|\nabla|w||^p + p\left(\tfrac{p-1}{p}\right)^{p-1}|w|^{p-1}(\nabla|w| \cdot \nabla\delta) \\
&\quad = c_p\delta^{p-1}|\nabla|w||^p + \left(\tfrac{p-1}{p}\right)^{p-1}(\nabla|w|^p \cdot \nabla\delta),
\end{aligned}$$

since $|\nabla|u|| \leq |\nabla u|$. Therefore

$$\begin{aligned}
&c_p^{-1}\int_\Omega[|\nabla u|^p - \left(\tfrac{p-1}{p}\right)^p \delta^{-p}|u|^p]d\mathbf{x} \\
&\quad \geq \int_\Omega[\delta^{p-1}|\nabla|w||^p + (\nabla|w|^p \cdot \nabla\delta)]d\mathbf{x} \\
&\quad = \int_\Omega[\delta^{p-1}|\nabla|w||^p + (-\Delta\delta)|w|^p]d\mathbf{x}.
\end{aligned}$$

Theorem 4.4.3 can be shown to apply to $v = |w|$ by a standard density argument. On making this application, the theorem follows. □

A corollary to Theorem 4.4.4 follows from Lemma 4.4.1.

**Corollary 4.4.5** *Let $\Omega$ be a weakly mean convex domain in $\mathbb{R}^n$.*

(1) *If $\Omega$ is bounded, then, for any $p \in [2,n)$, there exists a constant $C = C(n,p,H_0,D)$ such that (4.4.10) holds for $H_0 := \sup_{\mathbf{y}\in\partial\Omega}|H(\mathbf{y})|$.*
(2) *Alternatively, if $B := \sup_{\mathbf{y}\in\partial\Omega}\sum_{j=1}^{n-1}|\kappa_j(\mathbf{y})| < \infty$, then, for any $p \in [2,n)$, there exists a constant $C = C(n,p,BD)$ such that (4.4.10) holds.*

## 4.5 Exterior Domains

In [69], Gkikas established HSM inequalities on domains which are unbounded, these being **exterior** domains in the sense that they are open, connected subsets of $\mathbb{R}^n$ whose complements $\Omega^c := \mathbb{R}^n \setminus \Omega$ are connected and compact, and contain a neighborhood of the origin. It is also assumed in [69] that $\Omega$ has a $C^2$ boundary. Theorem 4.3.1, which is given in terms of the mean distance function $\delta_M$, is applicable for exterior domains, and, as we noted in Sect. 4.3, yields the prototypical Maz'ya's inequality's (4.1.2) when $\Omega = \mathbb{R}^n_+$, since then the mean distance function is equivalent to the distance function, by (3.3.9). Outside such examples, Gkikas provides an alternative approach.

To prove HSM inequalities for exterior domains, Gkikas [69] makes the assumption that

$$F(\mathbf{x}) := -\Delta\delta(\mathbf{x}) + (n-1)\frac{\nabla\delta(\mathbf{x})\cdot\mathbf{x}}{|\mathbf{x}|^2} \geq 0, \tag{4.5.1}$$

holds in the distributional sense

$$\int_\Omega F(\mathbf{x})\varphi(\mathbf{x})d\mathbf{x} \geq 0, \qquad \forall\varphi \in C_0^\infty(\Omega), \quad \varphi \geq 0.$$

This implies that (4.5.1) must hold in the pointwise sense for all $\mathbf{x} \in G(\Omega)$.

In the following lemma, $\Omega$ is not an exterior domain in part (ii), and is not necessarily so in part (iii).

**Lemma 4.5.1** *Condition (4.5.1) holds when*

(i) $\Omega = B_R(0)^c$,
(ii) $\Omega = \mathbb{R}^n_+ = \{(\mathbf{x}', x_n) \in \mathbb{R}^n : x_n > 0\}$, *and*
(iii) $\Omega$ *is a bounded, mean convex domain.*

*Proof* It is straightforward to show that (i) and (ii) imply (4.5.1). To see that (iii) does, first recall from Proposition 2.5.5, that $H(\mathbf{y})$ is continuous on $\partial\Omega$. Therefore, $H(\mathbf{y})$ assumes its maximum at some $\mathbf{y}_0 \in \partial\Omega$ with $H(\mathbf{y}_0) < 0$. From the proof of Proposition 2.5.4, for $\{\mathbf{y}\} = N(\mathbf{x})$

$$-\Delta\delta(\mathbf{x}) \geq -(n-1)H(\mathbf{y}) \geq -(n-1)H(\mathbf{y}_0) \geq (n-1)\varepsilon > 0, \qquad \mathbf{x} \in G(\Omega),$$

for some $\varepsilon > 0$. Proceeding as in the discussion after (2.1) in [69], we choose $\mathbf{x}_0 \in \Omega^c$ such that $|\mathbf{x}_0 - \mathbf{x}| > 1/\varepsilon$ for all $\mathbf{x} \in \Omega$. It follows that

$$-\Delta\delta(\mathbf{x}) + (n-1)\frac{\nabla\delta(\mathbf{x})\cdot(\mathbf{x}-\mathbf{x}_0)}{|\mathbf{x}-\mathbf{x}_0|^2} \geq 0, \qquad \mathbf{x} \in \Omega.$$

The change of variables $\mathbf{z} = \mathbf{x} - \mathbf{x}_0$ completes the proof. □

It is shown in Lemma 2.1 of [69] that, for a compact set $K$ with a $C^2$ boundary,

$$F(\mathbf{x}) \equiv 0 \quad \text{on} \quad \partial K$$

implies that $K$ is a closed ball centred at 0. However, if $\Omega$ is the exterior of an ellipse in $\mathbb{R}^2$, (4.5.1) does not hold. Indeed, using the representation $\Delta\delta(\mathbf{x}) = \kappa/(1+\kappa\delta)$ where $\kappa$ is the curvature at the near point $N(\mathbf{x})$ of $\mathbf{x}$, it can be shown that for some $\mathbf{x}$ in the exterior of the ellipse

$$F(\mathbf{x}) \leq -\frac{\kappa}{1+\kappa\delta} + \frac{1}{|\mathbf{x}|} < 0.$$

The following theorem is Theorem 2.2 in Gkikas [69].

**Theorem 4.5.2** *Let $\Omega \subset \mathbb{R}^n$, $n \geq 4$, be an exterior domain in $\mathbb{R}^n$ not containing the origin, which satisfies (4.5.1), and has a $C^2$ boundary $\partial\Omega$. Then there is a constant $C_n(\Omega) > 0$ depending only upon $n$ and $\Omega$ such that, for all $u \in C_0^\infty(\Omega)$,*

$$\int_\Omega |\nabla u|^2 d\mathbf{x} - \frac{1}{4}\int_\Omega \frac{|u(\mathbf{x})|^2}{\delta(\mathbf{x})^2} d\mathbf{x} \geq C_n(\Omega)\left(\int_\Omega |u(\mathbf{x})|^{\frac{2n}{n-2}} d\mathbf{x}\right)^{\frac{n-2}{n}}. \tag{4.5.2}$$

*Here $C_n(\Omega) = C(\rho, \rho', n)$, where*

$$\rho := \sup_{\mathbf{x}\in\partial\Omega} |\mathbf{x}|, \quad \rho' := \inf_{\mathbf{x}\in\partial\Omega} |\mathbf{x}|.$$

*Proof* Let

$$u(\mathbf{x}) = |\mathbf{x}|^{-\frac{n-1}{2}} \delta(\mathbf{x})^{\frac{1}{2}} v(\mathbf{x}).$$

Then

$$\nabla u = -\frac{n-1}{2}|\mathbf{x}|^{-\frac{n+1}{2}}\delta^{\frac{1}{2}} v \frac{\mathbf{x}}{|\mathbf{x}|} + \frac{1}{2}|\mathbf{x}|^{-\frac{n-1}{2}}\delta^{-\frac{1}{2}}(\nabla\delta)v + |\mathbf{x}|^{-\frac{n-1}{2}}\delta^{\frac{1}{2}}\nabla v$$

and this leads to

$$\begin{aligned}\int_\Omega |\nabla u|^2 d\mathbf{x} &= \int_\Omega \frac{\delta(\mathbf{x})|\nabla v|^2}{|\mathbf{x}|^{n-1}} d\mathbf{x} + \frac{1}{4}\int_\Omega \frac{|v(\mathbf{x})|^2}{\delta|\mathbf{x}|^{n-1}} d\mathbf{x} \\ &+ \frac{(n-1)^2}{4}\int_\Omega \frac{\delta(\mathbf{x})|v|^2}{|\mathbf{x}|^{n+1}} d\mathbf{x} - \frac{n-1}{2}\int_\Omega \frac{|v(\mathbf{x})|^2}{|\mathbf{x}|^{n+1}}(\mathbf{x}\cdot\nabla\delta) d\mathbf{x} \\ &- \frac{n-1}{2}\int_\Omega \frac{\delta(\mathbf{x})}{|\mathbf{x}|^{n+1}}[\mathbf{x}\cdot\nabla|v|^2] d\mathbf{x} + \frac{1}{2}\int_\Omega \frac{\nabla\delta\cdot\nabla|v|^2}{|\mathbf{x}|^{n-1}} d\mathbf{x}.\end{aligned}$$

Integration by parts gives

$$\int_\Omega \frac{\delta(\mathbf{x})}{|\mathbf{x}|^{n+1}}[\mathbf{x}\cdot\nabla|v|^2]d\mathbf{x} = -\int_\Omega \frac{\mathbf{x}\cdot\nabla\delta}{|\mathbf{x}|^{n+1}}|v|^2 d\mathbf{x} + \int_\Omega \frac{\delta|v|^2}{|\mathbf{x}|^{n+1}}d\mathbf{x}$$

and, on using (4.5.1),

$$\int_\Omega \frac{\nabla\delta\cdot\nabla|v|^2}{|\mathbf{x}|^{n-1}}d\mathbf{x} = \int_\Omega \Big[-\Delta\delta + (n-1)\frac{\mathbf{x}\cdot(\nabla\delta)}{|\mathbf{x}|^2}\Big]\frac{|v|^2}{|\mathbf{x}|^{n-1}}d\mathbf{x} \geq 0.$$

Since

$$\int_\Omega \frac{|v(\mathbf{x})|^2}{\delta(\mathbf{x})|\mathbf{x}|^{n-1}}d\mathbf{x} = \int_\Omega \frac{|u(\mathbf{x})|^2}{\delta^2(\mathbf{x})}d\mathbf{x},$$

then

$$\begin{aligned}
&\int_\Omega |\nabla u(\mathbf{x})|^2 d\mathbf{x} - \frac{1}{4}\int_\Omega \frac{|u(\mathbf{x})|^2}{\delta^2(\mathbf{x})}d\mathbf{x}\\
&\geq \int_\Omega \frac{\delta(\mathbf{x})|\nabla v(\mathbf{x})|^2}{|\mathbf{x}|^{n-1}}d\mathbf{x} + \Big[\frac{(n-1)^2}{4} - \frac{n-1}{2}\Big]\int_\Omega \frac{\delta(\mathbf{x})|v(\mathbf{x})|^2}{|\mathbf{x}|^{n+1}}d\mathbf{x}\\
&= \int_\Omega \frac{\delta(\mathbf{x})|\nabla v(\mathbf{x})|^2}{|\mathbf{x}|^{n-1}}d\mathbf{x} + \frac{(n-1)(n-3)}{4}\int_\Omega \frac{\delta(\mathbf{x})|v(\mathbf{x})|^2}{|\mathbf{x}|^{n+1}}d\mathbf{x}.
\end{aligned}$$

It is therefore sufficient to prove that

$$\begin{aligned}
&\int_\Omega \frac{\delta(\mathbf{x})|\nabla v(\mathbf{x})|^2}{|\mathbf{x}|^{n-1}}d\mathbf{x} + \frac{(n-1)(n-3)}{4}\int_\Omega \frac{\delta(\mathbf{x})|v(\mathbf{x})|^2}{|\mathbf{x}|^{n+1}}d\mathbf{x}\\
&\qquad \geq \left(\int_\Omega \frac{\delta(\mathbf{x})^{\frac{n}{n-2}}|v(\mathbf{x})|^{\frac{2n}{n-2}}}{|\mathbf{x}|^{\frac{n(n-1)}{n-2}}}d\mathbf{x}\right)^{\frac{n-2}{n}}, \qquad (4.5.3)
\end{aligned}$$

for all $v \in C_0^\infty(\Omega)$.

We may choose $\varepsilon$ sufficiently small such that $\Omega_\varepsilon := \{\mathbf{x}\in\Omega : \delta(\mathbf{x}) \leq \varepsilon\} \subset G(\Omega)$ by Remark 2.4.6, and we have that

$$\frac{\varepsilon}{\rho+\varepsilon} \leq \frac{\delta(\mathbf{x})}{|\mathbf{x}|} \leq 1, \quad \text{for } \mathbf{x} \in \Omega_\varepsilon^c := \Omega \setminus \Omega_\varepsilon, \qquad (4.5.4)$$

and

$$\rho' \leq |\mathbf{x}| \leq \rho + \varepsilon, \quad \text{for } \mathbf{x} \in \Omega_\varepsilon. \qquad (4.5.5)$$

The next step is to define cut-off functions supported near the boundary. Let $f \in C^\infty([0,\infty))$ be a non-increasing function satisfying

$$f(t) = \begin{cases} 1, & \text{for } t \in [0, 1/2), \\ 0 & \text{for } t \geq 1 \end{cases}$$

and $|f'(t)| \leq C_0$. Set $\varphi_\varepsilon(\mathbf{x}) := f(\delta(\mathbf{x})/\varepsilon)$; thus, since $\Omega_\varepsilon \subset G$, we have that $\varphi_\varepsilon \in C^2(\Omega_\varepsilon)$ and $|\nabla\varphi_\varepsilon(\mathbf{x})| = |f'(\delta(\mathbf{x})/\varepsilon)||\nabla\delta(\mathbf{x})|/\varepsilon \leq C_0/\varepsilon$; also $\varphi_\varepsilon = 1$ on $\Omega_{\varepsilon/2}$ and 0 on $\Omega_\varepsilon^c$.

From (4.5.4),

$$\begin{aligned} &\int_{\Omega^c_{\varepsilon/2}} \frac{\delta(\mathbf{x})|\nabla([1-\varphi_\varepsilon(\mathbf{x})]v(\mathbf{x}))|^2}{|\mathbf{x}|^{n-1}} d\mathbf{x} + \frac{(n-1)(n-3)}{4} \int_{\Omega^c_{\varepsilon/2}} \frac{\delta(\mathbf{x})|[1-\varphi_\varepsilon(\mathbf{x})]v(\mathbf{x})|^2}{|\mathbf{x}|^{n+1}} d\mathbf{x} \\ &\geq C(\varepsilon,\rho)\left(\int_{\Omega^c_{\varepsilon/2}} \frac{|\nabla([1-\varphi_\varepsilon(\mathbf{x})]v(\mathbf{x}))|^2}{|\mathbf{x}|^{n-2}} d\mathbf{x} + \frac{(n-1)(n-3)}{4} \int_{\Omega^c_{\varepsilon/2}} \frac{|[1-\varphi_\varepsilon(\mathbf{x})]v(\mathbf{x})|^2}{|\mathbf{x}|^{n}} d\mathbf{x}\right) \\ &\geq C(\varepsilon,\rho)\int_{\Omega^c_{\varepsilon/2}} \left(|\nabla w(\mathbf{x})|^2 + |w(\mathbf{x})|^2\right) d\mathbf{x}, \end{aligned} \tag{4.5.6}$$

where $w(\mathbf{x}) = (1-\varphi_\varepsilon(\mathbf{x}))v(\mathbf{x})/|\mathbf{x}|^{(n-2)/2}$ and we have used the assumption that $n \geq 4$. Hence $w \in W_0^{1,2}(\Omega^c_{\varepsilon/2})$, and we may therefore invoke the Sobolev embedding theorem to obtain

$$\begin{aligned} &\int_{\Omega^c_{\varepsilon/2}} \frac{\delta(\mathbf{x})|\nabla([1-\varphi_\varepsilon(\mathbf{x})]v(\mathbf{x}))|^2}{|\mathbf{x}|^{n-1}} d\mathbf{x} + \frac{(n-1)(n-3)}{4} \int_{\Omega^c_{\varepsilon/2}} \frac{\delta(\mathbf{x})|[1-\varphi_\varepsilon(\mathbf{x})]v(\mathbf{x})|^2}{|\mathbf{x}|^{n+1}} d\mathbf{x} \\ &\geq C(\varepsilon,\rho)\left(\int_{\Omega^c_{\varepsilon/2}} \frac{|(1-\varphi_\varepsilon(\mathbf{x}))v(\mathbf{x})|^{\frac{2n}{n-2}}}{|\mathbf{x}|^n} d\mathbf{x}\right)^{\frac{n-2}{2}} \\ &\geq C(\varepsilon,\rho)\left(\int_{\Omega^c_\varepsilon} \frac{\delta(\mathbf{x})^{\frac{n}{n-2}}|(1-\varphi_\varepsilon(\mathbf{x}))v(\mathbf{x})|^{\frac{2n}{n-2}}}{|\mathbf{x}|^{\frac{n(n-1)}{n-2}}} d\mathbf{x}\right)^{\frac{n-2}{2}} \end{aligned} \tag{4.5.7}$$

since $\Omega^c_\varepsilon \subset \Omega^c_{\varepsilon/2}$ and on using (4.5.4).

We next derive a similar estimate for the corresponding integrals involving $\varphi_\varepsilon$. From (4.4.1) applied to $v^s$, where $s = 2(n-1)/(n-2)$ and with $a = (n-1)/(n-2)$, we obtain for small enough $\varepsilon$,

$$\begin{aligned} C\left(\int_{\Omega_\varepsilon} \delta^{\frac{n}{n-2}}|v|^{\frac{2n}{n-2}} d\mathbf{x}\right)^{\frac{n-1}{n}} &\leq s\int_{\Omega_\varepsilon} \delta^{\frac{n-1}{n-2}}|v|^{\frac{n}{n-2}}|\nabla v| d\mathbf{x} \\ &\leq s\left(\int_{\Omega_\varepsilon} \delta^{\frac{n}{n-2}}|v|^{\frac{2n}{n-2}} d\mathbf{x}\right)^{\frac{1}{2}}\left(\int_{\Omega_\varepsilon} \delta|\nabla v|^2 d\mathbf{x}\right)^{\frac{1}{2}} \\ &\leq s\theta\left(\int_{\Omega_\varepsilon} \delta^{\frac{n}{n-2}}|v|^{\frac{2n}{n-2}} d\mathbf{x}\right)^{\frac{1}{2}} + C_\theta\left(\int_{\Omega_\varepsilon} \delta|\nabla v|^2 d\mathbf{x}\right)^{\frac{n-1}{n-2}} \end{aligned} \tag{4.5.8}$$

on using the inequality $ab \le \theta a^{\frac{2(n-1)}{n}} + C_\theta b^{\frac{2(n-1)}{n-2}}$, with arbitrary $\theta > 0$. It follows from(4.5.5)that

$$\int_{\Omega_\varepsilon} \frac{\delta(\mathbf{x})|\nabla(\varphi_\varepsilon(\mathbf{x})v(\mathbf{x}))|^2}{|\mathbf{x}|^{n-1}} d\mathbf{x} + \frac{(n-1)(n-3)}{4} \int_{\Omega_\varepsilon} \frac{\delta(\mathbf{x})|\varphi_\varepsilon(\mathbf{x})v(\mathbf{x})|^2}{|\mathbf{x}|^{n+1}} d\mathbf{x}$$
$$\ge C(\varepsilon, \rho, \rho') \left( \int_{\Omega_\varepsilon} \frac{\delta(\mathbf{x})^{\frac{n}{n-2}} |\varphi_\varepsilon(\mathbf{x})v(\mathbf{x})|^{\frac{2n}{n-2}}}{|\mathbf{x}|^{\frac{n(n-1)}{n-2}}} d\mathbf{x} \right)^{\frac{n-2}{2}}. \tag{4.5.9}$$

The addition of (4.5.7) and (4.5.9) gives

$$\begin{aligned} C(\varepsilon, \rho, \rho') &\left( \int_{\Omega_\varepsilon} \frac{\delta^{\frac{n}{n-2}} |\varphi_\varepsilon v|^{\frac{2n}{n-2}}}{|\mathbf{x}|^{\frac{n(n-1)}{n-2}}} d\mathbf{x} \right)^{\frac{n-2}{2}} \\ +C(\varepsilon, \rho) &\left( \int_{\Omega_\varepsilon^c} \frac{\delta^{\frac{n}{n-2}} |(1-\varphi_\varepsilon)v|^{\frac{2n}{n-2}}}{|\mathbf{x}|^{\frac{n(n-1)}{n-2}}} d\mathbf{x} \right)^{\frac{n-2}{2}} \\ &\le \int_{\Omega_\varepsilon} \frac{\delta|\nabla(\varphi_\varepsilon v)|^2}{|\mathbf{x}|^{n-1}} d\mathbf{x} \\ + \int_{\Omega_{\varepsilon/2}^c} &\frac{\delta|\nabla[(1-\varphi_\varepsilon)v]|^2}{|\mathbf{x}|^{n-1}} d\mathbf{x} + 2\frac{(n-1)(n-3)}{4} \int_\Omega \frac{\delta|v|^2}{|\mathbf{x}|^{n+1}} d\mathbf{x} \\ &\le C(\varepsilon) \left( \int_{\Omega_\varepsilon \setminus \Omega_{\varepsilon/2}} \frac{\delta|v|^2}{|\mathbf{x}|^{n-1}} d\mathbf{x} \right) \\ +C(n) &\left( \int_\Omega \frac{\delta|\nabla v|^2}{|\mathbf{x}|^{n-1}} d\mathbf{x} + \frac{(n-1)(n-3)}{4} \int_\Omega \frac{\delta|v|^2}{|\mathbf{x}|^{n+1}} d\mathbf{x} \right), \end{aligned} \tag{4.5.10}$$

where in the last inequality we have used the fact that $\nabla\varphi_\varepsilon$ is supported in $\Omega_\varepsilon \setminus \Omega_{\varepsilon/2}$ and $|\nabla\varphi_\varepsilon| \le C_0/\varepsilon$. Finally, by (4.5.5),

$$\int_{\Omega_\varepsilon \setminus \Omega_{\varepsilon/2}} \frac{\delta|v|^2}{|\mathbf{x}|^{n-1}} d\mathbf{x} \le (\rho + \varepsilon)^2 \left( \int_{\Omega_\varepsilon \setminus \Omega_{\varepsilon/2}} \frac{\delta|v|^2}{|\mathbf{x}|^{n+1}} d\mathbf{x} \right).$$

From this and (4.5.10), the inequality (4.5.3) follows and the theorem is proved. □

Gkikas also establishes the following theorem for the case $n = 3$.

**Theorem 4.5.3** *Let $\Omega$ be an exterior domain in $\mathbb{R}^3$ with a $C^2$ boundary, which satisfies (4.5.1) with strict inequality, i.e.,*

$$-\Delta\delta(\mathbf{x}) + 2\frac{\nabla\delta(\mathbf{x}) \cdot \mathbf{x}}{|\mathbf{x}|^2} > 0.$$

*Then, for all* $u \in C_0^\infty(\Omega)$,

$$\int_\Omega |\nabla u|^2 d\mathbf{x} - \frac{1}{4}\int_\Omega \frac{|u|^2}{\delta^2} d\mathbf{x} \geq C(\Omega)\left(\int_\Omega X^4(\frac{|\mathbf{x}|}{D})|u|^6 d\mathbf{x}\right)^{\frac{1}{3}}, \tag{4.5.11}$$

*where* $X(t) = (1 + \log t)^{-1}$ *and* $0 < D < \inf\{|\mathbf{x}| : \mathbf{x} \in \partial\Omega\}$. *The power* 4 *of* $X$ *can not be replaced by a smaller power.*

## 4.6 Equivalence of HSM and CLR Inequalities

In Sect. 1.5.2, we discussed the equivalence of the Sobolev and CLR inequalities implied by the Li-Yau proof of the CLR inequality in [109] and briefly described an abstract extension due to Levin and Solomyak which applies to

$$t[u] = q[u] - \int_\Omega V(\mathbf{x})|u|^2 d\mathbf{x} \tag{4.6.1}$$

for all quadratic forms $q$ associated with Markov generators in a sigma-finite measure space $(\Omega, d\mathbf{x})$. In [64], Frank, Lieb and Seiringer develop a more general theory for examining the equivalence of Sobolev and Lieb-Thirring (in particular CLR) inequalities, and this is shown in [62] to be capable of dealing with the example

$$q[u] = \int_\Omega |\nabla u|^2 d\mathbf{x} - \frac{(n-2)^2}{4}\int_\Omega \frac{|u|^2}{\delta(\mathbf{x})^2} d\mathbf{x}, \tag{4.6.2}$$

$$T := -\Delta - \frac{(n-2)^2}{4\delta(\mathbf{x})^2} - V, \tag{4.6.3}$$

when $\Omega$ is convex. In [64], the Beurling-Deny conditions on the quadratic form $q$ (with domain $\mathcal{H}_1(q)$) in [103] are generalised to the following:

(a) $q[u + iv] = q[u] + q[v]$ for real $u, v \in \mathcal{H}_1(q)$;
(b) if $u \in \mathcal{H}_1(q)$ is real, then $|u| \in \mathcal{H}_1(q)$ and $q[|u|] \leq q[u]$;
(c) there is a measurable function $\omega$ which is positive a.e. and is such that if $u \in \mathcal{H}_1(q)$ is non-negative, then $\min(u, \omega) \in \mathcal{H}_1(q)$ and $q[\min(u, \omega)] \leq q[u]$. Moreover, there is a dense subspace $D$ of $\mathcal{H}_1(q)$ (a core of $q$) such that $\omega^{-1}D$ is dense in $L^2(\Omega, \omega^{2\kappa/(\kappa-1)})$.

Thus the Beurling-Deny conditions correspond to the special case $\omega = 1$. The main theorem in [64] is

**Theorem 4.6.1** *Let $q$ satisfy the above generalised Beurling-Deny conditions for some $\kappa > 1$, and let $T$ be the self-adjoint operator in $L^2(\Omega)$ associated with the quadratic form $t$ in (4.6.1). Then the following are equivalent:*

*(i) there is a positive constant $S$ such that for all $u \in \mathcal{H}_1(q)$,*

$$t[u] \geq S\left(\int_\Omega |u|^s d\mathbf{x}\right)^{2/s}, \quad s := 2\kappa/(\kappa - 1); \tag{4.6.4}$$

*(ii) there is a positive constant $L$ such that for all $0 \leq V \in L^\kappa(\Omega)$,*

$$N(T - V) \leq L \int_\Omega V^\kappa d\mathbf{x}. \tag{4.6.5}$$

*Moreover,*

$$S^{-\kappa} \leq L \leq e^{\kappa-1} S^{-\kappa}.$$

In the case of $q$ being the quadratic form in (4.6.2), parts (a) and (b) of the generalised Beurling-Deny conditions above are clearly satisfied. Following [62], we demonstrate that the theorem applies when $\Omega \subset \mathbb{R}^3$ and is convex, with the choices $\omega = \delta^{1/2}$ and $D = C_0^\infty(\Omega)$ in part (c). For $u = \omega v$,

$$\begin{aligned} q[u] &= \int_\Omega \left(|\nabla u|^2 - \frac{|u|^2}{4\delta^2}\right) d\mathbf{x} \\ &= \int_\Omega \left(\delta|\nabla v|^2 + \frac{1}{2}[(\nabla\delta)\cdot(\nabla|v|^2)] + \frac{1}{4\delta}|\nabla\delta||v|^2 - \frac{1}{4\delta}|v|^2\right) d\mathbf{x} \\ &= \int_\Omega \left(|\nabla v|^2 - \frac{\Delta\delta}{2\delta}|v|^2\right) \delta d\mathbf{x} =: \tilde{q}[v]. \end{aligned}$$

Note that in $\tilde{q}$, $-\Delta\delta \geq 0$ in the distributional sense by Theorem 2.3.2, and so $\tilde{q}[v] \geq 0$. The map $M : u \mapsto u/\omega$ is an isometry of $L^2(\Omega)$ onto the weighted space $L^2(\Omega; \delta)$, and in view of the denseness of $C_0^\infty(\Omega)$ in the form domain of $q$, extends by continuity to an isometry of $\mathcal{H}_1(q)$ onto the form domain $\mathcal{H}_1(\tilde{q})$ of $\tilde{q}$ in $L^2(\Omega, \delta)$, equipped with the norm

$$\left(\tilde{q}[v] + \int_\Omega |v|^2 \delta d\mathbf{x}\right)^{1/2}.$$

Let $0 \leq u \in \mathcal{H}_1(q)$ and $u = \omega v$. Then $\min(u, \omega) = \omega \min(v, 1)$ and

$$\nabla[\min(v, 1)] = \begin{cases} \nabla v, & \text{if } v < 1, \\ 0, & \text{if } v > 1, \end{cases}$$

which implies that

$$|\nabla[\min(v, 1)]| \leq |\nabla v|.$$

Hence

$$\begin{aligned} q[\min(u, \omega)] &= \tilde{q}[\min(v, 1)] \\ &\leq \int_\Omega \left( |\nabla v|^2 - \left[ \frac{\Delta \delta}{2\delta} \right] |v|^2 \right) \delta d\mathbf{x} \\ &= \tilde{q}[v] = q[u]. \end{aligned}$$

Since multiplication by $\omega^{-1}$ is an isometry of $L^2(\Omega, \omega^{4/(n-2)})$ onto $L^2(\Omega, \omega^{2n/(n-2)})$ and $C_0^\infty(\Omega)$ is dense in $L^2(\Omega, \omega^{4/(n-2)})$, part (c) of the generalised Beurling-Deny conditions is satisfied. Consequently Theorem 4.6.1 holds, and in view of Corollary 4.3.2, we conclude that

$$N(-\Delta - \frac{(n-2)^2}{4\delta(\mathbf{x})^2} - V) \leq \int_\Omega V^{n/2} d\mathbf{x}. \tag{4.6.6}$$

Note that the theorem and (4.6.6) are also satisfied if $\Omega$ is a bounded, weakly mean convex domain with a $C^2$ boundary and null cut locus, on account of Theorem 4.4.4.

For a general domain $\Omega \subseteq \mathbb{R}^n$, it is proved in [62] that the theorem continues to hold if $\delta$ is replaced by the mean distance $\delta_M$, but the proof is much more complicated and uses a modified form of the generalised Beurling-Deny conditions which still imply the main theorem in [64]. We refer to [62] for the details.

# Chapter 5
# Inequalities and Operators Involving Magnetic Fields

## 5.1 Introduction

In classical mechanics the motion of charged particles depends only on electric and magnetic fields **E**, **B** which are uniquely described by Maxwell's equations:

$$\nabla \cdot \mathbf{E} = 4\pi\rho,$$

$$\nabla \cdot \mathbf{B} = 0,$$

$$\nabla \times \mathbf{E} = -\frac{\partial \mathbf{B}}{\partial t},$$

$$\nabla \times \mathbf{B} = 4\pi\mathbf{J} + \frac{\partial \mathbf{E}}{\partial t}.$$

These equations determine the relationships between the classical **electric field** **E**, the **magnetic field B**, the **electric current density J** and the charge density $\rho$. Most of classical electrodynamic can be described by Maxwell's equations together with the Lorentz forces on the charged particles, namely,

$$\vec{F}_i = e_i(\mathbf{E} + \mathbf{v}_i \times \mathbf{B}),$$

when the $i$th particle has charge $e_i$ and velocity $\mathbf{v}_i$. We see from the second Maxwell equation that the magnetic field is always divergence free, and this implies that it has the form

$$\mathbf{B} = \nabla \times \mathbf{A}$$



A.A. Balinsky et al., *The Analysis and Geometry of Hardy's Inequality*,
Universitext, DOI 10.1007/978-3-319-22870-9_5

for some (not uniquely defined) field $\mathbf{A}$ which is called a magnetic vector potential. Similarly, we can always rewrite the third Maxwell equation in the form

$$\mathbf{E} = -\nabla\phi - \frac{\partial \mathbf{A}}{\partial t}$$

for some scalar potential $\phi$. The representation of electric and magnetic fields in terms of a scalar potential $\phi$ and magnetic potential $\mathbf{A}$ is very useful since we need only the four component of $(\phi, \mathbf{A})$ to describe the electromagnetic field instead of the six components of $(\mathbf{E}, \mathbf{B})$. Until the development of quantum mechanics, it was widely believed that potentials $(\phi, \mathbf{A})$ are only nice mathematical constructions to simplify calculations and representations, and that they do not have real physical significance. However, quantum mechanics brought the realisation that $(\phi, \mathbf{A})$ play important roles, since the Schrödinger equation contains potentials $(\phi, \mathbf{A})$ and not fields $(\mathbf{E}, \mathbf{B})$, as we shall see in subsequent sections.

In 1959, Yakir Aharonov and his doctoral advisor David Bohm in [3] proposed an experiment to understand the role and significance of potentials in quantum mechanics. They predicted that a wave function can acquire some additional observable phase when traveling through non-simply connected domains with no electromagnetic fields ( i.e., $\mathbf{E} = 0, \mathbf{B} = 0$) but with non-zero potentials $(\phi, \mathbf{A})$. Such a phase shift was confirmed experimentally by R.G. Chambers in 1960, see [36] and also [25]. This is the famous **Aharonov-Bohm** effect, sometimes called the **Ehrenberg-Siday-Aharonov-Bohm** effect; in their 1961 paper [4], Aharonov and Bohm acknowledge the work of Werner Ehrenberg and Raymond E. Siday who had been the first to predict the effect in their 1949 paper [50]. The Aharonov-Bohm effect is so fundamental to our understanding of quantum physics that it was chosen by the *New Scientist* magazine as one of the " seven wonders of the quantum world". We refer the reader to the seminar by Kregar, published in [89], for a clear and comprehensive account.

## 5.2 The Magnetic Gradient and Magnetic Laplacian

The introduction of magnetic fields calls for changes to be made to the expressions which, in appropriate units, represent the momentum $\mathbf{p} = (1/i)\nabla$ and kinetic energy (the free Hamiltonian) $-\Delta$ of a charged particle. The new expressions are

$$\mathbf{p}_{\mathbf{A}} = \frac{1}{i}\nabla_{\mathbf{A}} := \frac{1}{i}(\partial_1 + iA_1, \cdots, \partial_3 + iA_3)$$

$$-\Delta_{\mathbf{A}} := -\sum_{j=1}^{3}(\partial_j + iA_j)^2,$$

where $\mathbf{A} = (A_1, A_2, A_3) : \mathbb{R}^3 \to \mathbb{R}^3$ is the vector potential, which determines the magnetic field $\mathbf{B}$ by the equation

$$\mathbf{B}(\mathbf{x}) = \text{curl } \mathbf{A}(\mathbf{x}). \tag{5.2.1}$$

In general $n$-dimensions we shall consider $\mathbf{A} = (A_1, A_2, \cdots, A_n) : \mathbb{R}^n \to \mathbb{R}^n$ and write

$$\nabla_{\mathbf{A}} = (\partial_1 + iA_1, \cdots, \partial_n + iA_n)$$

for the **covariant derivative** with respect to $\mathbf{A}$. We refer to $\nabla_{\mathbf{A}}$ as the **magnetic gradient**. In the language of **differential forms**, the magnetic potential is a 1-**form**, $\mathbf{A} = \sum_{j=1}^n A_j dx_j$, and the corresponding magnetic field $\mathbf{B}$ is a 2-**form**, $d\mathbf{A}$, which is the **exterior derivative** of $\mathbf{A}$ given by

$$d\mathbf{A} = \sum_{j=1}^n dA_j \wedge dx_j = \sum_{i,j=1}^n \frac{\partial A_j}{\partial x_i} dx_i \wedge dx_j = \sum_{i<j} \left( \frac{\partial A_j}{\partial x_i} - \frac{\partial A_i}{\partial x_j} \right) dx_i \wedge dx_j.$$

The changes make it necessary to introduce new Sobolev spaces which depend on the vector potential $\mathbf{A}$.

If $A_j \in L^2_{loc}(\Omega), j = 1, \cdots, n$, for $\Omega \in \mathbb{R}^n, n \geq 2$, then

$$(f, g)_{H^1_{0,\mathbf{A}}(\Omega)} := \sum_{j=1}^n \int_\Omega [(\partial_j + iA_j)f] \overline{[(\partial_j + iA_j)g]} d\mathbf{x} + \int_\Omega f\bar{g} d\mathbf{x} \tag{5.2.2}$$

is an inner product on $C_0^\infty(\Omega)$. We denote the completion of $C_0^\infty(\Omega)$ with respect to the norm $\|\cdot\|_{H^1_{0,\mathbf{A}}(\Omega)} := (\cdot,\cdot)^{1/2}_{H^1_{0,\mathbf{A}}(\Omega)}$ by $H^1_{0,\mathbf{A}}(\Omega)$; when $\Omega = \mathbb{R}^n$ we shall write $H^1_{\mathbf{A}}$. The kinetic energy operator $-\Delta_{\mathbf{A}}$ is the self-adjoint operator in $L^2(\Omega)$ determined by the quadratic form

$$\sum_{j=1}^n \int_\Omega [(\partial_j + iA_j)f] \overline{[(\partial_j + iA_j)g]} d\mathbf{x}$$

on $H^1_{0,\mathbf{A}}(\Omega)$; the expectation value of $-\Delta_{\mathbf{A}} f$ for $f$ in the domain of $-\Delta_{\mathbf{A}}$ is given by

$$(f, -\Delta_{\mathbf{A}} f)_{L^2(\Omega)} = \sum_{j=1}^n \int_\Omega |(\partial_j + iA_j)f|^2 d\mathbf{x}.$$

More generally, for $\Omega \subseteq \mathbb{R}^n$, $A_j \in L^p_{loc}(\Omega)$ and $1 < p < \infty$, we may define spaces $H^{1,p}_{0,\mathbf{A}}(\Omega)$ as the completion of $C_0^\infty(\Omega)$ with respect to the norms determined by

$$\|f\|^p_{H^{1,p}_{0,\mathbf{A}}(\Omega)} := \sum_{j=1}^n \int_{\mathbb{R}^n} \left|(\partial_j + iA_j)f\right|^p d\mathbf{x} + \int_{\mathbb{R}^n} |f|^p d\mathbf{x}. \tag{5.2.3}$$

## 5.3 The Diamagnetic (Kato's Distributional) Inequality

The inequality (5.3.2) in the following theorem is known as the **diamagnetic inequality**. It has a prominent role in problems involving magnetic fields.

**Theorem 5.3.1** *Let* $\mathbf{A} = (A_1, A_2, \cdots, A_n) : \Omega \to \mathbb{R}^n$ *be such that* $A_j \in L^p_{loc}(\Omega)$, *and let* $f \in H^{1,p}_{0,\mathbf{A}}(\Omega)$. *Then* $|f| \in H_0^{1,p}(\Omega)$ $(= W_0^{1,p}(\Omega))$, *and for* $j = 1, 2, \cdots, n$,

$$\partial_j |f|(\mathbf{x}) = \mathrm{Re}[\mathrm{sgn}(\bar{f})(\mathbf{x})(\partial_j + iA_j)f(\mathbf{x})], \quad a.e.\ \mathbf{x} \in \Omega, \tag{5.3.1}$$

*where*

$$\mathrm{sgn}(\bar{f})(\mathbf{x}) = \begin{cases} \frac{\bar{f}(\mathbf{x})}{f(\mathbf{x})}, & \text{if } f(\mathbf{x}) \neq 0, \\ 0, & \text{if } f(\mathbf{x}) = 0. \end{cases}$$

*Hence* $|\partial_j |f(\mathbf{x})|| \le |(\partial_j + iA_j)f(\mathbf{x})|$, *a.e.* $\mathbf{x} \in \Omega$, *and* $f \mapsto |f|$ *maps* $H^{1,p}_{0,\mathbf{A}}(\Omega)$ *continuously into* $H_0^{1,p}(\Omega)$ *with norm* $\le 1$. *Thus, when* $\Omega = \mathbb{R}^n$ *and* $p = 2$,

$$|\nabla |f(\mathbf{x})|| \le |\nabla_{\mathbf{A}} f(\mathbf{x})|, \quad a.e.\ \mathbf{x} \in \mathbb{R}^n. \tag{5.3.2}$$

*Proof* The diamagnetic inequality is equivalent to **Kato's distributional inequality**

$$\Delta |f(\mathbf{x})| \ge \mathrm{Re}[\mathrm{sgn}(\bar{f})(\mathbf{x}) \Delta_{\mathbf{A}} f(\mathbf{x})], \quad a.e.\ \mathbf{x} \in \mathbb{R}^n, \tag{5.3.3}$$

(see [84], Lemma A), and the following proof is a straightforward adaptation of Kato's $L^2$ proof to the more general result stated.

We begin by supposing that $f \in C_0^\infty(\mathbb{R}^n)$. Then, with $\varepsilon > 0$ and $f_\varepsilon := (|f|^p + \varepsilon)^{1/p}$, we have

$$f_\varepsilon^{p-1}(\partial_j f_\varepsilon) = |f|^{p-2}\mathrm{Re}[\bar{f}\partial_j f] = |f|^{p-2}\mathrm{Re}[\bar{f}(\partial_j + iA_j)f],$$

and hence

$$\partial_j f_\varepsilon = \left(\frac{|f|}{f_\varepsilon}\right)^{p-2} \mathrm{Re}[\frac{\bar{f}}{f_\varepsilon}(\partial_j + iA_j)f] =: \left(\frac{|f|}{f_\varepsilon}\right)^{p-2} \mathrm{Re}[F_\varepsilon(f)], \tag{5.3.4}$$

say.

The next step is to show that (5.3.4) holds for any $f \in H^{1,p}_{0,\mathbf{A}}(\Omega)$. With this in mind, let $f^{(m)} \in C_0^\infty(\Omega)$ be such that $f^{(m)} \to f$ in $H^{1,p}_{0,\mathbf{A}}(\Omega)$, and also pointwise a.e.; note that such a sequence $(f^{(m)})$ exists since $H^{1,p}_{0,\mathbf{A}}(\Omega) \hookrightarrow L^p(\Omega)$. We have

$$\begin{aligned} F_\varepsilon(f) - F_\varepsilon(f^{(m)}) &= \frac{\overline{f^{(m)}}}{f_\varepsilon^{(m)}} \left\{(\partial_j + iA_j)f - (\partial_j + iA_j)f^{(m)}\right\} \\ &+ \left(\frac{\bar{f}}{f_\varepsilon} - \frac{\overline{f^{(m)}}}{f_\varepsilon^{(m)}}\right)(\partial_j + iA_j)f. \end{aligned}$$

As $m \to \infty$, the first term on the right-hand side tends to zero in $L^p(\Omega)$ since $(\partial_j + iA_j)f^{(m)} \to (\partial_j + iA_j)f$ in $L^p(\Omega)$ and $|f^{(m)}/f_\varepsilon^{(m)}| \le 1$. The same is true for the second term, by the dominated convergence theorem and since the bracketed term converges a.e. to zero pointwise. Hence $F_\varepsilon(f) - F_\varepsilon(f^{(m)}) \to 0$ in $L^p(\Omega)$. Moreover, with

$$G_\varepsilon(f) := \left(\frac{|f|}{f_\varepsilon}\right)^{p-2} \mathrm{Re}[F_\varepsilon(f)],$$

it follows that $G_\varepsilon(f) - G_\varepsilon(f^{(m)}) \to 0$ in $L^p(\Omega)$. We therefore have that, for all $\varphi \in C_0^\infty(\Omega), j = 1, 2, \cdots, n$, and $m \to \infty$,

$$\int_\Omega f_\varepsilon \partial_j \varphi d\mathbf{x} = -\int_\Omega \varphi G_\varepsilon(f) d\mathbf{x}; \tag{5.3.5}$$

thus (5.3.4) is established for $f \in H^{1,p}_{0,\mathbf{A}}(\Omega)$. We now let $\varepsilon \to 0$ in (5.3.5). Since $f_\varepsilon \to |f|$ uniformly, $|\bar{f}/f_\varepsilon| \le 1$ and $\bar{f}/f_\varepsilon \to \mathrm{sgn}(f)$, we conclude that

$$\int_\Omega |f|\partial_j\varphi d\mathbf{x} = -\int_\Omega \varphi G_0(f) d\mathbf{x} = -\int_\Omega \varphi \mathrm{Re}\left[\mathrm{sgn}(\bar{f})(\partial_j + iA_j)f\right] d\mathbf{x}$$

and

$$\partial_j|f|(\mathbf{x}) = \mathrm{Re}[\mathrm{sgn}(\bar{f})(\mathbf{x})(\partial_j + iA_j)f(\mathbf{x})]$$

for a.e $\mathbf{x} \in \Omega$. This completes the proof. □

In [111], Sect. 7.20, $H^1_{\mathbf{A}}(\mathbb{R}^n)$ is defined as the space of functions $f : \mathbb{R}^n \to \mathbb{C}$ which are such that

$$f, \ (\partial_j + iA_j)f \in L^2(\mathbb{R}^n), \ \text{ for } j = 1, 2, \cdots, n,$$

with inner product (5.2.2) and $p = 2$, and then it is proved in Theorem 7.22 that $C_0^\infty(\mathbb{R}^n)$ is a dense subspace. Therefore this definition agrees with that given above. Note that for $f \in H^1_{\mathbf{A}}(\mathbb{R}^n)$, $\nabla_{\mathbf{A}} f \in L^2(\mathbb{R}^n)$, but $\nabla f$ and $Af$ need not separately be in $L^2(\mathbb{R}^n)$.

## 5.4 Schrödinger Operators with Magnetic Fields

### *5.4.1 The Free Magnetic Hamiltonian*

In the presence of the magnetic field **B** given in terms of the magnetic potential **A** in (5.2.1) (or, more generally, as the 2-form $d\mathbf{A}$), the free Hamiltonian $H_{\mathbf{A}} := -\Delta_{\mathbf{A}}$ in $L^2(\mathbb{R}^n)$, is the non-negative, self-adjoint operator associated with the closure $h_{\mathbf{A}}$ of the quadratic form

$$(H_{\mathbf{A}}\varphi, \varphi) = \int_{\mathbb{R}^n} |\nabla_{\mathbf{A}}\varphi|^2 d\mathbf{x}, \quad \varphi \in C_0^\infty(\mathbb{R}^n),$$

this being defined if $A_j \in L^2_{loc}(\mathbb{R}^n), j = 1, \cdots, n$. The form domain of $H_{\mathbf{A}}$ (i.e., the domain of $h_{\mathbf{A}}$) is therefore $H^1_{0,\mathbf{A}}(\mathbb{R}^n)$, and its domain is

$$\mathcal{D}(H_{\mathbf{A}}) := \{f : f \in H^1_{0,\mathbf{A}}(\mathbb{R}^n), -\Delta_{\mathbf{A}} f \in L^2(\mathbb{R}^n)\},$$

this being a dense subspace of $H^1_{0,\mathbf{A}}(\mathbb{R}^n)$. In view of the diamagnetic and Hardy inequalities,

$$h_{\mathbf{A}}[f] \geq \int_{\mathbb{R}^n} |\nabla |f|(\mathbf{x})|^2 \, d\mathbf{x} \geq C(n, \mathbb{R}^n) \int_{\mathbb{R}^n} \frac{|f(\mathbf{x})|^2}{|\mathbf{x}|^2} d\mathbf{x}, \tag{5.4.1}$$

with the optimal constant $C(n, \mathbb{R}^n) = \left|\frac{n-2}{2}\right|^2$ for $n > 2$; we saw in Sect. 1.2.5 that for $n = 2$, there is no Hardy inequality and we only deduce from (5.4.1) that $h_{\mathbf{A}} \geq 0$. However, the presence of a magnetic field can improve the situation significantly, as we shall now demonstrate. Let $n = 2$ and consider the symmetric operators

$$L_1 := -i\partial_1 + A_1, \quad L_2 := -i\partial_1 + A_2$$

on $C_0^\infty(\mathbb{R}^2)$. Then,

$$\begin{aligned} 0 &\le (L_1 \pm iL_2)(L_1 \pm iL_2)^* \\ &= (L_1 \pm iL_2)(L_1 \mp iL_2) \\ &= L_1^2 + L_2^2 \mp i(L_1L_2 - L_2L_1) \\ &= L_1^2 + L_2^2 \mp (\partial_1 A_2 - \partial_2 A_1) \\ &= |\nabla_{\mathbf{A}}|^2 \mp B \end{aligned}$$

where $B = \text{curl}\mathbf{A}$. Consequently

$$\int_{\mathbb{R}^2} |\nabla_{\mathbf{A}}\varphi|^2\, d\mathbf{x} \ge \pm \int_{\mathbb{R}^2} B|\varphi|^2 d\mathbf{x}, \quad \varphi \in C_0^\infty(\mathbb{R}^2), \tag{5.4.2}$$

and this holds for both signs. If $B$ is of one sign and $|B|$ is big, then (5.4.2) implies a significant lower bound on $h_{\mathbf{A}}$, but in general, it may not be of much use if $B$ has variable sign. The free Hamiltonian $H_{\mathbf{A}}$ in $\mathbb{R}^3$ associated with a constant magnetic field $\mathbf{B} = (0, 0, B),\ B \ge 0$, and magnetic vector potential

$$\mathbf{A} = \frac{1}{2}(-Bx_2, Bx_1, 0).$$

is known to have a spectrum which has least point $B$; see Sect. 3 of [13], and so

$$\|H_{\mathbf{A}}\varphi\| \ge B\|\varphi\|, \quad \varphi \in C_0^\infty(\mathbb{R}^3).$$

With $\mathbf{A} = (A_1, A_2)$ and $B = \partial_1 A_2 - \partial_2 A_1$, $H_{\mathbf{A}}$ can be regarded as an operator in $L^2(\mathbb{R}^2)$ defined by the quadratic form

$$h_{\mathbf{A}}[\varphi] = \int_{\mathbb{R}^2} |[\nabla + i\mathbf{A}(\mathbf{x})]\varphi(\mathbf{x})|^2\, d\mathbf{x}.$$

This is called the **Landau Hamiltonian** associated with the magnetic field of magnitude $B$ . Its spectrum consists of discrete eigenvalues

$$(2k+1)B : k = 0, 1, 2, \ldots$$

of infinite multiplicity called the **Landau levels**; see p. 171 in [95].

In section below, we shall discuss work of Laptev and Weidl in [99] in which they consider Aharonov-Bohm type magnetic fields for which it is possible for the angular part of $h_{\mathbf{A}}$ to have a positive lower bound and a Hardy inequality for $h_{\mathbf{A}}$ is available for $n = 2$ and $B = 0$ in $\mathbb{R}^2 \setminus \{0\}$.

### 5.4.2 Gauge Invariance

Let $L$ be any linear differential operator in $L^2(\Omega)$, $\Omega \subseteq \mathbb{R}^n$, and $\phi : \Omega \to U(1) := \{z \in \mathbb{C} : |z| = 1\}$. The transformation

$$u \mapsto \phi^{-1}u, \quad L_\phi := \phi^{-1}L\phi,$$

is called a **gauge transformation**. It preserves the pointwise norm of $u$ and $P : u \mapsto \phi^{-1}u$ is a unitary map on $L^2(\Omega)$. Thus $L$ and $L_\phi = PLP^{-1}$ are unitarily equivalent.

If $L = \nabla_{\mathbf{A}}$, then

$$L_\phi = \nabla + i\mathbf{A} + \phi^{-1}\nabla\phi =: \nabla_{\tilde{\mathbf{A}}},$$

and so a new magnetic gradient $\nabla_{\tilde{\mathbf{A}}}$ is formed with magnetic potential

$$\tilde{\mathbf{A}} = \mathbf{A} + \frac{1}{i}\phi^{-1}\nabla\phi. \tag{5.4.3}$$

However, the magnetic field remains unchanged:

$$\mathbf{B} = \mathrm{curl}\mathbf{A} = \mathrm{curl}\tilde{\mathbf{A}}.$$

Moreover, the operators $H_{\mathbf{A}}$ and $H_{\tilde{\mathbf{A}}}$ are unitarily equivalent and hence have the same spectral properties. The magnetic potentials $\mathbf{A}$ and $\tilde{\mathbf{A}}$ are said to be **gauge equivalent**

In a simply connected domain $\Omega$, any continuous function $\phi : \Omega \to U(1)$ can be written as $\phi(\mathbf{x}) = e^{if(\mathbf{x})}$, for some continuous $f : \Omega \to \mathbb{R}$, and from (5.4.3),

$$\tilde{\mathbf{A}} - \mathbf{A} = \nabla f. \tag{5.4.4}$$

If $\Omega$ is not simply-connected, there are gauge transformations which can not be represented in the form $\phi(\mathbf{x}) = e^{if(\mathbf{x})}$. An example in $\mathbb{R}^2$, which will be of particular interest in the next section, is

$$\phi(\mathbf{x}) = (z/|z|)^k, k \in \mathbb{Z} \setminus \{0\}, \quad z = x + iy \in \mathbb{C} \setminus \{0\}, \tag{5.4.5}$$

where $\mathbf{x} = (x, y) \in \mathbb{R}^2 \setminus \{0\}$. Then

$$\frac{1}{i}\phi^{-1}\nabla\phi = k\frac{1}{(x^2 + y^2)}(-y, x). \tag{5.4.6}$$

In $\mathbb{R}^2 \setminus \{0\}$, let $(r, \theta)$ be polar co-ordinates, and consider the vector potential

$$\mathbf{A}(r, \theta) := \frac{\psi(\theta)}{r}\mathbf{e}_\theta, \tag{5.4.7}$$

where $\mathbf{e}_\theta$ is the unit vector $(-\sin\theta, \cos\theta)$, which is orthogonal to the radial vector $\mathbf{e}_r = \mathbf{x}/r = (\cos\theta, \sin\theta)$. Set

$$\Psi := \frac{1}{2\pi}\int_0^{2\pi} \psi(\theta)d\theta.$$

We have that curl$\mathbf{A} = 0$ in $\mathbb{R}^2 \setminus \{0\}$, and $\mathbf{A}$ is in the so-called **transversal** (or **Poincaré**) gauge characterized by $\mathbf{A}\cdot\mathbf{e}_r = 0$. This involves no loss of generality as any vector potential is gauge equivalent to one in Poincaré gauge (see [141], Sect. 8.4.2). What is assumed in (5.4.7) is that the **flux** (or **circulation**) of $\mathbf{A}$ about the origin, namely the integral

$$\Psi(\mathbf{A}) := \frac{1}{2\pi}\int_0^{2\pi} \mathbf{A}(r,\theta)\cdot\mathbf{e}_\theta r d\theta,$$

is independent of $r$, and equal to the constant $\Psi$. Let $\mathbf{A}'$ be any vector potential in $\mathbb{R}^2 \setminus \{0\}$ which is such that curl$\mathbf{A}' = 0$ and $\Psi(\mathbf{A}') = \Psi$, and define

$$\tilde{\mathbf{A}}' := \mathbf{A}' - \frac{\Psi}{r}\mathbf{e}_\theta.$$

Then curl$\tilde{\mathbf{A}}' = 0$ and $\Psi(\tilde{\mathbf{A}}') = 0$. Therefore, for any $\mathbf{x}_0, \mathbf{x} \in \mathbb{R}^2 \setminus \{0\}$, the line integral

$$f(\mathbf{x}) = \int_{\mathbf{x}_0}^{\mathbf{x}} \tilde{\mathbf{A}}'(\mathbf{z})\cdot d\mathbf{z}$$

is independent of the path from $\mathbf{x}_0$ to $\mathbf{x}$. It follows that

$$\tilde{\mathbf{A}}' = \nabla f$$

and so

$$\mathbf{A}'(r,\theta) - \frac{\Psi}{r}\mathbf{e}_\theta = \nabla f(r,\theta). \tag{5.4.8}$$

We therefore infer that $\mathbf{A}'$ is gauge equivalent to $(\Psi/r)\mathbf{e}_\theta$. In particular, this is true for $\mathbf{A}' = \mathbf{A}$. We also note from (5.4.3) and (5.4.6) that $\mathbf{A}$ and

$$\tilde{\mathbf{A}} = \mathbf{A} + \frac{k}{r}\mathbf{e}_\theta, \quad k \in \mathbb{Z},$$

are gauge equivalent, and

$$\Psi(\tilde{\mathbf{A}}) = \Psi(\mathbf{A}) + k, \quad \Psi(\mathbf{A}) = \Psi.$$

Thus under the gauge transformation (5.4.3), the flux of $\mathbf{A}$ is transformed to $\Psi + k$.

## 5.5 The Aharonov-Bohm Magnetic Field

### 5.5.1 *The Laptev-Weidl Inequality*

We know that there is no Hardy inequality in $\mathbb{R}^2$, and (5.4.1) gives only trivial information. A natural question is if there are magnetic potentials $\mathbf{A}$ for which there is a constant $c > 0$ such that

$$\int_{\mathbb{R}^2} |\nabla_{\mathbf{A}} u|^2 d\mathbf{x} \geq c \int_{\mathbb{R}^2} \frac{|u|^2}{|\mathbf{x}|^2} d\mathbf{x}, \qquad u \in C_0^\infty(\mathbb{R}^2 \setminus \{0\}),$$

and the corresponding magnetic field $\mathbf{B}$ is bounded on $\mathbb{R}^2 \setminus \{0\}$; note that by (5.4.2), the inequality is satisfied for $B(\mathbf{x}) = c/|\mathbf{x}|^2$, which is not bounded on $\mathbb{R}^2 \setminus \{0\}$. The question was answered in the affirmative by Laptev and Weidl in [99], where the following theorem is proved for magnetic potentials of Aharonov-Bohm type.

**Theorem 5.5.1** *Let*

$$\mathbf{A} = \frac{\psi(\theta)}{r} \mathbf{e}_\theta, \tag{5.5.1}$$

*where* $\psi \in L^1(0, 2\pi)$, *and*

$$\Psi(\mathbf{A}) = \Psi = \frac{1}{2\pi} \int_0^{2\pi} \psi(\theta) d\theta. \tag{5.5.2}$$

*Then, for all* $u \in C_0^\infty(\mathbb{R}^2 \setminus \{0\})$,

$$\int_{\mathbb{R}^2} |(\nabla + i\mathbf{A}(\mathbf{x}))u(\mathbf{x})|^2 \, d\mathbf{x} \geq (\min_{k \in \mathbb{Z}} |k + \Psi|)^2 \int_{\mathbb{R}^2} \frac{|u(\mathbf{x})|^2}{|\mathbf{x}|^2} d\mathbf{x}. \tag{5.5.3}$$

*The constant* $(\min_{k\in\mathbb{Z}} |k + \Psi|^2)$ *is sharp.*

*Proof* From the discussion in Sect. 5.4.2, it follows that the magnetic potential (5.5.1) is gauge equivalent to

$$\mathbf{A}(r, \theta) = \Psi r^{-1}(-\sin\theta, \cos\theta), \quad \Psi := \Psi(\mathbf{A}). \tag{5.5.4}$$

Therefore, we may, and shall, assume that $\mathbf{A}$ is given by (5.5.4).

In polar co-ordinates, the form

$$\int_{\mathbb{R}^2} |(\nabla + i\mathbf{A}(\mathbf{x}))u(\mathbf{x})|^2 \, d\mathbf{x}$$

becomes

$$h_{\mathbf{A}} := \int_0^\infty \int_0^\infty \left( \left| \frac{\partial u}{\partial r} \right|^2 + r^{-2} |K_\theta u|^2 \right) r dr d\theta,$$

where $K_\theta := i\frac{\partial}{\partial\theta} + \Psi$. The operator $K_\theta$ has domain $H^1(\mathbb{S}^1)$ in $L^2(\mathbb{S}^1)$, eigenvalues $\lambda_k = k + \Psi,\ k \in \mathbb{Z}$, and corresponding eigenfunctions

$$\varphi_k(\theta) = \frac{1}{\sqrt{2\pi}} \exp(-ik\theta).$$

The sequence $\{\varphi_k\}$ is an orthonormal basis of $L^2(\mathbb{S}^1)$ and hence any $u \in L^2(\mathbb{S}^1)$ has a representation

$$u(r,\theta) = \sum_{k\in\mathbb{Z}} u_k(r)\varphi_k(\theta),$$

where

$$u_k(r) = \int_0^{2\pi} u(r,\theta)\overline{\varphi_k(\theta)} d\theta.$$

For any $u \in H^1(\mathbb{S}^1)$,

$$h_{\mathbf{A}}[u] = \sum_{k\in\mathbb{Z}} \int_0^\infty \left( |u_k'(r)|^2 + \frac{\lambda_k^2}{r^2} |u_k(r)|^2 \right) r dr$$

and so

$$\begin{aligned}
\int_{\mathbb{R}^2} \frac{|u(\mathbf{x})|^2}{|\mathbf{x}|^2} d\mathbf{x} &= \sum_{k\in\mathbb{Z}} \int_0^\infty \frac{|u_k(r)|^2}{r^2} r dr \\
&\leq \sum_{k\in\mathbb{Z}} \frac{1}{\min_{k\in\mathbb{Z}} \lambda_k^2} \int_0^\infty \lambda_k^2 \frac{|u_k(r)|^2}{r^2} r dr \\
&\leq \frac{1}{(\min_{k\in\mathbb{Z}} |k+\Psi|)^2} h_{\mathbf{A}}[u],
\end{aligned}$$

which proves (5.5.3).

To verify that the constant is sharp, suppose that the minimum is attained at $k = k_0$, and let $u(r,\theta) = v(r)\varphi_{k_0}(\theta),\ v \in C_0^\infty(0,\infty)$. Then

$$h_{\mathbf{A}}[u] - (k_0 + \Psi)^2 \int_0^\infty \int_0^{2\pi} \frac{|u(r,\theta)|^2}{r^2} r dr d\theta = \int_0^\infty v'(r)^2 r dr. \tag{5.5.5}$$

The constant will have been proved to be sharp if we can show that, for any $\varepsilon > 0$, there exists $v \in C_0^\infty(0,\infty)$ such that

$$\int_0^\infty v'(r)^2 r dr \le \varepsilon \int_0^\infty v^2(r)/r dr.$$

On making the substitution $r = e^x$, $\tilde{v}(x) = v(e^x)$, this becomes

$$\int_{\mathbb{R}} |\tilde{v}'(x)|^2 dx \le \varepsilon \int_{\mathbb{R}} |\tilde{v}(x)|^2 dx$$

which is satisfied for some $\tilde{v} \in C_0^\infty(\mathbb{R})$ since the Laplacian $-\frac{d^2}{dx^2}$ on $\mathbb{R}$ has spectrum $[0,\infty)$. □

*Remark 5.5.2* If the flux $\Psi$ is an integer, then the operator $H_{\mathbf{A}}$ defined by the form $h_{\mathbf{A}}$ in Theorem 5.5.1 is unitarily equivalent to $-\Delta$ and there is no Hardy inequality. For suppose that $\Psi(\mathbf{A}) = k \in \mathbb{Z}$. Then from (5.4.6) and (5.4.8),

$$\mathbf{A} = \frac{1}{i}\phi^{-1}\nabla\phi + \nabla f,$$

for some $f \in \mathbb{R}^2 \setminus \{0\}$. Hence $\mathbf{A}$ is gauge equivalent to $\nabla f$ and hence to 0.

### 5.5.2 An Inequality of Sobolev Type

Closely associated with the Hardy-type inequality in Theorem 5.5.1 is a Sobolev inequality featuring the Aharonov-Bohm potential $\mathbf{A}$ defined in (5.5.1). The inequality is expressed in terms of the space

$$X := L^\infty(\mathbb{R}^+; L^2(\mathbb{S}^1); rdr) \equiv L^\infty(\mathbb{R}^+; rdr) \otimes L^2(\mathbb{S}^1), \tag{5.5.6}$$

with norm

$$\|u\|_X := \text{ess sup}_{r>0}\left\{\left(\int_0^{2\pi} |u(r,\theta)|^2 d\theta\right)^{1/2}\right\}. \tag{5.5.7}$$

We shall denote $H^1_{\mathbf{A}}(\mathbb{R}^2 \setminus \{0\})$ by $H^1_{\mathbf{A}}$ throughout this section.

**Theorem 5.5.3** *Let* $\mathbf{A}$ *be given by (5.5.1) and suppose that its flux* $\Psi(\mathbf{A}) = \Psi \notin \mathbb{Z}$. *Then, for all* $u \in H^1_{\mathbf{A}}$,

$$\|(\nabla + i\mathbf{A})u\|^2 \ge (\min_{k\in\mathbb{Z}} |k + \Psi|)\|u\|_X^2. \tag{5.5.8}$$

*Proof* We may again assume, without loss of generality, that $\mathbf{A}$ is given by (5.5.4). In the proof of Theorem 5.5.1, we used polar co-ordinates to write $h_{\mathbf{A}}$ as

$$h_{\mathbf{A}}[u] = \sum_{k\in\mathbb{Z}} \left\{ \int_0^\infty \left( |u_k'(r)|^2 + \frac{\lambda_k^2}{r^2}|u_k(r)|^2 \right) rdr \right\},$$

where the $u_k(r)$ are Fourier coefficients of $u(r,\cdot)$ and $\lambda_k = k+\Psi$. For any $t \in (0,\infty)$,

$$\begin{aligned} |u_k(t)|^2 &= 2\mathrm{Re} \int_0^t \overline{u}_k(r)u_k'(r)dr \\ &\leq 2\left(\int_0^t |u_k'(r)|^2 rdr\right)^{1/2} \left(\int_0^t |u_k(r)|^2 \frac{dr}{r}\right)^{1/2} \\ &= \frac{2}{|\lambda_k|}\left(\int_0^t |u_k'(r)|^2 rdr\right)^{1/2} \left(\lambda_k^2 \int_0^t |u_k(r)|^2 \frac{dr}{r}\right)^{1/2} \\ &\leq \frac{1}{|\lambda_k|}\left\{ \int_0^\infty \left( |u_k'(r)|^2 + \frac{\lambda_k^2}{r^2}|u_k(r)|^2 \right) rdr \right\}. \end{aligned}$$

Hence

$$\begin{aligned} \int_0^{2\pi} |u(t,\theta)|^2 d\theta &= \sum_{k\in\mathbb{Z}} |u_k(t)|^2 \\ &\leq \frac{1}{\min_{k\in\mathbb{Z}} |\lambda_k|} \|(\nabla + i\mathbf{A})u\|^2, \end{aligned}$$

whence (5.5.8). □

*Remark 5.5.4* It follows from the theorem that the set of radial functions in $H^1_{\mathbf{A}}$ is continuously embedded in $L^\infty(\mathbb{R}^2)$. This is not true for $H^1(\mathbb{R}^2)$.

## 5.6 A CLR Inequality

In view of the close relationship between the Hardy, Sobolev and CLR inequalities described in Chap. 1, it is only to be expected that a CLR inequality exists for an operator $H_{\mathbf{A}} + V$, when $\mathbf{A}$ satisfies the conditions of Theorem 5.5.1 with non-integer flux, and $V$ is a real-valued function which satisfies some appropriate conditions. To establish such a result, we need some preliminaries. The first concerns a compactness property of the operator of multiplication by $V$, when $V$ is a member of the space

$$Y := L^1(\mathbb{R}^+; L^\infty(\mathbb{S}^1); rdr) = L^1(\mathbb{R}^+; rdr) \otimes L^\infty(\mathbb{S}^1) \tag{5.6.1}$$

endowed with the norm

$$\|u\|_Y := \int_0^\infty \left( \text{ess} \sup_{\theta \in (0,2\pi)} |u(r,\theta)| \right) rdr. \tag{5.6.2}$$

The inequality (5.5.8) implies that $H_{\mathbf{A}}$ has no eigenvalue at zero, and hence $H_{\mathbf{A}}^{1/2}$ is injective and its domain $\mathcal{D}(H_{\mathbf{A}}^{1/2})$ and range $\mathcal{R}(H_{\mathbf{A}}^{1/2})$ are dense in $L^2(\mathbb{R}^2)$. Let $D_{\mathbf{A}}^1$ denote the completion of $\mathcal{D}(H_{\mathbf{A}}^{1/2})$ with respect to the norm

$$\|\varphi\|_{D_{\mathbf{A}}^1} := \|H_{\mathbf{A}}^{1/2}\varphi\| = \|(\nabla + i\mathbf{A})\varphi\|, \tag{5.6.3}$$

where $\|\cdot\|$ is the standard $L^2(\mathbb{R}^n)$ norm. Note that $D_{\mathbf{A}}^1$ is not a subspace of $L^2(\mathbb{R}^2)$, but it lies in the weighted space $L^2(\mathbb{R}^2; |\mathbf{x}|^{-2}d\mathbf{x})$ on account of Theorem 5.5.1.

**Lemma 5.6.1** *Let* $\mathbf{A}$ *be given by (5.5.1) with* $\Psi(\mathbf{A}) = \Psi \notin \mathbb{Z}$, *and let* $V \in Y$. *Then the operator* $P := H_{\mathbf{A}}^{-1/2}|V|H_{\mathbf{A}}^{-1/2}$ *is compact in* $L^2(\mathbb{R}^2)$. *Hence,* $V : D_{\mathbf{A}}^1 :\to L^2(\mathbb{R}^2)$ *is compact.*

*Proof* It is sufficient to prove that

$$T := |V|^{1/2}H_{\mathbf{A}}^{-1/2} \quad \text{is compact,}$$

since $P = T^*T$.

Given $\varepsilon > 0$, choose $W \in C_0^\infty(\mathbb{R}^+; L^\infty(\mathbb{S}^1))$ such that $\|V - W\|_Y < \varepsilon$, with support in

$$\Omega_\varepsilon = B(0, k_\varepsilon) \setminus B(0, 1/k_\varepsilon)$$

for some constant $k_\varepsilon > 0$, and such that $\|W\|_{L^\infty(\mathbb{R}^2)} \le k_\varepsilon$. Let $\varphi_n \rightharpoonup 0$ in $L^2(\mathbb{R}^2)$. Then, with $\psi_n = H_{\mathbf{A}}^{-1/2}\varphi_n$, we have that $\psi_n \rightharpoonup 0$ in $D_{\mathbf{A}}^1$ and

$$\begin{aligned}
\|T\varphi_n\|^2 = \||V|^{1/2}\psi_n\|^2 &\le \||W|^{1/2}\psi_n\|^2 + \||V - W|^{1/2}\psi_n\|^2 \\
&\le k_\varepsilon \int_{\Omega_\varepsilon} |\psi_n|^2 d\mathbf{x} + \|V - W\|_Y \|\psi_n\|_X^2 \\
&\le k_\varepsilon \int_{\Omega_\varepsilon} |\psi_n|^2 d\mathbf{x} + C\varepsilon \|H_{\mathbf{A}}^{1/2}\psi_n\|^2,
\end{aligned} \tag{5.6.4}$$

by (5.5.8). For any $\psi \in C_0^\infty(\mathbb{R}^2 \setminus \{0\})$, there exists a constant $C(\varepsilon)$, depending on $\varepsilon$, such that

$$\|\psi\|^2_{L^2(\Omega_\varepsilon)} \leq C(\varepsilon)\|\psi\|^2_X \leq C(\varepsilon)\|(\nabla + i\mathbf{A})\psi\|^2,$$

by (5.5.8), and

$$\begin{aligned}\|\nabla\psi\|_{L^2(\Omega_\varepsilon)} &\leq \|(\nabla + i\mathbf{A})\psi\|^2 + C(\varepsilon)\|\psi\|^2_{L^2(\Omega_\varepsilon)}\\ &\leq C(\varepsilon)\|(\nabla + i\mathbf{A})\psi\|^2.\end{aligned}$$

This implies that $D^1_{\mathbf{A}}$ is continuously embedded in the standard Sobolev space $H^1(\Omega_\varepsilon)$. Since $H^1(\Omega_\varepsilon)$ is compactly embedded in $L^2(\Omega_\varepsilon)$ by the Rellich-Kondrachov theorem, it follows that $\psi_n \to 0$ in $L^2(\Omega_\varepsilon)$ as $n \to \infty$. Hence, from (5.6.4),

$$\limsup_{n\to\infty} \|T\varphi_n\|^2 \leq C\varepsilon\|H^{1/2}_{\mathbf{A}}\psi_n\|^2 = C\varepsilon\|\varphi_n\|^2.$$

As $\varepsilon$ is arbitrary, this means that $\|T\varphi_n\| \to 0$ and the lemma is proved. □

**Lemma 5.6.2** *Let* $\mathbf{A}$ *satisfy (5.5.1). Then* $\sigma(H_{\mathbf{A}}) = [0, \infty)$.

*Proof* Take $\psi(\theta) = (2\pi - \alpha)^{-1}[1 - \chi(0, \alpha)]\Psi$, $0 < \alpha < \pi/2$, where $\chi(0, \alpha)$ denotes the characteristic function of the interval $(0, \alpha)$ . In the sector

$$S_\alpha := \{(r, \theta) : 0 < r < \infty,\ 0 < \theta < \alpha\}$$

the Laplace operator with Dirichlet boundary conditions, has essential spectrum $[0, \infty)$, and there exists a Weyl singular sequence $\{\psi_n\}$ for any point $\lambda \in (0, \infty)$, with each $\psi_n$ supported in $S_\alpha$; see [49], Theorem X.6.5. Since $\mathbf{A}\psi_n = 0$ in $S_\alpha$ for each $n$, $\{\psi_n\}$ is also a singular sequence at $\lambda$ for $H_{\mathbf{A}}$ and the lemma follows. □

Lemma 5.6.1 asserts that the multiplication operator $V$ is compact relative to the quadratic form $\|\nabla_{\mathbf{A}} \cdot\|$ (see Sect. 1.5.1), with the result that $H_{\mathbf{A}} - V$ is defined as a form sum with form domain $\mathcal{Q}(H_{\mathbf{A}}) = H^1_{\mathbf{A}}$, and $H_{\mathbf{A}} - V$ and $H_{\mathbf{A}}$ have the same essential spectra, namely $[0, \infty)$ by Lemma 5.6.2. The following theorem establishes an analogue of the CLR inequality for $H_{\mathbf{A}} - V$.

**Theorem 5.6.3** *Let* $\mathbf{A}$ *be given by (5.5.1) and suppose that its flux* $\Psi = \Psi(\mathbf{A}) \notin \mathbb{Z}$. *Let* $V$ *be a real-valued function in* $L^1_{loc}(\mathbb{R}^2 \setminus \{0\})$ *which satisfies*

$$V_+ := \max(V, 0) \in Y = L^1(\mathbb{R}^+; L^\infty(\mathbb{S}^1); rdr).$$

*Then $N(T_{\mathbf{A}}(V))$, the number of negative eigenvalues of $T_{\mathbf{A}}(V) := H_{\mathbf{A}} - V$ is finite and*

$$N(T_{\mathbf{A}}(V)) \leq \sum_{m\in\mathbb{Z}}{}' \frac{1}{4\pi|m+\Psi|}\|V_+\|_Y,$$

*where $\sum'_{m\in\mathbb{Z}}$ indicates that all summands less that* 1 *are omitted.*

*Proof* The operator $T_{\mathbf{A}}(V) := H_{\mathbf{A}} - V$ is the self-adjoint operator associated with the lower semi-bounded quadratic form

$$q[u] := \int_{\mathbb{R}^2} \{|(\nabla + i\mathbf{A})u|^2 - V|u|^2\}d\mathbf{x}, \quad u \in C_0^\infty(\mathbb{R}^2 \setminus \{0\}). \tag{5.6.5}$$

The form domain of $T_{\mathbf{A}}(V)$ is the domain of the closure of $q$ (which we continue to denote by $q$), and this coincides with $H^1_{\mathbf{A}}$.

As in the proof of Theorem 5.5.1, we take $\mathbf{A}$ to be given by (5.5.4), namely

$$\mathbf{A} = \frac{\Psi}{r}(-\sin\theta, \cos\theta).$$

Let

$$W(r) := \|V_+(r,\cdot)\|_{L^\infty(\mathbb{S}^1)}, \tag{5.6.6}$$

so that

$$\begin{aligned} \|W\|_{L^1(\mathbb{R}^+, rdr)} &= \|W\|_{L^1(\mathbb{R}^+;L^\infty(\mathbb{S}^1);rdr)} \\ &= \|V_+\|_{L^1(\mathbb{R}^+;L^\infty(\mathbb{S}^1);rdr)} < \infty. \end{aligned} \tag{5.6.7}$$

Thus Lemma 5.6.1 also implies that $T_{\mathbf{A}}(W)$ is lower semi-bounded, self-adjoint and has essential spectrum $[0,\infty)$. Since $T_{\mathbf{A}}(V) \geq T_{\mathbf{A}}(W)$, we have that

$$N(T_{\mathbf{A}}(V)) \leq N(T_{\mathbf{A}}(W)),$$

and so the theorem will follow if we prove it with $V$ replaced by $W$.

Let $u \in H^1_{\mathbf{A}} = \mathcal{Q}(T_{\mathbf{A}}(W))$, and let $u(r,\cdot)$ have Fourier coefficients $u_m(r)$, $m \in \mathbb{Z}$. Then, as we saw in the proof of Theorem 5.5.1,

$$\begin{aligned} &\textstyle\int_{\mathbb{R}^2} \left(|(\nabla + i\mathbf{A})u|^2 - W|u|^2\right) d\mathbf{x} \\ &= \textstyle\sum_{m\in\mathbb{Z}} \int_0^\infty \left(|u_m'|^2 + \frac{(m+\Psi)^2}{r^2}|u_m|^2 - W|u_m|^2\right) rdr. \end{aligned}$$

It follows that

$$T_{\mathbf{A}}(W) = \bigoplus_{m\in\mathbb{Z}} \{(D_m - W) \otimes 1_m\}, \tag{5.6.8}$$

where $D_m$ is the (Friedrichs) operator in $L^2((0,\infty); rdr)$ associated with the form

$$h_m[u] := \int_0^\infty \left( |u_m'(r)|^2 + \frac{(m+\Psi)^2}{r^2} |u_m(r)|^2 \right) rdr.$$

Furthermore, the negative spectrum of $T_{\mathbf{A}}(W)$ is the aggregate of the negative eigenvalues of the operators $D_m - W$. To complete the proof, we use the Bargmann estimate

$$N(D_m - W) \le \frac{1}{2|m+\Psi|} \int_0^\infty W(r) rdr,$$

from [24]; see also [132], where it is proved that the inequality is sharp. In view of (5.6.8), this yields

$$N(T_{\mathbf{A}}(W)) \le \sum_{m\in\mathbb{Z}}{}' \frac{1}{2|m+\Psi|} \int_0^\infty W(r) rdr,$$

and the theorem follows since $2\pi \int_0^\infty W(r) rdr = \|V_+\|_Y$. □

The following theorem is also obtained in [18], on applying a result from [97].

**Theorem 5.6.4** *Suppose the hypothesis of Theorem 5.6.3 is satisfied. Then*

$$N(T_{\mathbf{A}}(V)) \le c(\Psi) \|V_+\|_Y, \tag{5.6.9}$$

*where $c(\Psi)$ is a constant depending only on $\Psi$.*

*Proof* For any $n \in \mathbb{Z}$, the gauge function $f(r,\theta) \mapsto e^{in\theta} f(r,\theta)$ in (5.4.8) takes $\Psi = \Psi(\mathbf{A})$ into $\Psi + n$ and gives rise to unitarily equivalent operators. Therefore, we may assume that $\Psi \in (0,1)$. For $m \ge 0$,

$$D_m \ge -\frac{1}{r}\frac{d}{dr}\left(r\frac{d}{dr}\right) + \frac{\Psi^2}{r^2} + \frac{m^2}{r^2}, \tag{5.6.10}$$

and, for $m < 0$,

$$D_m \ge -\frac{1}{r}\frac{d}{dr}\left(r\frac{d}{dr}\right) + \frac{(1-\Psi)^2}{r^2} + \frac{(m+1)^2}{r^2}. \tag{5.6.11}$$

The two-dimensional case of the operator considered by Laptev and Netrusov is of the form

$$L(b;W) \equiv -\Delta + \frac{b}{|\mathbf{x}|^2} - W = \bigoplus_{m\in\mathbb{Z}} \{(L_m(b) - W) \otimes 1_m\}, \quad b > 0, \tag{5.6.12}$$

in

$$\bigoplus_{m\in\mathbb{Z}} \left\{ L^2(\mathbb{R}^+; rdr) \otimes \left[ \frac{e^{im\phi}}{\sqrt{2\pi}} \right] \right\},$$

where $[\cdot]$ denotes the linear span; in (5.6.12),

$$L_m(b) := -\frac{1}{r}\frac{d}{dr}\left(r\frac{d}{dr}\right) + \frac{b}{r^2} + \frac{m^2}{r^2}$$

and $L_m(b) - W$ is defined by the associated quadratic form in $L^2(\mathbb{R}^+; rdr)$. In Theorem 1.2 in [97], it is proved that

$$N(L(b;W)) \le C(b)\|W\|_{L^1(\mathbb{R}^+;rdr)}. \tag{5.6.13}$$

By (5.6.10) and (5.6.11),

$$\begin{aligned} N(\bigoplus_{m\ge 0}\{(D_m - W) \otimes 1_m\}) &\le N(\bigoplus_{m\ge 0}\{(L_m(\Psi^2) - W) \otimes 1_m\}) \\ &\le N(L_m(\Psi^2; W)) \end{aligned} \tag{5.6.14}$$

and

$$\begin{aligned} N(\bigoplus_{m<0}\{(D_m - W) \otimes 1_m\}) &\le N(\bigoplus_{m<0}\{(L_m([1-\Psi^2]) - W) \otimes 1_m\}) \\ &\le N(L_m([1-\Psi^2]; W)). \end{aligned} \tag{5.6.15}$$

Therefore, from (5.6.13),

$$N(T_{\mathbf{A}}(W)) \le c(\Psi)\|W\|_{L^1(\mathbb{R}^+;rdr)} = c(\Psi)\|V_+\|_Y$$

and the theorem is proved. □

*Remark 5.6.5* For $V(\mathbf{x}) = V(|\mathbf{x}|)$, Laptev obtains in [96], Sect. 3.4, the inequality

$$N(T_A(V)) \le \frac{R(\Psi)}{4\pi} \int_{\mathbb{R}^2} V(\mathbf{x}) d\mathbf{x},$$

with the sharp constant

$$R(\Psi) = \sup_{k\in\mathbb{Z}} \left\{ \nu^{-1/2} \left( \sharp\{k : -\nu + (k-\Psi)^2 < 0\} \right) \right\}.$$

## 5.7 Hardy-Type Inequalities for Aharonov-Bohm Magnetic Potentials with Multiple Singularities

In this section we are interested in Hardy type inequalities for magnetic Dirichlet forms with Aharonov-Bohm vector potentials that have multiple singularities.

Let $P_1 = (x_1, y_1), \ldots, P_n = (x_n, y_n)$ be $n$ different points in $\mathbb{R}^2$. We can identify $\mathbb{R}^2$ with $\mathbb{C}$ by the correspondence $(x, y) \mapsto z = x + iy$ and the points $P_1, \ldots, P_n$ then correspond to the complex numbers $z_1 = x_1 + iy_1, \ldots, z_n = x_n + iy_n$.

Consider a smooth vector potential $\mathbf{A} = (A_1(x, y), A_2(x, y))$ in the punctured plane $M = \mathbb{R}^2 \setminus \{P_1, \ldots, P_n\}$ with zero magnetic field (5.7.1):

$$\mathbf{B} := \operatorname{curl} \mathbf{A} = 0. \tag{5.7.1}$$

If we denote by $\omega_{\mathbf{A}}$ the differential 1-form $A_1(x, y)dx + A_2(x, y)dy$, then (5.7.1) says that $\omega_{\mathbf{A}}$ is a **closed differential form** in $M$, i.e. $d\omega_{\mathbf{A}} = 0$, where $d$ is the exterior derivative. Such a vector potential $\mathbf{A}$ is known as a **magnetic vector potential of Aharonov-Bohm type with multiple singularities**. The condition (5.7.1) implies that in any simply connected, open subset of $M$, there exists a gauge function $f$ such that $\mathbf{A} = \nabla f$, as we saw in Sect. 5.4.2.

For each point $P_k$ $(k = 1, \ldots, n)$ let us define a *circulation* of $\mathbf{A}$ round $P_k$ as

$$\Phi_k = \frac{1}{2\pi} \oint_{\gamma_k} A_1(x, y)dx + A_2(x, y)dy, \tag{5.7.2}$$

where $\gamma_k$ is a small circle in $M$ which winds once around $P_k$ in an anticlockwise direction; see Fig. 5.1. Condition (5.7.1) implies that (5.7.2) is invariant under

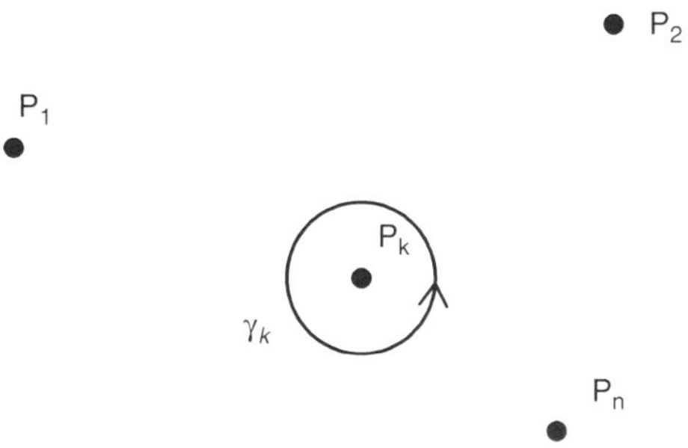


**Fig. 5.1** Small circle

continuous deformations of $\gamma_k$ inside the punctured plane $M = \mathbb{R}^2 \setminus \{P_1, \dots, P_n\}$. Furthermore, if the circulations $\Phi = (\Phi_1, \dots, \Phi_n)$ of two distinct Aharonov-Bohm type vector potentials $\mathbf{A}$ and $\mathbf{A}'$ on the punctured plane $M$ are equal modulo $\mathbb{Z}^n$ then $\mathbf{A}$ and $\mathbf{A}'$ are equivalent under some gauge transformation $\phi : M \to U(1) = \{z \in \mathbb{C} : |z| = 1\}$, i.e. $\mathbf{A}' = \mathbf{A} + \frac{1}{i}\phi^{-1}\nabla\phi$.

We now introduce the special magnetic potentials

$$\mathbf{A}^{(j)} := \frac{\Phi_j}{r_j^2} \cdot (-y + y_j, x - x_j), \quad j = 1, \dots, n,$$

where $r_j^2 = (x - x_j)^2 + (y - y_j)^2$ and $\Phi_j$ is the circulation of $\mathbf{A}$ round $P_j$. Each $\mathbf{A}^{(j)}$ satisfies (5.7.1) on $\mathbb{R}^2 \setminus \{P_j\}$ and has the circulation $\Phi_j$ round $P_j$ and the circulations zero round $P_i$, $i \neq j$. Then $\mathbf{A} - \sum_j \mathbf{A}^{(j)}$ is a magnetic potential with zero magnetic field and zero circulations on the punctured plane $M$. Therefore, for any magnetic vector potential $\mathbf{A}$ satisfying (5.7.1) in $M$ there exists a gauge function $f$ such that

$$\mathbf{A}(x, y) - \sum_{j=1}^{n} \frac{\Phi_j}{r_j^2} \cdot (-y + y_j, x - x_j) = (\nabla f)(x, y),$$

Given the vector potential $\mathbf{A}$, we define the corresponding **magnetic Dirichlet** form on $C_0^\infty(M)$ by

$$\mathcal{Q}_{\mathbf{A}}[u] = \int_M |(\nabla + i\mathbf{A})u|^2 dxdy, \quad u \in C_0^\infty(M). \tag{5.7.3}$$

Our main goal in this section is to find an estimate from below for (5.7.3) by a *Hardy-type expression*

$$\mathcal{Q}_{\mathbf{A}}[u] \geq \int_M H(x, y)|u(x, y)|^2 dxdy, \quad u \in C_0^\infty(M) \tag{5.7.4}$$

with a suitable nonnegative function $H(x, y)$ on $M$.

Under a gauge transformation $u \mapsto \phi \cdot u$ with an arbitrary smooth function $\phi : M \to U(1)$, the Dirichlet form $\mathcal{Q}_{\mathbf{A}}[u]$ becomes $\mathcal{Q}_{\mathbf{A}'}[u]$ with $\mathbf{A}' = \mathbf{A} + \frac{1}{i}\phi^{-1}\nabla\phi$. The right hand side of (5.7.4) is invariant under this gauge transform. Hence, it is sufficient to establish (5.7.4) for any $\mathbf{A}$ from a given gauge equivalent class of magnetic vector potentials.

For any real number $\Psi$, we denote by $p(\Psi)$ the distance from $\Psi$ to the set of integers $\mathbb{Z}$, i.e.

$$p(\Psi) := \min_{k \in \mathbb{Z}} |k - \Psi|. \tag{5.7.5}$$

There may be many functions $H(x, y)$ that give the inequality (5.7.4). By analogy with the Laptev-Weidl inequality, we are interested in finding those $H(x, y)$ that satisfy the following conditions.

1 $H(x, y)$ depends on $\mathbf{A}$ only through the circulations $\Phi_1, \Phi_2, \ldots, \Phi_n$ and the coordinates of $P_j, j = 1, \ldots, n$. That is, we would like to find functions $H(x, y)$ that are the same for equivalent magnetic potentials.
2 $H(x, y)$ behaves like

$$\frac{(p(\Phi_j))^2}{(x - x_j)^2 + (y - y_j)^2}$$

near each point $P_j, j = 1, 2, \ldots, n$, and $H(x, y)$ behaves like

$$\frac{(p(\Phi_1 + \Phi_2 + \ldots + \Phi_n))^2}{x^2 + y^2}$$

near infinity.

Such a class of functions $H(x, y)$ was discovered in [14]; we shall now describe the main properties. To be specific, we show that any analytic function $F(z)$ on $\mathbb{C}$ with zero set $\{P_1, P_2, \ldots, P_n\}$ and $F(\infty) = \infty$ generates a function $H(x, y)$ with the properties 1 and 2 above.

Before going into the general description, for the reader's convenience, we present an example of $H(x, y)$ in the case of two points $P_1 = -1$ and $P_2 = 1$ in $\mathbb{C}$ with the circulations $c_1 \equiv \Phi_1$ and $c_2 \equiv \Phi_2$, respectively.

*Example 5.7.1* Let $P_1 = (-1, 0)$, $P_2 = (1, 0)$ be two points in $\mathbb{R}^2$, $M = \mathbb{R}^2 \setminus \{P_1, P_2\}$ and suppose that $\mathbf{A}$ is a magnetic vector potential of Aharonov-Bohm type in $M$ with the circulations $c_j$ round $P_j$, $j = 1, 2$. Then the inequality (5.7.4) holds with

$$H(x, y) = C(x, y) \cdot \left| \frac{2z}{z^2 - 1} \right|^2, \quad z = x + iy,$$

where $C(x, y)$ is the piecewise constant function on $\mathbb{R}^2$ shown in Fig. 5.2.

In Fig. 5.2, $C$ is the curve $(x^2 - y^2 - 1) + 4x^2y^2 = 1$ which divides the plane $\mathbb{R}^2$ into three regions $\Omega_1$, $\Omega_2$ and $\Omega_\infty$, where $P_1 \in \Omega_1$ and $P_2 \in \Omega_2$; $C(x, y)$ equals $(p(c_1))^2$ in $\Omega_1$, $(p(c_2))^2$ in $\Omega_2$ and $(p(c_1 + c_2))^2/4$ in $\Omega_\infty$.

The general case of (5.7.4) will be made clearer by first considering a special case of magnetic potentials with zero magnetic fields in doubly connected domains in $\mathbb{R}^2$.

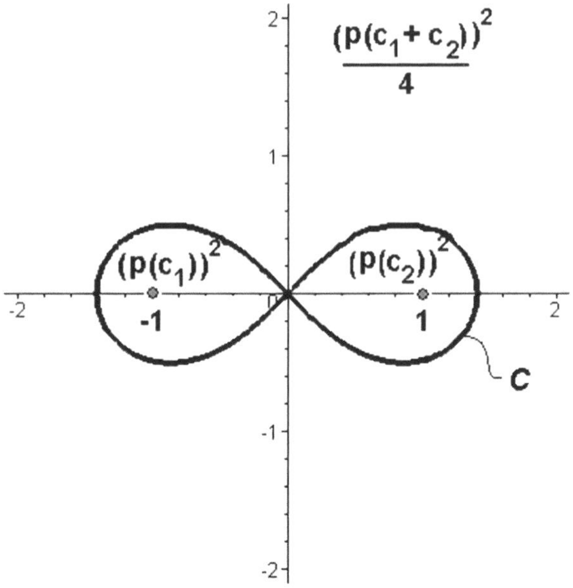


**Fig. 5.2** Function $C(x, y)$

### 5.7.1 *Inequality for Doubly Connected Domains*

Let $\Omega$ denote a bounded *doubly connected domain* (i.e. the boundary of $\Omega$ is a disjoint union of two closed simple curves) with a smooth boundary in the plane $\mathbb{R}^2 = \mathbb{C}$; $\Omega$ is homeomorphic to an open annulus.

Let $\Omega_{r,R}$ $(r < R)$ be an annulus in $\mathbb{C}$ with internal radius $r$, external radius $R$ and with centre at the origin, thus,

$$\Omega_{r,R} = \{z \in \mathbb{C} \mid r < |z| < R\}.$$

From the theory of functions of one complex variable we know (see [149, Theorem 1.2]) that any doubly connected domain can be conformally mapped onto an annulus $\Omega_{r,R}$ for some $r$ and $R$, as illustrated in Fig. 5.3.

For any such conformal mapping $F : \Omega \to \Omega_{r,R}$ we define a function $\mathcal{B}_{\Omega,F}(x, y)$ on $\Omega$ by

$$\mathcal{B}_{\Omega,F}(x, y) = \left| \frac{F'_z(z)}{F(z)} \right|^2, \tag{5.7.6}$$

where $z = x + iy$ and $F'_z$ denote the complex derivative of $F$.

**Lemma 5.7.2** *The function $\mathcal{B}_{\Omega,F}$ defined by (5.7.6) does not depend on the choice of the conformal mapping $F$.*

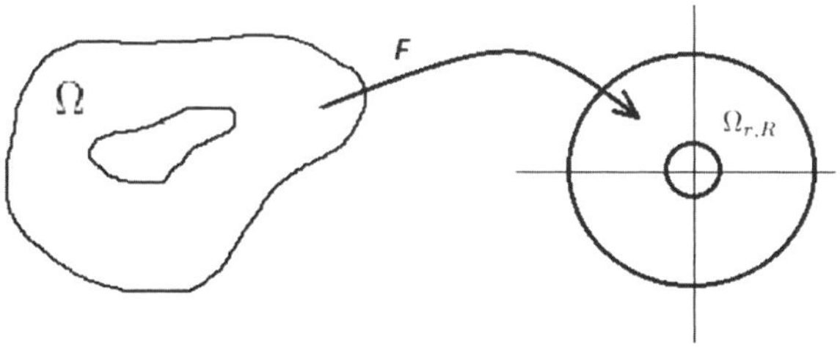


**Fig. 5.3** Conformal mapping

*Proof* Consider any other conformal mapping $\tilde{F}$ from $\Omega$ onto $\Omega_{\tilde{r},\tilde{R}}$. From Theorem 1.3 [149] we know that

$$\frac{R}{r} = \frac{\tilde{R}}{\tilde{r}}.$$

Hence, since the right-hand side of (5.7.6) is invariant under scaling $F \mapsto const \cdot F$, we can assume that $r = \tilde{r}$ and $R = \tilde{R}$. The mapping $\tilde{F} \circ F^{-1}$ is a conformal automorphism of $\Omega_{r,R}$. Since any holomorphic automorphism of $\Omega_{r,R}$ is a composition of rotations and reflections (see p. 133 in [88]), we have to check that the right-hand side of (5.7.6) is invariant under $F \mapsto \mu \cdot F$ (for $\mu$ a unimodular constant) and under $F \mapsto \frac{r \cdot R}{F}$. This is clear and hence the proof is completed. □

We shall use the notation $\mathcal{B}_\Omega$ instead of $\mathcal{B}_{\Omega,F}$.

Let $\mathbf{A} = (A_1(x,y), A_2(x,y))$ be a smooth magnetic vector potential in $\overline{\Omega}$ with zero magnetic field (5.7.1). Recall that a circulation $\Phi$ of $\mathbf{A}$ in the doubly connected domain $\Omega$ is

$$\Phi = \frac{1}{2\pi} \oint_\sigma A_1(x,y)dx + A_2(x,y)dy,$$

where $\sigma$ is a closed path which parameterizes the "internal" component of the boundary of $\Omega$. The last integral is invariant under continuous deformations of $\sigma$.

The following theorem explains the importance of the function $\mathcal{B}_\Omega$ for establishing a Hardy type inequality for a bounded doubly connected domain in $\mathbb{R}^2$:

**Theorem 5.7.3** *Let $\Omega$ be a bounded doubly connected domain in $\mathbb{R}^2$ with a smooth boundary. For any smooth function $f \in C^\infty(\overline{\Omega})$ we have*

$$\int_\Omega |(\nabla + i\mathbf{A})f|^2 dxdy \geq (p(\Phi))^2 \int_\Omega \mathcal{B}_\Omega(x,y)|f(x,y)|^2 dxdy, \tag{5.7.7}$$

*where $\mathcal{B}_\Omega(x,y)$ is defined by (5.7.6) and $p(\Phi)$ is defined by (5.7.5).*

*Proof* First we prove the following inequality for $\Omega_{r,R}$ and $\tilde{\mathbf{A}} = \frac{\Phi}{x^2+y^2}(-y, x)$:

$$\int_{\Omega_{r,R}} |(\nabla + i\tilde{A})f|^2 dxdy \geq (p(\Phi))^2 \int_{\Omega_{r,R}} \frac{|f(x,y)|^2}{x^2+y^2} dxdy, \tag{5.7.8}$$

for any $f \in C^\infty(\overline{\Omega_{r,R}})$ (see [99] for more general results). The left-hand and right-hand sides are both invariant under rotation of $\mathbb{R}^2$ around the origin. So, it is sufficient to establish (5.7.8) for spherical functions $f(r)e^{in\theta}$, $n \in \mathbb{Z}$ and $r = \sqrt{x^2+y^2}$. For such functions

$$\int_{\Omega_{r,R}} |(\nabla + i\tilde{\mathbf{A}})f(r)e^{in\theta}|^2 dxdy = \int_{\Omega_{r,R}} (|f_r'|^2 + \frac{1}{r^2}|f(r)|^2 \cdot (n+\Phi)^2) dxdy$$

$$\geq \int_{\Omega_{r,R}} \frac{1}{r^2}|f(r)|^2 \cdot (n+\Phi)^2 dxdy \geq (p(\Phi))^2 \int_{\Omega_{r,R}} \frac{|f(r)e^{in\theta}|^2}{r^2} dxdy.$$

Now, let $F : \Omega \to \Omega_{r,R}$ be a conformal mapping, $F(x,y) = (u(x,y), v(x,y))$. Denote by $\mathbf{A}^F(u,v) = (A_1^F(u,v), A_2^F(u,v))$ a magnetic vector potential in $\Omega_{r,R}$ such that $F^*(\omega_{\mathbf{A}^F}) = \omega_{\mathbf{A}}$, i.e.,

$$A_1^F(u,v)du + A_2^F(u,v)dv = A_1(x,y)dx + A_2(x,y)dy.$$

The magnetic vector potential $\mathbf{A}^F$ also has zero magnetic field and the same circulation $\Phi$ as $\mathbf{A}$ since the integral of a differential form and the property of being closed are invariant under diffeomorphisms.

Since $F$ is a conformal mapping, the reader will have no difficulty in showing that, for any $f \in C^\infty(\overline{\Omega_{r,R}})$,

$$\int_{\Omega_{r,R}} (|(\nabla_u + iA_1^F(u,v))f(u,v)|^2 + |(\nabla_v + iA_2^F(u,v))f(u,v)|^2) dudv =$$

$$\int_{\Omega} (|(\nabla_x + iA_1(x,y))f(u(x,y), v(x,y))|^2 + |(\nabla_y + iA_2(x,y))f(u(x,y), v(x,y))|^2) dxdy. \tag{5.7.9}$$

Since $\mathbf{A}^F$ is gauge equivalent to the magnetic vector potential $\frac{\Phi}{u^2+v^2}(-v, u)$ and the inequality (5.7.8) is also invariant under gauge transformations, we have that

$$\int_{\Omega} |(\nabla + i\mathbf{A})f(u(x,y), v(x,y))|^2 dxdy \geq (p(\Phi))^2 \int_{\Omega_{r,R}} \frac{|f(u,v)|^2}{u^2+v^2} dudv. \tag{5.7.10}$$

Taking into account that

$$\int_{\Omega_{r,R}} \frac{|f(u,v)|^2}{u^2+v^2} dudv = \int_{\Omega} |f(u(x,y), v(x,y))|^2 \frac{|F'_z|^2}{|F|^2} dxdy,$$

we obtain from (5.7.10) the inequality (5.7.7). This completes the proof. □

## 5.7.2 *Inequality for Punctured Planes*

In this section we establish a Hardy type inequality for a punctured plane. Theorem 5.7.3 gives us a Hardy-type inequality for a magnetic vector potential of Aharonov-Bohm type in a bounded doubly connected domain. For a more general domain $\Omega$, e.g. a Riemann surface or multiply connected domain, our strategy will be:

(a) find a decomposition, up to a zero measure set, of the domain $\Omega$ into doubly connected domains;
(b) find conformal mappings of these doubly connected domains into annuli and apply Theorem 5.7.3.

The most general tool for decomposing a manifold into simple parts is the classical Morse theory. We don't need the full power of the Morse theory here, the simple version that we now present being adequate.

Let $\Omega$ be a two-dimensional manifold and $f : \Omega \to (0, \infty)$. The map $f$ is said to be *proper* if the pre-image under $f$ of a compact set is compact, and $\omega \in \Omega$ is a *critical point* of $f$ if $\nabla f(\omega) = 0$, $f(\omega)$ being called a *critical value*. We assume that $f$ has a finite number of critical points in $\Omega$, and that there is no critical value in $[a,b] \subset (0,\infty)$. Then the pre-image $f^{-1}([a,b]) \subset \Omega$ is diffeomorphic to $f^{-1}(a) \times [a,b]$. To show this, we consider a regular vector field in $f^{-1}([a,b])$ defined by $\nabla f / \|\nabla f\|$. This is a well-defined vector field of length one which is orthogonal to level sets. Integral curves of this vector field will define the required diffeomorphism between $f^{-1}([a,b])$ and $f^{-1}(a) \times [a,b]$. Since $a$ is not a critical point, the pre-image $f^{-1}(a)$ is a one-dimensional compact manifold, i.e., a union of closed simple curves. Thus, $f^{-1}([a,b])$ is a union of doubly connected domains.

In general, finding a conformal mapping from a doubly connected domain into $\Omega_{r,R}$ is a difficult problem. The idea in [14] is to use a function $|F|$ for constructing a Morse complex, where $F : \Omega \to \mathbb{C}$ is an holomorphic function. In this case, the function $|F|$ provides a decomposition and $F$ provides the conformal mapping. We apply this idea in the case of the punctured plane $M = \mathbb{C} \setminus \{P_1, P_2, \ldots, P_n\}$ with a smooth magnetic vector potential $A$ of Aharonov-Bohm type.

Let $F : \mathbb{C} \to \mathbb{C}$ be an analytic function with zero set $\{P_1, P_2, \ldots, P_n\}$, i.e., $F^{-1}(0) = \{P_1, P_2, \ldots, P_n\}$, and $F(\infty) = \infty$. Consider the strictly positive function $f = |F| : M \to (0, \infty)$. Then the critical points of $f$ coincide with the zero set of the complex derivative $F'_z$ of the function $F$. This can be shown as follows. We have that $\nabla f(x,y) = 0$ iff $\nabla(f^2)(x,y) = 0$ and $f^2 = F \cdot \bar{F}$. If

$$\partial_x(F \cdot \bar{F}) = F_x \cdot \bar{F} + F \cdot \bar{F}_x = 0,$$

and

$$\partial_y(F \cdot \bar{F}) = F_y \cdot \bar{F} + F \cdot \bar{F}_y = 0,$$

then

$$(F_x - iF_y) \cdot \bar{F} + F \cdot \overline{F_x + iF_y} = 0.$$

Since the function $F$ is an analytic function, $F_x + iF_y = 0$ and this implies that $F_x - iF_y = 0$, i.e., the complex derivative $F'_z$ of the function $F$ is equal to zero.

Denote by $ord_{P_j}F$, the order of zero of $F$ at $P_j$. Let $\{Q_1, Q_2, \ldots, Q_l\}$ be the zero set of the complex derivative $F'_z$ of the function $F$, i.e., $\{Q_1, Q_2, \ldots, Q_l\} = (F'_z)^{-1}(0)$, and denote by $Crit_F$ the following subset of $\mathbb{R}_+ = \{x \in \mathbb{R} : x \geq 0\}$:

$$Crit_F = \{0, |F(Q_1)|, \ldots, |F(Q_l)|\}.$$

Under the map $|F| : \mathbb{C} \to \mathbb{R}_+$ the pre-image of $Crit_F$ is a zero measure set $\mathcal{F}_c$.

We shall now define a piecewise constant function $C_F$ on $\mathbb{R}^2$. For any $(x,y) \in \mathbb{R}^2$, $x + iy \notin \mathcal{F}_c$, the set $|F|^{-1}(|F|(x+iy))$ is a disjoint union of smooth simple curves in $\mathbb{C}$; let $\gamma_{(x,y)}$ denote one of them that goes through the point $(x,y)$. This $\gamma_{(x,y)}$ divides $\mathbb{C}$ into two domains, a bounded domain $\Omega_{int}(\gamma_{(x,y)})$ and an unbounded domain $\Omega_{ext}(\gamma_{(x,y)})$. Then

$$C_F(x,y) := \frac{\left(p\Big(\sum\limits_{P_k \in \Omega_{int}(\gamma_{(x,y)})} \Phi_k\Big)\right)^2}{(ord_{\gamma_{(x,y)}}F)^2}, \tag{5.7.11}$$

where $\Phi_k$ is a circulation of $\mathbf{A}$ round $P_k$ and

$$ord_{\gamma_{(x,y)}}F = \sum_{P_k \in \Omega_{int}(\gamma_{(x,y)})} ord_{P_k}F. \tag{5.7.12}$$

We can now state our main result.

**Theorem 5.7.4** *Let $C_F$ be defined in (5.7.11) for the analytic function $F$. For any $u \in C_0^\infty(M)$ the following inequality holds*

$$\int_M |(\nabla + i\mathbf{A})u|^2 dxdy \geq \int_M C_F(x,y) \left| \frac{F'_z(x+iy)}{F(x+iy)} \right|^2 |u(x,y)|^2 dxdy. \tag{5.7.13}$$

*Proof* Let $\mathbb{R}_+ \setminus Crit_F = \bigcup_{m\in\mathbb{Z}} [a_m, b_m]$ such that $(a_m, b_m) \cap (a_{m'}, b_{m'}) = \emptyset$ for $m \neq m'$. From Morse theory we know that $|F|^{-1}((a_m, b_m))$ is a disjoint union of doubly connected domains. Let $\Omega_0$ be any connected component of $|F|^{-1}((a_m, b_m))$. Then $F|_{\Omega_0} : \Omega_0 \to \Omega_{a_m,b_m}$ is a holomorphic function from $\Omega_0$ onto an annulus $\Omega_{a_m,b_m}$. Since we are away from the critical set of $F$, the holomorphic map $F|_{\Omega_0}$ is a covering map from $\Omega_0$ onto the annulus $\Omega_{a_m,b_m}$. From the Argument Principle (see, e.g. [5]) the degree of $F|_{\Omega_0}$ equals to $ord_{\gamma_{(x,y)}} F$ defined in (5.7.12), where $(x, y) \in \Omega_0$. Therefore the function $(F|_{\Omega_0})^{1/ord_{\gamma_{(x,y)}}F}$ is well defined and is a conformal mapping from $\Omega_0$ onto an annulus $\Omega_{r_m,R_m}$, where

$$r_m = (a_m)^{1/ord_{\gamma_{(x,y)}}F} \quad \text{and} \quad R_m = (b_m)^{1/ord_{\gamma_{(x,y)}}F}.$$

From Theorem 5.7.3

$$\int_{\Omega_0} |(\nabla + i\mathbf{A})u|^2 dxdy \geq \int_{\Omega_0} \Bigg(p\Big(\sum_{P_k\in\Omega_{int}(\gamma_{(x,y)})} \Phi_k\Big)\Bigg)^2 \left|\frac{((F|_{\Omega_0})^{1/ord_{\gamma_{(x,y)}}F})'_z}{(F|_{\Omega_0})^{1/ord_{\gamma_{(x,y)}}F}}\right|^2 |u|^2 dxdy$$

$$= \int_{\Omega_0} C_F(x, y) \left|\frac{F'_z(x+iy)}{F(x+iy)}\right|^2 |u(x, y)|^2 dxdy. \qquad (5.7.14)$$

Summing (5.7.14) over all connected component of $|F|^{-1}((a_m, b_m))$ and over all $m \in \mathbb{Z}$, we obtain Theorem 5.7.4. This conclude the proof. □

*Remark 5.7.5* Choosing $F(z) = \prod_{j=1}^{n} (z - z_j)$ we obtain a function

$$H_F(x, y) = C_F(x, y) \left|\frac{F'_z(x+iy)}{F(x+iy)}\right|^2$$

which satisfies the conditions 1 and 2 from the introduction to this section.

Another interesting choice is

$$F(z) = \frac{1}{\sum_{j=1}^{n} \frac{p(\Phi_j)}{(z-z_j)}},$$

which yields $H_F(x, y)$.

## 5.8 Generalised Hardy Inequality For Magnetic Dirichlet Forms

In this section we present some results from [22], where lower bounds for the magnetic Dirichlet form

$$h[u] = \int |(\nabla + i\mathbf{A})u|^2 d\mathbf{x}$$

on $C_0^\infty(\mathbb{R}^n)$, $n \geq 2$, were obtained.

For $n = 2$, the results generalise a well known lower bound by the *magnetic field strength*, where the actual magnetic field $\mathbf{B}$ is replaced by an non-vanishing *effective field* (the precise definition will be given below) which decays outside the support of $\mathbf{B}$ as $\text{dist}(\mathbf{x}, \text{supp}\,\mathbf{B})^{-2}$. In the case $d \geq 3$ the magnetic form is bounded from below by the magnetic field strength if one assumes that the field does not vanish and its direction is slowly varying.

We consider separately two cases: $n = 2$ and $n \geq 3$. To derive a meaningful estimate for $n = 2$ we exploit two elementary ideas. The first of them is the standard lower bound established in (5.4.2),

$$h[u] \geq \int \pm B|u|^2 d\mathbf{x},$$

which holds with either of the signs $\pm$; with $\mathbf{A} = (A_1, A_2)$, the magnetic field $\mathbf{B}$ is identified with the scalar $B = \partial_1 A_2 - \partial_2 A_1$. The second ingredient is the Hardy inequality for domains with Lipschitz boundaries. Put together, they yield a bound of the form

$$h[u] \geq c \int \tilde{B}|u|^2 d\mathbf{x},$$

with an *effective* magnetic field $\tilde{B}$, which coincides with $\pm B$ on its support, and decays outside the support as $\text{dist}\{\mathbf{x}, \text{ supp}\,B\}^{-2}$. The constant $c$ in the above estimate depends only on the support of $B$; see Theorem 5.8.1 for the precise statement. The constant $c$ is explicit and does not depend on the flux.

In the case $n \geq 3$ the problem becomes more complicated, as the magnetic field $\mathbf{B} = d\mathbf{A} = \text{curl}\mathbf{A}$ may now change its direction; see Sect. 5.8.5 for the precise definition of this notion. Assuming that the field never vanishes, and under some extra conditions on the smoothness of $\mathbf{B}$, we prove the Sobolev-type bound

$$h[u] \geq c \int |\mathbf{B}||u|^2 d\mathbf{x},$$

in Theorem 5.8.6.

### 5.8.1 Magnetic Forms

Let $\mathbf{A} = (A_1, A_2, \ldots A_n)$ be a real-valued vector function with $A_j \in L^2_{\text{loc}}(\mathbb{R}^n)$ for $j = 1, \cdots, n$ . Then the symmetric quadratic form

$$h[u] = \int |(\nabla + i\mathbf{A})u|^2 d\mathbf{x},$$

is closable on $C_0^\infty(\mathbb{R}^d)$, and the magnetic Laplacian $H_{\mathbf{A}} = -\Delta_{\mathbf{A}}$ is the unique self-adjoint operator associated with its closure. Assume that the magnetic field $\mathbf{B} = d\mathbf{A}$ exists in the sense of distributions and it is measurable on $\mathbb{R}^n$. We shall need the notation

$$B_{jk} = \partial_j A_k - \partial_k A_j, \quad k, j = 1, 2, \ldots, n.$$

Since the two-form $\mathbf{B}$ is antisymmetric, it is fully determined by the components $B_{jk}$ with $j < k$; the number of these components is

$$\varkappa_n = n(n-1)/2.$$

We measure the strength of the field by the quantity

$$|\mathbf{B}| = \sqrt{\sum_{j<k} B_{jk}^2}.$$

In the two- and three-dimensional cases, this quantity coincides with the length of the magnetic field vector. If $n = 2$, the only non-zero components of $\mathbf{B}$ are $B_{12}$ and $B_{21} = -B_{12}$, i.e., in our previous notation, $B$ and $-B$..

In the next two subsections we state our results separately for two cases: $n = 2$ and $n \geq 3$. They have much in common but due to the simplicity of the magnetic field structure for $n = 2$, our results in this case are obtained under more general assumptions on the field than for $n \geq 3$. For both cases we need to introduce a positive continuous functions $\ell$ which plays the role of a slowly varying spatial scale reflecting variations of the magnetic field. We associate with the function $\ell$ the open ball

$$\mathcal{K}(\mathbf{x}) = \{\mathbf{y} \in \mathbb{R}^2 : |\mathbf{x} - \mathbf{y}| < \ell(\mathbf{x})\}.$$

The precise conditions on the function $\ell$ for $n = 2$ and $n \geq 3$ are slightly different and will be specified in each case separately.

### 5.8.2 Case $n = 2$

We revert to our original notation and denote by $B(\mathbf{x})$ the component $B_{12}(\mathbf{x})$. The scale $\ell$ is assumed to satisfy the conditions

$$\ell \in C_0^1(\mathbb{R}^2); \quad |\nabla \ell(\mathbf{x})| \leq 1, \ \ \ell(\mathbf{x}) > 0, \ \forall \mathbf{x} \in \mathbb{R}^2. \tag{5.8.1}$$

To specify further conditions on $B$ we need to divide $\mathbb{R}^2$ into sets relevant to the strength of the field. For a (measurable) set $\mathcal{C} \subset \mathbb{R}^2$ define

$$\mathcal{C}^{\uparrow} = \cup_{\mathbf{x} \in \mathcal{C}} \mathcal{K}(\mathbf{x}). \tag{5.8.2}$$

With the field $B$ we associate two open sets $\Omega, \Lambda \subset \mathbb{R}^2$, such that $\Omega^{\uparrow} \subset \Lambda$ and $(\mathbb{R}^2 \setminus \Lambda)^{\uparrow} \cap \Omega^{\uparrow} = \emptyset$; the case $\Lambda = \mathbb{R}^2$ is not excluded. Let $ł_0 > 0$ be the lowest eigenvalue of the Laplace operator $-\Delta$ on the unit disk, with Dirichlet boundary conditions. Put

$$A_0 = \frac{5(2 + 4\sqrt{ł_0})}{\sqrt{2}}, \tag{5.8.3}$$

and assume that

$$|B(\mathbf{x})| \ell(\mathbf{x})^2 \geq 2A_0^2, \ \ \text{a.e. } \mathbf{x} \in \Lambda. \tag{5.8.4}$$

The physical meaning of the sets $\Omega$, $\Lambda$, is that on $\Omega$ the field $B$ is "large", on $\mathbb{R}^2 \setminus \Lambda$ the field $B$ is negligibly small, and the set $\Lambda \setminus \Omega$ is a "transition zone".

Before stating the main result we remind the reader about an important constant depending on $\Omega$. Suppose that the boundary of $\Omega$ is Lipschitz, and let $\delta(\mathbf{x})$ denote the distance from $\mathbf{x} \in \mathbb{R}^2$ to $\Omega$. Then there exists a positive constant $\mu \leq 1/4$ such that for any $u \in \mathsf{H}_0^1(\Omega')$, $\Omega' = \mathbb{R}^2 \setminus \Omega$, one has the Hardy inequality

$$\int_{\Omega'} |\nabla u(\mathbf{x})|^2 d\mathbf{x} \geq \mu \int_{\Omega'} \frac{|u(\mathbf{x})|^2}{\delta(\mathbf{x})^2} d\mathbf{x}; \tag{5.8.5}$$

see Theorem 3.2.1 and comment (vi) in Sect. 3.1. If $\Omega'$ is a union of convex connected components, one has $\mu = 1/4$. In view of the diamagnetic inequality

$$\int |(\nabla + i\mathbf{A})u|^2 d\mathbf{x} \geq \int |\nabla |u||^2 d\mathbf{x}, \ \forall u \in C_0^1(\mathbb{R}^2), \tag{5.8.6}$$

we immediately infer from (5.8.5) that

$$\int_{\Omega'} |(\nabla + i\mathbf{A})u(\mathbf{x})|^2 d\mathbf{x} \geq \mu \int_{\Omega'} \frac{|u(\mathbf{x})|^2}{\delta(\mathbf{x})^2} d\mathbf{x}, \ \forall u \in C_0^1(\Omega'). \tag{5.8.7}$$

**Theorem 5.8.1** *Let $\Omega$ be an open set with Lipschitz boundary. Let the function $\ell$ be as specified in* (5.8.1), *and let the field $B$ satisfy* (5.8.4). *Suppose also that the field $B$ is either non-negative or non-positive for a.e.* $\mathbf{x} \in \mathbb{R}^2$. *Then*

$$h[u] \geq \frac{\mu}{2} \int \frac{|u(\mathbf{x})|^2}{\ell(\mathbf{x})^2 + \delta(\mathbf{x})^2} d\mathbf{x}$$

*for all $u \in \mathcal{D}[h]$, the domain of $h$.*

To apply Theorem 5.8.1, the first step is to make an appropriate choice of the function $\ell$ for a given magnetic field. Below we illustrate how it can be done in the case $n = 2$ for two special cases. Both examples are deliberately made strongly radially asymmetric in order to guarantee that the separation of variables is not applicable.

*Example 5.8.2* The first example is a compactly supported magnetic field. We denote by $D_R(x, y)$ the open disk in $\mathbb{R}^2$ of radius $R$, centred at $(x, y) \in \mathbb{R}^2$; let $x_0 > R > 0$ and $\Lambda = D_R(x_0, 0) \cup D_R(-x_0, 0)$. Assume that

$$B(\mathbf{x}) = 0, \;\; \mathbf{x} \notin \Lambda, \quad B \geq B_0, \;\; \mathbf{x} \in \Lambda \tag{5.8.8}$$

with some positive constant $B_0$, and define the function $\ell(\mathbf{x})$ by

$$\ell(\mathbf{x}) \equiv \ell_0 = \sqrt{2A_0^2/B_0}, \;\; \mathbf{x} \in \mathbb{R}^2. \tag{5.8.9}$$

Clearly, $\ell$ satisfies the conditions (5.8.1) and (5.8.4) on the set $\Lambda$. Moreover, $(\mathbb{R}^2 \setminus \Lambda)^{\uparrow} \cap \Omega^{\uparrow} = \emptyset$. If $2\ell_0 < R$, then $\Omega$ can be chosen as follows: $\Omega = D_{R-2\ell_0}(x_0, 0) \cup D_{R-2\ell_0}(-x_0, 0)$. Now Theorem 5.8.1 leads to the inequality

$$h[u] \geq \frac{\mu}{2} \int \frac{|u(\mathbf{x})|^2}{\ell_0^2 + \delta(\mathbf{x})^2} d\mathbf{x}, \;\; u \in D[h], \tag{5.8.10}$$

where, with $\mathbf{x} = (x, y)$,

$$\delta(x, y) = \begin{cases} 0, & \text{if } (x, y) \in \Omega, \\ \sqrt{(x - x_0)^2 + y^2} - (R - 2\ell_0), & \text{if } x \geq 0 \text{ and } (x, y) \notin \Omega, \\ \sqrt{(x + x_0)^2 + y^2} - (R - 2\ell_0), & \text{if } x \leq 0 \text{ and } (x, y) \notin \Omega. \end{cases}$$

*Example 5.8.3* Next, we consider an "opposite" example: a *magnetic field with holes in its support*. Suppose as above, that $R < x_0$, assume again that $B$ satisfies (5.8.8) with the set

$$\Lambda = \mathbb{R} \setminus \overline{D_R(x_0, 0) \cup D_R(-x_0, 0)},$$

and define $\ell$ by (5.8.9). If $2\ell_0 < x_0 - R$, define $\Omega$ by

$$\Omega = \mathbb{R}^2 \setminus \overline{D_{R+2\ell_0}(x_0, 0) \cup D_{R+2\ell_0}(-x_0, 0)}.$$

Then Theorem 5.8.1 yields again (5.8.10) with the distance function $\delta(\mathbf{x}) = \delta(x, y)$ given by

$$\delta(x, y) = \begin{cases} 0, & \text{if } (x, y) \in \Omega, \\ (R + 2\ell_0) - \sqrt{(x - x_0)^2 + y^2}, & \text{if } (x, y) \in D_{R+2\ell_0}(x_0, 0), \\ (R + 2\ell_0) - \sqrt{(x + x_0)^2 + y^2}, & \text{if } (x, y) \in D_{R+2\ell_0}(-x_0, 0). \end{cases}$$

Moreover, since $\Omega' = \mathbb{R}^2 \setminus \Omega$ is a union of two convex sets (namely disks), one has $\mu = 1/4$; see the comment after formula (5.8.5).

Note that in both cases the effective field

$$\tilde{B}(\mathbf{x}) = \frac{1}{\ell_0^2 + \delta(\mathbf{x})^2}$$

in (5.8.10), shows the following behaviour in the strong field regime, that is when $B_0 \to \infty$: if $\mathbf{x} \in \Omega$, then $\tilde{B} \to \infty$ as well. For $\mathbf{x} \in \mathbb{R}^2 \setminus \Lambda$ the function $\tilde{B}$ behaves like $\delta(\mathbf{x})$, and thus, effectively, it "does not feel" the magnetic field, irrespective of its strength. The set $\Lambda \setminus \Omega$, which consists of two rings of width $B_0^{-1/2}$, is a transition region.

Obviously, both examples can be generalised to any number of disks.

Before presenting proof of Theorem 5.8.1, we describe a very useful partition of unity in two-dimensions.

### 5.8.3 A Partition of Unity

Let $\Omega, \Lambda \subset \mathbb{R}^2$ be the sets introduced in the previous subsection , and let $\Omega^\uparrow$ be as defined in (5.8.2). As was previously mentioned, Theorem 5.8.1 trivially follows from (5.8.4) and (5.8.24) if $\Lambda = \mathbb{R}^2$. Henceforth we assume that $\mathbb{R}^2 \setminus \Lambda \neq \emptyset$.

Let $\Upsilon \in C_0^1(\mathbb{R}^2)$ be a non-negative function such that $\Upsilon(\mathbf{x}) = 0$ for $|\mathbf{x}| \geq 1$ and $\int \Upsilon(\mathbf{x})^2 d\mathbf{x} = 1$; set

$$\lambda = \lambda(\Upsilon) = \int |\nabla \Upsilon(\mathbf{x})|^2 d\mathbf{x}. \tag{5.8.11}$$

Then, by the min-max variational principle, the lowest eigenvalue $\lambda_0$ of the Dirichlet Laplacian on the unit disk is given by

$$\lambda_0 = \inf_{\Upsilon} \lambda. \tag{5.8.12}$$

**Lemma 5.8.4** *Suppose that* $\ell$ *satisfies* (5.8.1). *Then the function*

$$\phi(\mathbf{x}) = \frac{1}{\ell(\mathbf{x})^2} \int_{\Omega^\uparrow} \Upsilon\left(\frac{\mathbf{x}-\mathbf{y}}{\ell(\mathbf{x})}\right)^2 d\mathbf{y}$$

*possesses the following properties:*

(i) $\phi \in C^1(\mathbb{R}^2)$, *and* $|\nabla \phi(\mathbf{x})| \leq (2 + 4\sqrt{\lambda})\ell(\mathbf{x})^{-1}$;
(ii) $\phi(\mathbf{x}) = 1$ *for* $\mathbf{x} \in \Omega$, $\phi(\mathbf{x}) = 0$ *for* $\mathbf{x} \in \mathbb{R}^2 \setminus \Lambda$, *and* $0 \leq \phi(\mathbf{x}) \leq 1$.

*Proof* The inclusion $\phi \in C^1(\mathbb{R}^2)$ is obvious, since $\ell \in C^1(\mathbb{R}^2)$. The estimate for $\nabla\phi$ is checked by a direct calculation:

$$\begin{aligned}
|\nabla \phi(\mathbf{x})| &\leq \frac{2|\nabla \ell(\mathbf{x})|}{\ell(\mathbf{x})^3} \int_{\Omega^\uparrow} \Upsilon(\mathbf{y})^2 d\mathbf{y} \\
&\quad + \frac{2}{\ell(\mathbf{x})^3} \int_{\Omega^\uparrow} |\Upsilon(\mathbf{y})||\nabla \Upsilon(\mathbf{y})| \left[1 + \frac{|\mathbf{x}-\mathbf{y}|}{\ell(\mathbf{x})} |\nabla \ell(\mathbf{x})|\right] d\mathbf{y} \\
&\leq \frac{2}{\ell(\mathbf{x})} + \frac{4}{\ell(\mathbf{x})^3} \int_{\Omega^\uparrow} |\Upsilon(\mathbf{y})||\nabla \Upsilon(\mathbf{y})| d\mathbf{y} \\
&\leq \frac{2}{\ell(\mathbf{x})} + \frac{4}{\ell(\mathbf{x})} \left[\int |\nabla \Upsilon(\mathbf{x})|^2 d\mathbf{x}\right]^{\frac{1}{2}} = \frac{2 + 4\sqrt{\lambda}}{\ell(\mathbf{x})}.
\end{aligned}$$

Here we have taken into account that $|\nabla \ell(\mathbf{x})| \leq 1$.

In view of the formula $\int \Upsilon(\mathbf{x})^2 d\mathbf{x} = 1$, we always have $\phi(\mathbf{x}) \leq 1$. Furthermore, if $\mathbf{x} \in \Omega$, then by definition, $\mathcal{K}(\mathbf{x}) \subset \Omega^\uparrow$, and hence

$$\phi(\mathbf{x}) = \frac{1}{\ell(\mathbf{x})^2} \int_{\mathbb{R}^2} \Upsilon\left(\frac{\mathbf{x}-\mathbf{y}}{\ell(\mathbf{x})}\right)^2 d\mathbf{y} = 1,$$

as required. Otherwise, if $\mathbf{x} \in \mathbb{R}^2 \setminus \Lambda$, then by definition, $\Upsilon\big((\mathbf{x}-\mathbf{y})\ell(\mathbf{x})^{-1}\big) = 0$ for all $\mathbf{y} \in \Omega^\uparrow$, and therefore $\phi(\mathbf{x}) = 0$. □

This Lemma allows one to introduce a convenient partition of unity:

**Lemma 5.8.5** *Let the domains* $\Omega$ *and* $\Lambda$ *be as in Theorem 5.8.1, and let* $\mathbb{R}\setminus\Lambda \neq \emptyset$. *Then there exist two functions* $\zeta, \eta \in C^1(\mathbb{R}^2)$ *such that*

(i) $0 \le \zeta \le 1,\ \ 0 \le \eta \le 1$;
(ii) $\zeta(\mathbf{x}) = 1$ *for* $\mathbf{x} \in \Omega$, $\eta(\mathbf{x}) = 1$ *for* $\mathbf{x} \in \mathbb{R}^2 \setminus \Lambda$;
(iii) $\zeta^2 + \eta^2 = 1$;
(iv) $|\nabla\zeta| \le A\ell^{-1}$, $|\nabla\eta| \le A\ell^{-1}$ *with*

$$A = \frac{5(2+4\sqrt{\lambda})}{\sqrt{2}}. \tag{5.8.13}$$

*Proof* Let $\phi$ be the function constructed in Lemma 5.8.4, and let $\psi = 1 - \phi$. It is straightforward to check that $\phi^2 + \psi^2 = 2(\phi - 1/2)^2 + 1/2 \ge 1/2$. Define

$$\zeta = \frac{\phi}{\sqrt{\phi^2+\psi^2}}, \quad \eta = \frac{\psi}{\sqrt{\phi^2+\psi^2}}.$$

These functions, obviously satisfy properties (i), (ii), (iii). To prove (iv) note that

$$|\nabla\zeta| \le \frac{5}{2\sqrt{\phi^2+\psi^2}}|\nabla\phi| \le \frac{5}{\sqrt{2}}|\nabla\phi|,$$

and a similar bound holds for $\nabla\eta$. The required estimate now follows from Lemma 5.8.4. □

## 5.8.4 *Proof of Theorem 5.8.1*

We are now ready to present a proof of Theorem 5.8.1. Suppose that the conditions of Theorem 5.8.1 are fulfilled. Our next step is to split the magnetic form $h[u]$ into two parts that will be estimated in two different ways. Let $\zeta$, $\eta$ be the functions from Lemma 5.8.5. Since $\zeta^2 + \eta^2 = 1$, we have for any $u \in C_0^1(\mathbb{R}^2)$:

$$\begin{aligned} h[u] &= \int |\zeta(\nabla + i\mathbf{A})u|^2 d\mathbf{x} + \int |\eta(\nabla + i\mathbf{A})u|^2 d\mathbf{x} \\ &= h[\zeta u] + h[\eta u] - \int (|\nabla\zeta|^2 + |\nabla\eta|^2)|u|^2 d\mathbf{x}. \end{aligned}$$

We use the following decomposition:

$$h[u] = \frac{1}{2}\left[h[u] - \int (|\nabla\zeta|^2 + |\nabla\eta|^2)|u|^2 d\mathbf{x}\right] + \frac{1}{2}h[\zeta u] + \frac{1}{2}h[\eta u]. \tag{5.8.14}$$

Since $B$ does not change sign, it follows from (5.8.24) that $h[u] \geq (|B|u, u)$. Let us estimate from below the first term on the right-hand side of (5.8.14), bearing in mind that $\nabla\zeta$ and $\nabla\eta$ are supported on the set $\Lambda$:

$$h[u] - \int (|\nabla\zeta|^2 + |\nabla\eta|^2)|u|^2 d\mathbf{x} \geq \int_\Lambda \Big[|B| - (|\nabla\zeta|^2 + |\nabla\eta|^2)\Big]|u|^2 d\mathbf{x}.$$

In view of the condition (5.8.4) and of the fact that $|\nabla\zeta|^2 + |\nabla\eta|^2 \leq 2A^2\ell^{-2}$, the right-hand side is bounded from below by $-\nu E(u)$ with

$$E(u) = \int_\Lambda \frac{1}{\ell(\mathbf{x})^2}|u(\mathbf{x})|^2 d\mathbf{x}, \quad \nu = 2(A^2 - A_0^2) \geq 0.$$

Next we estimate the remaining two terms in (5.8.14). For the term with $\zeta$ use (5.8.4) again, keeping in mind that $\zeta$ is supported on $\Lambda$:

$$h[\zeta u] \geq \int |B(\mathbf{x})|\zeta(\mathbf{x})^2|u(\mathbf{x})|^2 d\mathbf{x} \geq 2A_0^2 \int \frac{1}{\ell(\mathbf{x})^2}\zeta(\mathbf{x})^2|u|^2 d\mathbf{x}.$$

For the term with $\eta$ we use Hardy's inequality (5.8.7):

$$h[\eta u] \geq \mu \int \eta(\mathbf{x})^2 \frac{|u(\mathbf{x})|^2}{\delta(\mathbf{x})^2} d\mathbf{x}.$$

Collecting all the estimates, we obtain the lower bound

$$2h[u] \geq 2A_0^2 \int \zeta^2 \frac{|u|^2}{\ell^2} d\mathbf{x} + \mu \int \eta^2 \frac{|u|^2}{\delta^2} d\mathbf{x} - 2\nu E(u).$$

Since $2A_0^2 \geq 100$ and $\mu \leq 1/4$, one can write

$$2h[u] \geq \mu \int \frac{|u|^2}{\delta^2 + \ell^2} d\mathbf{x} - 2\nu E(u).$$

Neither the right-hand side nor the left-hand side depends on the function $\Upsilon$. Therefore we can take the sup of both sides over all admissible $\Upsilon$. In view of definitions (5.8.3) and (5.8.13), the equality (5.8.12) yields $\sup_\Upsilon(-\nu) = \inf_\Upsilon \nu = 0$. This leads to the required bound from below and thus completes the proof of Theorem 5.8.1.

### 5.8.5 Results for $n \geq 3$

In this case our conditions on **B** are more restrictive. To state the precise conditions, let us begin with the function $\ell$. We assume that

$$|\ell(\mathbf{x}) - \ell(\mathbf{y})| \leq \varrho|\mathbf{x} - \mathbf{y}|,\ 0 \leq \varrho < 1,\ \ \ell(\mathbf{x}) > 0,\ \forall \mathbf{x}, \mathbf{y} \in \mathbb{R}^n. \tag{5.8.15}$$

Assume that for some $\Phi > 0$

$$|\mathbf{B}(\mathbf{x})|\ell(\mathbf{x})^2 \geq \Phi,\ \ \text{a.e.}\ \ \mathbf{x} \in \mathbb{R}^n. \tag{5.8.16}$$

This assumption guarantees that the field **B** never vanishes. Denote by $\mathbf{n} = \{n_{jk}\}_{j,k=1}^n$ the matrix with the components

$$n_{jk}(\mathbf{x}) = \frac{B_{jk}(\mathbf{x})}{|\mathbf{B}(\mathbf{x})|},\ j, k = 1, 2, \dots, n.$$

One may loosely call **n** the *direction matrix* for the field **B**. Our second assumption on **B** is that for all $k, l = 1, 2, \dots, n$ and $\mathbf{z} \in \mathbb{R}^n$

$$|\mathbf{n}(\mathbf{x}) - \mathbf{n}(\mathbf{y})| \leq \alpha, \forall \mathbf{x}, \mathbf{y} \in \mathcal{K}(\mathbf{z}), \tag{5.8.17}$$

with some $0 \leq \alpha \leq \sqrt{\varkappa_d^{-1}}/4$. This assumption implies that the direction **n** of the field varies slowly with **x**. Note that this condition is automatically fulfilled in the case $n = 2$ with $\alpha = 0$.

**Theorem 5.8.6** *Let* $n \geq 3$. *Let the function* $\ell$ *be as specified in* (5.8.15), *and let the field* **B** *be a continuous function satisfying* (5.8.16) *and* (5.8.17). *Then for a sufficiently large* $\Phi > 0$ *in* (5.8.16) *we have*

$$h[u] \geq c \int |\mathbf{B}(\mathbf{x})||u(\mathbf{x})|^2 d\mathbf{x} \tag{5.8.18}$$

*for all* $u \in \mathcal{D}[h]$, *with some positive constant* $c$ *depending on* $\varrho$ *and* $\Phi$.

Theorem 5.8.6 holds for the case $n = 2$ as well, but it is a trivial corollary of Theorem 5.8.1.

Note that in contrast to Theorem 5.8.1 we do not specify the constant $c$ in the inequality (5.8.18), neither do we provide any precise estimates on the value of $\Phi$ sufficient for (5.8.18) to hold. In fact, as a careful examination of the proof will show, one can always control the constants in all the estimates, but their values will hardly be optimal.

### 5.8.6 Proof of Theorem 5.8.6

Suppose that the condition of Theorem 5.8.6 are fulfilled, and in particular, the function $\ell$ satisfies (5.8.15).

The keystone of the proof is the following partition of unity associated with the scale function $\ell(\mathbf{x})$.

**Lemma 5.8.7** *Let $\ell(\mathbf{x})$ (resp. $\ell(x)$) be a continuous function satisfying* (5.8.15). *Then there exists a set of points $\mathbf{x}_j \in \mathbb{R}^n$, $j \in \mathbb{N}$ such that the open balls $\mathcal{K}_j = \mathcal{K}(\mathbf{x}_j)$ form a covering of $\mathbb{R}^n$ with the finite intersection property (i.e. each ball $\mathcal{K}_j$ intersects with no more than $\tilde{N} = \tilde{N}(\varrho) < \infty$ other balls). Moreover, there exists a set of non-negative functions $\phi_j \in C_0^\infty(K_j)$, $j \in \mathbb{N}$, such that*

$$\sum_j \phi_j^2 = 1, \tag{5.8.19}$$

*and*

$$|\partial^m \phi_j| \le C_m \ell^{-|m|}, \ \forall m, \tag{5.8.20}$$

*uniformly in j.*

We note that the square in (5.8.19) will be convenient for us, though the common definition of the partition of unity requires $\sum_j \phi_j = 1$. Proof of this lemma is analogous to that of Theorem 1.4.10 from [79] and we do not reproduce it here.

We rephrase the finite intersection property for balls $\mathcal{K}_j$ as follows: setting

$$\mathfrak{m}_j = \{k \in \mathbb{N} : \mathcal{K}_j \cap \mathcal{K}_k \ne \emptyset\},$$

then

$$\operatorname{card} \mathfrak{m}_j \le N(\varrho) := \tilde{N}(\varrho) + 1,$$

with the number $\tilde{N}(\varrho)$ defined in Lemma 5.8.7.

The next step is to use the partition of unity constructed in Lemma 5.8.7. For $u \in C_0^1(\mathbb{R}^n)$, a simple calculation, similar to that in the proof of Theorem 5.8.1, yields

$$h[u] = \sum_k h[\phi_k u] - \sum_k \int |\nabla \phi_k|^2 |u|^2 d\mathbf{x}. \tag{5.8.21}$$

The first term on the right-hand side satisfies

$$N \sum_k h[\phi_k u] \ge \sum_k \sum_{l \in \mathfrak{m}_k} h[\phi_l u], \ N = N(\varrho).$$

Let $k \in \mathbb{N}$ be fixed, and let $j, l \in [1, d]$ be a pair of integers such that $|B_{jl}(\mathbf{x}_k)| \geq \sqrt{\varkappa_d^{-1}}|\mathbf{B}(\mathbf{x}_k)|$; such a pair always exists. Assume, without loss of generality, that $B_{jl} > 0$. Then, in view of (5.8.17),

$$B_{jl}(\mathbf{x}) \geq \frac{3}{4\sqrt{\varkappa_d}}|\mathbf{B}(\mathbf{x})|, \ \mathbf{x} \in \mathcal{K}_k.$$

On using (5.8.17) again, we obtain

$$B_{jl}(\mathbf{x}) \geq \frac{1}{2\sqrt{\varkappa_d}}|\mathbf{B}(\mathbf{x})|, \ \mathbf{x} \in \cup_{s \in \mathfrak{m}_k} \mathcal{K}_s.$$

From (5.8.23), we have the lower estimate

$$\sum_{s \in \mathfrak{m}_k} h[\phi_s u] \geq \int B_{jl} \sum_{s \in \mathfrak{m}_k} \phi_s^2 |u|^2 d\mathbf{x} \geq \frac{1}{2\sqrt{\varkappa_d}} \int |\mathbf{B}| \sum_{s \in \mathfrak{m}_k} \phi_s^2 |u|^2 d\mathbf{x},$$

the last integral being bounded from below by

$$\frac{1}{2\sqrt{\varkappa_d}} \int_{\mathcal{K}_k} |\mathbf{B}||u|^2 d\mathbf{x}.$$

Here we have used the fact that $\sum_{s \in \mathfrak{m}_k} \phi_s(\mathbf{x})^2 = 1$ for all $\mathbf{x} \in \mathcal{K}_k$, which follows from the definition of $\mathfrak{m}_k$. Consequently

$$\sum_k h[\phi_k u] \geq \frac{1}{N} \sum_k \sum_{s \in \mathfrak{m}_k} h[\phi_s u] \geq \frac{1}{2N\sqrt{\varkappa_d}} \sum_k \int_{\mathcal{K}_k} |\mathbf{B}|\, |u|^2 d\mathbf{x}. \tag{5.8.22}$$

We estimate $h[u]$ from below using (5.8.21) and (5.8.22):

$$h[u] \geq \sum_k \int_{\mathcal{K}_k} \left[ \frac{1}{2N\sqrt{\varkappa_d}} |\mathbf{B}| - |\nabla \phi_k|^2 \right] |u|^2 d\mathbf{x},$$

where we have used the fact that $\nabla \phi_k$ is supported on $\mathcal{K}_k$. According to (5.8.16) and (5.8.20) we have

$$|\nabla \phi_k|^2 \leq c^2 \ell^{-2} \leq c^2 \Phi^{-1} |\mathbf{B}|,$$

so that

$$h[u] \geq \left[ \frac{1}{2N\sqrt{\varkappa_d}} - c^2 \Phi^{-1} \right] \sum_k \int_{\mathcal{K}_k} |\mathbf{B}|\, |u|^2 d\mathbf{x}.$$

If we assume that $\Phi$ is sufficiently large that the factor before the integral is positive, we then obtain

$$h[u] \geq c \int |\mathbf{B}|\, |u|^2 d\mathbf{x},\ u \in C_0^1(\mathbb{R}^n).$$

This is the required bound. The proof of Theorem 5.8.6 is now complete.

*Remark 5.8.8* In [22], Sect. 2.4, there is the following illuminating discussion of Theorems 5.8.1 and 5.8.6.

The simplest known source of lower bounds for the magnetic Schrödinger operator is the following representation for the quadratic form $h[u]$. Set $L_k = \partial_k + iA_k$. Then

$$\|L_k u\|^2 + \|L_l u\|^2 = \|(L_k \pm iL_l)u\|^2 \pm (B_{kl}u, u),\ u \in C_0^1(\mathbb{R}^n),$$

for any pair $k, l = 1, 2, \ldots, n$. This identity implies that

$$h[u] \geq \pm(B_{kl}u, u),\ \forall k, l = 1, 2, \ldots, n. \tag{5.8.23}$$

If one knows that, for some $k, l$, the quantity $B_{kl}$ preserves its sign, and $c|\mathbf{B}| \leq |B_{kl}|$, then the above inequality leads to the lower bound (5.8.18) in Theorem 5.8.6. The bound is especially useful in the case $n = 2$, when it can be rewritten as

$$h[u] \geq \pm(Bu, u),\ u \in C_0^1(\mathbb{R}^2), \tag{5.8.24}$$

as we saw in Sect. 5.4.1. In fact, Theorem 5.8.1 trivially follows from this estimate and (5.8.4) if one assumes that $\Lambda = \mathbb{R}^2$. In this case, assuming, for instance, that $B > 0$, one uses (5.8.24) with the "+" sign, which leads, in view of (5.8.4), to the bound

$$h[u] \geq 2A_0^2 \int \frac{|u(\mathbf{x})|^2}{\ell(\mathbf{x})^2} d\mathbf{x}.$$

Since $2A_0^2 \geq 100$ and $\mu \leq 1/4$, this implies the sought lower bound. If, on the other hand, $B \geq 0$ and the support of the field does not coincide with $\mathbb{R}^2$, then Theorem 5.8.1 yields a bound similar to (5.8.24), but with an *effective* magnetic field

$$\tilde{B}(\mathbf{x}) = \frac{1}{\ell(\mathbf{x})^2 + \delta(\mathbf{x})^2},$$

which, loosely speaking, coincides with $B$ inside the support, and decays away from it. It is important that this effective field does not vanish in contrast to $B$.

In the multi-dimensional situation the picture is different: the field **B** is allowed to change its direction. In these circumstances the estimate (5.8.23) is not very helpful as all the components $B_{kj}$ may change their signs. Theorem 5.8.6 is specifically designed to handle this situation. We need to assume however that **B** never vanishes.

A lower bound of a type similar to (5.8.18) was proved in [76]. Instead of the function $|\mathbf{B}|$ the inequality in [76] features a specific weight function, which coincides with $|\mathbf{B}|$ in the case of a polynomial magnetic field. Another instance when such an inequality is known to hold, is described in [135]. If the magnetic field is assumed to belong to a certain reverse Hölder class, then it is shown in [135] that $h[u] + \|u\|^2 \geq c(\ell^{-2}u, u)$ with some explicitly defined scale function $\ell$. Theorem 5.8.6 is close in the spirit to these results, but the proof in [22] is much more elementary, and is based on a natural partition of unity associated with the scale function $\ell$.

## 5.9 Pauli Operators in $\mathbb{R}^3$ with Magnetic Fields

In relativistic quantum mechanics, when electron spin is taken into account, the Schrödinger operator $H_{\mathbf{A}}$ discussed in Sect. 5.4 is replaced by the Pauli operator, which in $\mathbb{R}^3$, is formally given by

$$\mathbb{P}_{\mathbf{A}} = \left\{\boldsymbol{\sigma} \cdot \left(\frac{1}{i}\nabla + \mathbf{A}\right)\right\}^2 \equiv \sum_{j=1}^{3} \left\{\sigma_j \left(\frac{1}{i}\partial_j + A_j\right)\right\}^2, \tag{5.9.1}$$

where $\mathbf{A} = (A_1, A_2, A_3)$ is a vector potential associated with the magnetic field $\mathbf{B} = \text{curl}\mathbf{A}$, and $\boldsymbol{\sigma} = (\sigma_1, \sigma_2, \sigma_3)$ is the triple of Pauli matrices

$$\sigma_1 = \begin{pmatrix} 0 & 1 \\ 1 & 0 \end{pmatrix} \quad \sigma_2 = \begin{pmatrix} 0 & -i \\ i & 0 \end{pmatrix} \quad \sigma_3 = \begin{pmatrix} 1 & 0 \\ 0 & -1 \end{pmatrix}.$$

The expression (5.9.1) can also be written in the convenient form

$$\mathbb{P}_{\mathbf{A}} = \mathbb{S}_{\mathbf{A}} + \boldsymbol{\sigma} \cdot \mathbf{B}, \qquad \mathbf{B} = \text{curl}\mathbf{A}, \tag{5.9.2}$$

where we have written $\mathbb{S}_{\mathbf{A}}$ for the magnetic Schrödinger operator as an operator in $L^2(\mathbb{R}^3; \mathbb{C}^2)$, namely,

$$\mathbb{S}_{\mathbf{A}} = \left(\frac{1}{i}\nabla + \mathbf{A}\right)^2 \mathbb{I}_2 \equiv \sum_{j=1}^{3} \left(\frac{1}{i}\partial_j + A_j\right)^2 \mathbb{I}_2, \tag{5.9.3}$$

where $\mathbb{I}_2$ is the $2 \times 2$ identity matrix, and $\boldsymbol{\sigma} \cdot \mathbf{B}$ is called the **Zeeman** term. To simplify notation in this section, we denote $L^2(\mathbb{R}^3; \mathbb{C}^2)$ by $\mathcal{H}$ and its standard inner-product and norm by $(\cdot,\cdot)$ and $\|\cdot\|$ respectively: for $f = \{f_1, f_2\}, g = \{g_1, g_2\} \in \mathcal{H}$

$$(f,g) = \sum_{j=1}^{2} \int_{\mathbb{R}^3} f_j(\mathbf{x})\bar{g}_j(\mathbf{x})d\mathbf{x}, \quad \|f\|^2 = (f,f).$$

Suppose that $A_j \in L^2_{loc}(\mathbb{R}^3)$, $j = 1,2,3$. Then, as an operator in $\mathcal{H}$, $\mathbb{S}_{\mathbf{A}}$ is the Friedrichs operator associated with the form

$$\left\{ \left\| \left( \frac{1}{i}\nabla + \mathbf{A} \right) \varphi \right\|^2 + \|\varphi\|^2 \right\}^{\frac{1}{2}}. \tag{5.9.4}$$

It is non-negative and has no zero modes, i.e., no eigenvalue at zero. Its form domain is the completion of $C_0^\infty(\mathbb{R}^3;\mathbb{C}^2)$ with respect to the norm determined by (5.9.4), which we denote by $\|\cdot\|_{1,\mathbf{A}}$.

Our objective in this section is to establish Sobolev, Hardy and CLR type inequalities for the Pauli operator which are analogous to those of $\mathbb{S}_{\mathbf{A}}$. An obstacle is the fact that the Pauli operator may have zero modes. This means that in order to obtain the aforementioned inequalities, the zero modes must be avoided and the inequalities should reflect this. To achieve these aims,we need some technical preliminary results which lead to the introduction of a Birman-Schwinger type operator. Initially we shall assume that $|\mathbf{A}| \in L^3(\mathbb{R}^3)$, but later $|\mathbf{B}| \in L^{3/2}(\mathbb{R}^3)$ is also needed. However, it is proved in [66], Theorem A1 in Appendix A, that, given $|\mathbf{B}| \in L^{3/2}(\mathbb{R}^3)$, there is a unique magnetic potential $\mathbf{A}$ with the properties

$$|\mathbf{A}| \in L^3(\mathbb{R}^3), \quad \operatorname{curl} \mathbf{A} = \mathbf{B}, \quad \operatorname{div} \mathbf{A} = 0,$$

this being given by

$$\mathbf{A}(\mathbf{x}) = \frac{1}{4\pi} \int_{\mathbb{R}^3} \frac{(\mathbf{x}-\mathbf{y})}{|\mathbf{x}-\mathbf{y}|} \times \mathbf{B}(\mathbf{y})d\mathbf{y}.$$

We shall approach the problem through the Weyl-Dirac (or massless Dirac) operator

$$\mathbb{D}_{\mathbf{A}} : \boldsymbol{\sigma}\cdot\mathbf{p} + \boldsymbol{\sigma}\cdot\mathbf{A}, \quad \mathbf{p} := \frac{1}{i}\nabla,$$

where we have use the standard notation $\mathbf{p}$ for the momentum operator, but note that it now operates on $\mathbb{C}^2$-valued functions. The first lemma determines the domain of $\mathbb{D}_{\mathbf{A}}$.

**Lemma 5.9.1** *Let* $|\mathbf{A}| \in L^3(\mathbb{R}^3)$. *Then, given* $\varepsilon > 0$, *there exists* $k_\varepsilon > 0$ *such that for all* $\varphi \in H^1(\mathbb{R}^3; \mathbb{C}^2)$, *the Sobolev space* $H^1 = W^{1,2}$ *on* $\mathbb{R}^3$ *for* $\mathbb{C}^2$*-valued functions,*

$$\|(\boldsymbol{\sigma} \cdot \mathbf{A})\varphi\| \leq \varepsilon \|\nabla \varphi\| + k_\varepsilon \|\varphi\|. \tag{5.9.5}$$

*Proof* We may choose $|\mathbf{A}| = a_1 + a_2$, where $|a_1(\mathbf{x})| \leq k_\varepsilon$ and $\|a_2\|_{L^3(\mathbb{R}^3)} < \varepsilon$. Then, by Hölder's inequality and Sobolev's embedding theorem,

$$\begin{aligned} \|(\boldsymbol{\sigma} \cdot \mathbf{A})\varphi\| &\leq \|a_1\varphi\| + \|a_2\varphi\| \leq k_\varepsilon \|\varphi\| + \varepsilon \|\varphi\|_{L^6(\mathbb{R}^3)} \\ &\leq k_\varepsilon \|\varphi\| + \gamma\varepsilon \|\nabla\varphi\|, \end{aligned}$$

where $\gamma$ denotes the norm of the embedding $D^{1,2}(\mathbb{R}^n) \hookrightarrow L^6(\mathbb{R}^n)$. □

It follows from the lemma and a well-known result on relative bounded perturbations of self-adjoint operators (see [48], Corollary III.8.5) that $\mathbb{D}_\mathbf{A}$ is the operator sum

$$\mathbb{D}_\mathbf{A} = \boldsymbol{\sigma} \cdot \mathbf{p} + \boldsymbol{\sigma} \cdot \mathbf{A},$$

with domain

$$\mathcal{D}(\mathbb{D}_\mathbf{A}) = \mathcal{D}(\boldsymbol{\sigma} \cdot \mathbf{p}) = H^1(\mathbb{R}^3; \mathbb{C}^2).$$

The Pauli operator $\mathbb{P}_\mathbf{A}$ can now be defined as the non-negative self-adjoint operator associated with the form

$$\|\mathbb{D}_\mathbf{A}\varphi\|^2 + \|\varphi\|^2, \quad \varphi \in \mathcal{D}(\mathbb{D}_\mathbf{A}),$$

which is the square of the graph norm of $\mathbb{D}_\mathbf{A}$ on its domain; $\mathcal{D}(\mathbb{D}_\mathbf{A})$ is therefore the form domain of $\mathbb{P}_\mathbf{A}$ and

$$\mathcal{D}(\mathbb{P}_\mathbf{A}^{1/2}) = \mathcal{D}(\mathbb{D}_\mathbf{A}) = H^1(\mathbb{R}^3; \mathbb{C}^2). \tag{5.9.6}$$

Similarly, if $|\mathbf{A}| \in L^3(\mathbb{R}^3)$, $\mathbf{p} + \mathbf{A}$ is an operator sum and the form domain of the Schrödinger operator $\mathbb{S}_\mathbf{A}$ is

$$\mathcal{D}(\mathbb{S}_\mathbf{A}^{1/2}) = \mathcal{D}(\mathbf{p} + \mathbf{A}) = H^1(\mathbb{R}^3; \mathbb{C}^2).$$

To proceed, we need some more notation.

(i) The operator

$$\mathbb{P} = \mathbb{P}_\mathbf{A} + |\mathbf{B}|$$

is the non-negative self-adjoint operator associated with

$$p[\varphi] = p[\varphi, \varphi] = ([\mathbb{P}_{\mathbf{A}} + |\mathbf{B}|]\varphi, \varphi) = (\mathbb{P}\varphi, \varphi).$$

If $|\mathbf{B}| \in L^{3/2}(\mathbb{R}^3)$, then for all $\varphi \in H^1(\mathbb{R}^3; \mathbb{C}^2)$,

$$(|\mathbf{B}|\varphi, \varphi) \leq \|B\|_{L^{3/2}(\mathbb{R}^3)} \|\varphi\|^2_{L^6(\mathbb{R}^3)} \leq \gamma \|B\|_{L^{3/2}(\mathbb{R}^3)} \|\nabla\varphi\|^2.$$

Therefore $\mathbb{P}$ has the same form domain as $\mathbb{P}_{\mathbf{A}}$, namely $H^1(\mathbb{R}^3; \mathbb{C}^2)$ by (5.9.6). The operator $\mathbb{S}_{\mathbf{A}}$ does not have an eigenvalue at 0, and clearly, neither does $\mathbb{P}$. Therefore $\mathbb{S}_{\mathbf{A}}$ and $\mathbb{P}$ are injective and have dense domains and ranges in $\mathcal{H}$.

(ii) We denote by $H^1_{\mathbf{A}}$ the completion of $\mathcal{D}(\mathbb{S}_{\mathbf{A}}^{1/2})$ with respect to the norm

$$\|\varphi\|_{H^1_{\mathbf{A}}} := \|\mathbb{S}_{\mathbf{A}}^{1/2}\varphi\|. \tag{5.9.7}$$

Note that in this notation, $H^1_{\mathbf{0}}$ has norm $\|\nabla\varphi\|$, and is therefore the $\mathbb{C}^2$-vector version of the space denoted by $D_0^{1,2}$ in Sect. 1.3.1.

(iii) We shall also need the space $\mathbb{H}^1_{\mathbf{B}}$ which is the completion of $\mathcal{D}(\mathbb{P}^{1/2})$ with respect to

$$\|\varphi\|_{\mathbb{H}^1_{\mathbf{B}}} := \|\mathbb{P}^{1/2}\varphi\|. \tag{5.9.8}$$

(iv) The spaces $H^1_{\mathbf{A}}, \mathbb{H}^1_{\mathbf{B}}$ do not lie in $\mathcal{H}$, but they both contain $C_0^\infty(\mathbb{R}^3; \mathbb{C}^2)$ as dense subspaces, and since $\mathbb{P} \geq \mathbb{S}_{\mathbf{A}}$ by (5.9.2), we have the natural embedding

$$\mathbb{H}^1_{\mathbf{B}} \hookrightarrow H^1_{\mathbf{A}}. \tag{5.9.9}$$

The diamagnetic inequality asserts that $f \mapsto |f|$ maps $H^1_{\mathbf{A}}$ continuously into $H^1_{\mathbf{0}}$, and hence, by the Sobolev's embedding theorem, we have

$$f \mapsto |f| : H^1_{\mathbf{A}} \to H^1_{\mathbf{0}} \hookrightarrow L^6(\mathbb{R}^3, \mathbb{C}^2). \tag{5.9.10}$$

Furthermore, Hardy's inequality yields

$$H^1_{\mathbf{A}} \hookrightarrow L^2(\mathbb{R}^3; |\mathbf{x}|^{-2}d\mathbf{x}, \mathbb{C}^2).$$

This also implies that

$$\mathcal{D}(\mathbb{P}^{1/2}) = \mathbb{H}^1_{\mathbf{B}} \cap \mathcal{H}, \tag{5.9.11}$$

with norm

$$\{\|\mathbb{P}^{1/2}\varphi\|^2 + \|\varphi\|^2\}^{1/2};$$

the embedding $\mathbb{H}^1_{\mathbf{B}} \hookrightarrow L^6(\mathbb{R}^3;\mathbb{C}^2)$ guarantees the completeness to establish (5.9.11), since convergent sequences in $\mathbb{H}^1_{\mathbf{B}}$ therefore converge pointwise to their limits, a.e..

Let

$$b[\varphi];=(|\mathbf{B}|\varphi,\varphi). \tag{5.9.12}$$

Then

$$0\le b[\varphi]\le p[\varphi]$$

and it follows that there exists a bounded self-adjoint operator $\mathcal{B}$ on $\mathbb{H}^1_{\mathbf{B}}$ such that

$$b[\varphi]=(\mathcal{B}\varphi,\varphi)_{\mathbb{H}^1_{\mathbf{B}}},\quad \varphi\in\mathbb{H}^1_{\mathbf{B}}. \tag{5.9.13}$$

For $\varphi\in\mathcal{R}(\mathbb{P}^{1/2})$, the range of $\mathbb{P}^{1/2}$,

$$\|\mathbb{P}^{-1/2}\varphi\|_{\mathbb{H}^1_{\mathbf{B}}}=\|\varphi\|,$$

and hence, since $\mathcal{D}(\mathbb{P}^{1/2})$ and $\mathcal{R}(\mathbb{P}^{1/2})$ are dense subspaces of $\mathbb{H}^1_{\mathbf{B}}$ and $\mathcal{H}$, respectively, $\mathbb{P}^{-1/2}$ extends to a unitary map

$$U:\mathcal{H}\to\mathbb{H}^1_{\mathbf{B}},\quad U=\mathbb{P}^{-1/2}\ \text{on}\ \mathcal{R}(\mathbb{P}^{1/2}). \tag{5.9.14}$$

Define

$$\mathcal{S}:=|\mathbf{B}|^{1/2}U:\mathcal{H}\to\mathcal{H}. \tag{5.9.15}$$

Note that

$$\||\mathbf{B}|^{1/2}u\|^2\le\||\mathbf{B}|\|_{L^{3/2}(\mathbb{R}^3)}\|u\|^2_{L^6(\mathbb{R}^3;\mathbb{C}^2)}\le C\|u\|^2_{\mathbb{H}^1_{\mathbf{B}}}, \tag{5.9.16}$$

for some positive constant $C$, by (5.9.9) and (5.9.10).

The results we seek rely on the properties of the **Birman Schwinger** type operator $\mathcal{S}\mathcal{S}^*$. For $f\in\mathcal{R}(\mathbb{P}^{1/2}),\ g\in\mathcal{D}(\mathbb{P}^{1/2})$,

$$\begin{aligned}(f,\mathcal{S}^*f,g)&=(\mathcal{S}f,g)=(|\mathbf{B}|^{1/2}Uf,g)\\&=(Uf,|\mathbf{B}|^{1/2}g)\\&=(\mathbb{P}^{-1/2}f,|\mathbf{B}|^{1/2}g);\end{aligned}$$

note that $|\mathbf{B}|^{1/2}g \in \mathcal{H}$ by (5.9.16). Hence $|\mathbf{B}|^{1/2}g \in \mathcal{D}(\mathbb{P}^{-1/2})$ and $\mathbb{P}^{-1/2}|\mathbf{B}|^{1/2}g = \mathcal{S}^*g$. In other words

$$\mathcal{S}^* = \mathbb{P}^{-1/2}|\mathbf{B}|^{1/2} \quad \text{on } \mathcal{D}(\mathbb{P}^{1/2}), \tag{5.9.17}$$

Hence

$$\mathcal{S}\mathcal{S}^* = |\mathbf{B}|^{1/2}U^2|\mathbf{B}|^{1/2} \quad \text{on } \mathcal{D}(\mathbb{P}^{1/2}), \tag{5.9.18}$$

and this extends by continuity to a bounded operator on $\mathcal{H}$.

**Lemma 5.9.2** *If* $|\mathbf{B}| \in L^{3/2}(\mathbb{R}^3)$, *the operator* $\mathcal{S}\mathcal{S}^* : \mathcal{H} \to \mathcal{H}$ *is compact.*

*Proof* It is sufficient to prove that $|\mathbf{B}|^{1/2} : \mathbb{H}^1_{\mathbf{B}} \to \mathcal{H}$ is compact, for then it will follow that $\mathcal{S} = |\mathbf{B}|^{1/2}U : \mathcal{H} \to \mathcal{H}$ is compact. Let $\{\psi_n\}$ be a sequence which converges weakly to zero in $\mathbb{H}^1_{\mathbf{B}}$, and hence $|\psi_n| \rightharpoonup 0$ in $H^1_{\mathbf{0}}$ by (5.9.9) and (5.9.10). Then, in particular, $\||\psi_n|\|_{H^1_{\mathbf{0}}} \leq k$ for some constant $k$. Given $\varepsilon > 0$, set $|\mathbf{B}| = B_1 + B_2$, where $B_1 \in C_0^\infty(\mathbb{R}^3)$ with support $\Omega_\varepsilon$ and $|\mathbf{B}| \leq k_\varepsilon$ say, and $\|\mathbf{B}_2\|_{L^{3/2}(\mathbb{R}^3)} < \varepsilon$. Then

$$\begin{aligned}\||\mathbf{B}|^{1/2}\psi_n\|^2 &\leq k_\varepsilon\|\psi_n\|^2_{L^2(\Omega_\varepsilon;\mathbb{C}^2)} + \gamma^2\|\mathbf{B}_2\|_{L^{3/2}(\mathbb{R}^3)}\||\psi_n|\|^2_{H^1_{\mathbf{0}}} \\ &\leq k_\varepsilon\|\psi_n\|^2_{L^2(\Omega_\varepsilon;\mathbb{C}^2)} + \gamma^2\varepsilon\||\psi_n|\|^2_{H^1_{\mathbf{0}}}.\end{aligned}$$

The first term tends to zero as $n \to \infty$ by the Rellich-Kondrachov theorem. Hence

$$\limsup_{n\to\infty} \||\mathbf{B}|^{1/2}\psi_n\|^2 \leq k\gamma^2\varepsilon.$$

Since $\varepsilon$ is arbitrary, it follows that $|\mathbf{B}|^{1/2} : \mathbb{H}^1_{\mathbf{B}} \to \mathcal{H}$ is compact, and the lemma is proved. □

The last result we need before giving the main theorem in this section concerns the number of zero modes of the Pauli operator $\mathbb{P}_{\mathbf{A}}$, that is, the dimension of the kernel of $\mathbb{P}_{\mathbf{A}}$; this is denoted by nul $\mathbb{P}_{\mathbf{A}}$ where nul stands for *nullity*.

**Lemma 5.9.3**

$$\begin{aligned}\text{nul } \mathbb{P}_{\mathbf{A}} &= \dim\{u : \mathcal{B}u = u,\ u \in \mathbb{H}^1_{\mathbf{B}} \cap \mathcal{H}\} \\ &\leq \text{nul } F,\end{aligned}$$

*where* $F = 1 - \mathcal{S}\mathcal{S}^*$. *There is equality if and only if*

$$Fu = 0 \Rightarrow Uu \in \mathbb{H}^1_{\mathbf{B}} \cap \mathcal{H}.$$

*Proof* Let $u, v \in \mathcal{D}(\mathbb{P}^{1/2})$. Then

$$\begin{aligned} p_{\mathbf{A}}[u, v] &= p[u, v] - b[u, v] \\ &= (u - \mathcal{B}u, v)_{\mathbb{H}^1_{\mathbf{B}}}. \end{aligned}$$

Hence $u \in \ker \mathbb{P}_{\mathbf{A}} \subset \mathcal{D}(\mathbb{P}^{1/2})$, if and only if, $\mathcal{B}u = u$, with $u \in \mathcal{H}$. Moreover, for any $f, g \in \mathcal{H}$,

$$(\mathcal{S}f, \mathcal{S}g) = (\mathcal{B}Uf, Ug)_{\mathbb{H}^1_{\mathbf{B}}},$$

whence

$$([\mathcal{S}^*\mathcal{S} - 1]f, g) = ([\mathcal{B} - 1]Uf, Ug)_{\mathbb{H}^1_{\mathbf{B}}}.$$

Since nul $[\mathcal{S}^*\mathcal{S} - 1] =$ nul $[\mathcal{S}\mathcal{S}^* - 1]$, we see that

$$\mathbb{P}_{\mathbf{A}} u = 0 \quad \Leftrightarrow \quad F(U^{-1}u) = 0,$$

where it is understood that $u \in \mathcal{D}(\mathbb{P}_{\mathbf{A}}) \subset \mathbb{H}^1_{\mathbf{B}}$. The lemma follows. □

The following theorem is proved in [19]

**Theorem 5.9.4** *Let* $|\mathbf{A}| \in L^3(\mathbb{R}^3)$ *and* $|\mathbf{B}| \in L^{3/2}(\mathbb{R}^3)$. *Let* $\mathcal{S}$ *be given by (5.9.15) and* $F = 1 - \mathcal{S}\mathcal{S}^*$. *Suppose that for some* $\mathbf{B}$, *the operator* $F$ *has no zero mode, and set*

$$\delta(\mathbf{B}) := \inf_{\|f\|=1,\ Uf \in \mathbb{H}^1_{\mathbf{B}} \cap \mathcal{H}} \|[1 - \mathcal{S}^*\mathcal{S}]f\|^2. \tag{5.9.19}$$

*Then* $\delta(\mathbf{B}) > 0$ *and*

$$\mathbb{P}_{\mathbf{A}} \geq \delta(\mathbf{B}) \mathbb{S}_{\mathbf{A}}. \tag{5.9.20}$$

*The following hold for all* $\varphi \in \mathcal{D}(\mathbb{P}^{\frac{1}{2}}_{\mathbf{A}}) = H^1(\mathbb{R}^3, \mathbb{C}^2)$ :

(i) *(Sobolev-type inequality)*

$$\left\| \mathbb{P}^{\frac{1}{2}}_{\mathbf{A}} \varphi \right\|^2 \geq \frac{\delta(\mathbf{B})}{\gamma^2} \|\varphi\|^2_{[L^6(\mathbb{R}^3)]^2}, \tag{5.9.21}$$

*where* $\gamma$ *is the norm of the embedding* $H^1(\mathbb{R}^3) \hookrightarrow [L^6(\mathbb{R}^3)]^2$;

(ii) *(Hardy-type inequality)*

$$\|\mathbb{P}^{\frac{1}{2}}_{\mathbf{A}} \varphi\|^2 \geq \frac{\delta(\mathbf{B})}{4} \left\| \frac{\varphi}{|\cdot|} \right\|^2; \tag{5.9.22}$$

(iii) *(CLR-type inequality) for $V_- \in L^{\frac{3}{2}}(\mathbb{R}^3)$, the number $N(\mathbb{P}_{\mathbf{A}} + V)$ of negative eigenvalues $\{-\lambda_n\}$ of $\mathbb{P}_{\mathbf{A}} + V$ satisfies*

$$N(\mathbb{P}_{\mathbf{A}} + V) \leq \frac{c}{\delta(\mathbf{B})^{\frac{3}{2}}} \int_{\mathbb{R}^3} V_-^{\frac{3}{2}} d\mathbf{x}, \tag{5.9.23}$$

*where $c$ is the best constant in the CLR inequality for $\mathbb{S}_{\mathbf{A}}$;*

(iv) *(Lieb-Thirring-type inequalities)*

$$\sum_n \lambda_n^\nu \leq \frac{c}{\delta(\mathbf{B})^{\frac{3}{2}}} \int_{\mathbb{R}^3} V_-^{\nu+\frac{3}{2}} d\mathbf{x} \tag{5.9.24}$$

*for any $\nu \geq 0$.*

*Proof* If $F$ has no zero mode, the compact operator $\mathcal{S}\mathcal{S}^*$ on $\mathcal{H}$ does not have eigenvalue 1 and hence neither does $\mathcal{S}^*\mathcal{S}$, since

$$\sigma(\mathcal{S}\mathcal{S}^*) \setminus \{0\} = \sigma(\mathcal{S}^*\mathcal{S}) \setminus \{0\};$$

see [141], Sect. 5.2. Hence, $\delta(\mathbf{B}) > 0$, and for any $f \in \mathcal{H}$ with $Uf \in \mathcal{H} \cap \mathbb{H}^1_{\mathbf{B}}$,

$$\begin{aligned}\delta(\mathbf{B})\|f\|^2 &\leq \|(1 - \mathcal{S}^*\mathcal{S})f\|^2\\ &= \|f\|^2 - 2(\mathcal{S}^*\mathcal{S}f, f) + \|\mathcal{S}^*\mathcal{S}f\|^2.\end{aligned}$$

Let $f = \mathbb{P}^{\frac{1}{2}}\varphi$ for $\varphi \in \mathcal{H}$. Then $Uf = \varphi$ and $\mathcal{S}f = |\mathbf{B}|^{\frac{1}{2}}\varphi$ from (5.9.14). Consequently

$$\begin{aligned}\delta(\mathbf{B})\|\mathbb{P}^{\frac{1}{2}}\varphi\|^2 &\leq \|\mathbb{P}^{\frac{1}{2}}\varphi\|^2 - 2\||\mathbf{B}|^{\frac{1}{2}}\varphi\|^2 + \|\mathcal{S}^*|\mathbf{B}|^{\frac{1}{2}}\varphi\|^2\\ &= \|\mathbb{P}_{\mathbf{A}}{}^{\frac{1}{2}}\varphi\|^2 - \||\mathbf{B}|^{\frac{1}{2}}\varphi\|^2 + \|\mathcal{S}^*|\mathbf{B}|^{\frac{1}{2}}\varphi\|^2,\end{aligned} \tag{5.9.25}$$

since $\mathbb{P} = \mathbb{P}_{\mathbf{A}} + |\mathbf{B}|$ in the form sense. Also if $g \in \mathcal{R}(\mathbb{P}^{\frac{1}{2}})$,

$$\|\mathcal{S}g\| = \||\mathbf{B}|^{\frac{1}{2}}\mathbb{P}^{-\frac{1}{2}}g\| \leq \|g\|$$

as $\mathbb{P} \geq |\mathbf{B}|$. Since $\mathcal{R}(\mathbb{P}^{\frac{1}{2}})$ is dense in $\mathcal{H}$, we have that $\|\mathcal{S}^*\| = \|\mathcal{S}\| \leq 1$. It follows from (5.9.25) that

$$\delta(\mathbf{B})\|\mathbb{P}^{\frac{1}{2}}\varphi\|^2 \leq \|\mathbb{P}_{\mathbf{A}}{}^{\frac{1}{2}}\varphi\|^2,$$

whence $\mathbb{P}_{\mathbf{A}} \geq \delta(\mathbf{B})\mathbb{P} \geq \delta(\mathbf{B})\mathbb{S}_{\mathbf{A}}$.

The inequalities exhibited are now consequences of (5.9.20) and the corresponding standard inequalities featuring $-\Delta$, on using the diamagnetic inequality. □

*Remark 5.9.5* If any one of the inequalities (5.9.21)–(5.9.24) is satisfied, then $\mathbb{P}_{\mathbf{A}}$ has no zero mode. Whether or not nul $\mathbb{P}_{\mathbf{A}} = 0$ implies nul $F = 0$ is not clear. It is of interest to observe that in (5.9.19), the infimum is taken over the subspace of $\mathcal{H}$ in which $\mathbb{P}_{\mathbf{A}}$ and $F$ have common nullity.

In [27] it is proved that if $\mathbf{B} \in L^{3/2}(\mathbb{R}^3)$ is such that $\delta(\mathbf{B}) = 0$ and there exist $\beta > 0, C \geq 0$ and $r_0 \geq 0$ such that

$$|\mathbf{B}(\mathbf{x})| \leq C|\mathbf{x}|^{-2-\beta}$$

for all $|\mathbf{x}| \geq r_0$, then the associated Pauli operator $\mathbb{P}_{\mathbf{A}}$ has a zero mode.

# Chapter 6
# The Rellich Inequality

## 6.1 Introduction

In lectures delivered at New York University in 1953, and published posthumously in the proceedings [128] of the International Congress of Mathematicians held in Amsterdam in 1954, Rellich proved the following inequality which bears his name: for $n \neq 2$

$$\int_{\mathbb{R}^n} |\Delta u(\mathbf{x})|^2 d\mathbf{x} \geq \frac{n^2(n-4)^2}{16} \int_{\mathbb{R}^n} \frac{|u(\mathbf{x})|^2}{|\mathbf{x}|^4} d\mathbf{x}, \quad u \in C_0^\infty(\mathbb{R}^n \setminus \{0\}), \tag{6.1.1}$$

while for $n = 2$, the inequality continues to hold but for a restricted class of functions $u$; see Remark 6.4.4 below.

Since Rellich's proof, many versions of the inequality have been proven in various settings, and we present some of these in this chapter. First, we review methods due to Schminke [133] and Bennett [29] for proving the inequality in $L^2(\mathbb{R}^n)$, $n > 4$, which are different from those used in [128, 129]. These motivated the proof of a weighted $L^p(\mathbb{R}^n)$ version of the inequality for any $1 < p < \infty$, obtained by Davies and Hinz in [45]; we give a full account of the proof in [45], and also note an earlier paper by Okazawa [125] in which a more general inequality is proved and then applied to establishing the Rellich inequality and the accretiveness of Schrödinger operators in $L^p(\mathbb{R}^n)$, (see Lemmas 3.5 and 3.8 of [125]). Rellich-Sobolev inequalities derived by Frank (private communication, 2007) for convex domains are then discussed and shown to hold on weakly mean convex domains. Finally, a Rellich inequality in $L^2(\mathbb{R}^n)$ with magnetic potentials is proved by methods reminiscent of those in Rellich's original proof. These methods provide a path to studying the eigenvalues of biharmonic operators $-\Delta_{\mathbf{A}}^2 - V$, where $\mathbf{A}$ is a magnetic potential of Aharonov-Bohm type. Bounds for the number of negative eigenvalues are established which depend upon the magnetic flux $\tilde{\Psi}$ of $\mathbf{A}$.

 
A.A. Balinsky et al., *The Analysis and Geometry of Hardy's Inequality*,
Universitext, DOI 10.1007/978-3-319-22870-9_6

## 6.2 Rellich and Rellich-Sobolev Inequalities in $L^2$

### 6.2.1 The Rellich Inequality

The following proof of an inequality in $L^2(\mathbb{R}^n)$ associated with the Rellich inequality is much in the spirit of an elegant proof given by Schmincke [133], (and a generalisation by Bennett in [29]), which recovers Rellich's inequality for $n > 4$. It resembles the Hardy inequality in Theorem 1.2.8 .

**Theorem 6.2.1** *Let $\Omega \subseteq \mathbb{R}^n$ for $n \geq 2$. If a real-valued function $V \in C^2(\Omega)$ and $\Delta V(\mathbf{x}) < 0$, then for all $u \in C_0^\infty(\Omega)$ and any $\varepsilon > 0$,*

$$\int_\Omega \frac{|V|^2}{|\Delta V|}|\Delta u|^2 d\mathbf{x} \geq 2\varepsilon \int_\Omega V|\nabla u|^2 d\mathbf{x} + \varepsilon(1-\varepsilon)\int_\Omega |\Delta V||u|^2 d\mathbf{x}. \tag{6.2.1}$$

*Proof* Since $\Delta V(\mathbf{x}) < 0$ in $\Omega$,

$$\int_\Omega |\Delta V||u|^2 d\mathbf{x} = -\int_\Omega V\Delta|u|^2 d\mathbf{x} = -2\int_\Omega V[\mathrm{Re}(\overline{u}\Delta u) + |\nabla u|^2]d\mathbf{x}.$$

Then

$$\begin{aligned}\int_\Omega |\Delta V||u|^2 d\mathbf{x} &\leq 2[\frac{1}{\varepsilon}\int_\Omega \frac{|V|^2}{|\Delta V|}|\Delta u|^2 d\mathbf{x}]^{\frac{1}{2}}[\varepsilon\int_\Omega |\Delta V||u|^2 d\mathbf{x}]^{\frac{1}{2}} \\ &\quad -2\int_\Omega V|\nabla u|^2 d\mathbf{x} \\ &\leq \frac{1}{\varepsilon}\int_\Omega \frac{|V|^2}{|\Delta V|}|\Delta u|^2 d\mathbf{x} + \varepsilon\int_\Omega |\Delta V||u|^2 d\mathbf{x} \\ &\quad -2\int_\Omega V|\nabla u|^2 d\mathbf{x}\end{aligned} \tag{6.2.2}$$

and the conclusion follows. □

**Corollary 6.2.2** *If $n > \alpha + 4$ for some $\alpha > -2$, then*

$$\int_{\mathbb{R}^n\setminus\{0\}} \frac{|\Delta u|^2}{|\mathbf{x}|^\alpha} d\mathbf{x} \geq \frac{(n+\alpha)^2(n-\alpha-4)^2}{16}\int_{\mathbb{R}^n\setminus\{0\}} \frac{|u|^2}{|\mathbf{x}|^{\alpha+4}} d\mathbf{x} \tag{6.2.3}$$

*for all $u \in C_0^\infty(\mathbb{R}^n \setminus \{0\})$.*

*Proof* Let $V(\mathbf{x}) = |\mathbf{x}|^{-(\alpha+2)}$ in Theorem 6.2.1, and so

$$\nabla V = -(\alpha+2)|\mathbf{x}|^{-\alpha-4}\mathbf{x}, \quad \Delta V = -(\alpha+2)(n-\alpha-4)|\mathbf{x}|^{-(\alpha+4)}.$$

Set $c(n,\alpha) = (\alpha+2)(n-\alpha-4)$, and observe that $\Delta V = -c(n,\alpha)|\mathbf{x}|^{-(\alpha+4)} < 0$ for $|\mathbf{x}| \neq 0$. It follows from (6.2.1) that

$$\int_{\mathbb{R}^n\setminus\{0\}} \frac{|\Delta u|^2}{|\mathbf{x}|^\alpha} d\mathbf{x} \geq 2c(n,\alpha)\varepsilon \int_{\mathbb{R}^n\setminus\{0\}} \frac{|\nabla u|^2}{|\mathbf{x}|^{\alpha+2}} d\mathbf{x} + c(n,\alpha)^2 \varepsilon(1-\varepsilon) \int_{\mathbb{R}^n\setminus\{0\}} \frac{|u|^2}{|\mathbf{x}|^{\alpha+4}} d\mathbf{x}. \tag{6.2.4}$$

To derive (6.2.3), we first apply the Hardy inequality (1.2.19) with $V(\mathbf{x}) = |\mathbf{x}|^{-(\alpha+2)}$ to get that, for $\alpha \in (-2, n-4)$,

$$\int_{\mathbb{R}^n\setminus\{0\}} \frac{|\nabla u|^2}{|\mathbf{x}|^{\alpha+2}} d\mathbf{x} \geq \frac{(n-\alpha-4)^2}{4} \int_{\mathbb{R}^n\setminus\{0\}} \frac{|u|^2}{|\mathbf{x}|^{\alpha+4}} d\mathbf{x},$$

and then let $\varepsilon = (n+\alpha)/4(\alpha+2)$ in (6.2.4), which is the choice of $\varepsilon$ that gives the maximum right-hand side. □

*Remark 6.2.3* The case $\alpha = 0$ of (6.2.3) is, of course, the Rellich inequality (6.1.1) for $n > 4$. Following the original technique of Rellich, Allegretto proved (6.2.3) in [6], Corollary 3, but required $\alpha \geq 0$ and $n \geq 1 + \sqrt{(\alpha+1)(\alpha+3)}$. That result is a corollary of Theorem 6.4.1 below; see Remark 6.4.2.

*Remark 6.2.4* Note that Corollary 6.2.2 is valid for a different range of values $\alpha$ than Corollary 3 in [6]. For example, Corollary 6.2.2 holds for $n = 3$ or $n = 4$ if $\alpha \in (-2, n-4)$, whereas Allegretto's Corollary 3 does not apply. When $n \geq 5$, Corollary 6.2.2 is applicable for $\alpha \in (-2, n-4)$ and Allegretto requires $0 \leq \alpha \leq -2 + \sqrt{(n-1)^2+1}$. Thus, on combining these results we are able to conclude that for $n \geq 5$, (6.2.3) holds for $\alpha \in (-2, -2+\sqrt{(n-1)^2+1})$. The weighted Rellich inequality in $L^p(\mathbb{R}^n)$, $p \in (1,\infty)$, corresponding to Corollary 6.2.2 will be given in Corollary 6.3.4.

In [34], an inequality similar to (6.2.3) is obtained for all $u \in C_0^\infty(\Omega\setminus\{0\})$, where $\Omega$ is a cone $\{r\sigma : r > 0,\ \sigma \in \Sigma\}$, with $\Sigma$ a domain with $C^2$ boundary in the unit sphere. Their inequality holds for all real $\alpha$ and the constant is sharp.

*Remark 6.2.5* We shall return to the cases $n = 2, 3,$ and 4 of the Rellich inequality in Sect. 6.4.2 below. On page 91 in [129], it is proved that when $n = 2$, the Rellich inequality still holds, but only for functions $u \in C_0^\infty(\mathbb{R}^2 \setminus \{0\})$ which also satisfy

$$\int_0^{2\pi} u(r,\theta)\ \cos\ \theta d\theta = \int_0^{2\pi} u(r,\theta)\ \sin\ \theta d\theta = 0. \tag{6.2.5}$$

This is recovered within the general result proved in Sect. 6.4.

### 6.2.2 Rellich-Sobolev Inequalities

In a private communication, Rupert Frank showed how to prove certain refinements of the Rellich inequality. In order to present these here, we first need a corollary of the Hardy-type inequality (1.2.19): we recall, for convenience, that for $\Delta V \le 0$ in the distributional sense, (1.2.19) is

$$\int_\Omega |\Delta V||u|^2 d\mathbf{x} \le 4\int_\Omega \frac{|\nabla V|^2}{|\Delta V|}|\nabla u|^2 d\mathbf{x}, \quad u \in C_0^\infty(\Omega).$$

For $\alpha \neq 0$, make the substitution, $V(\mathbf{x}) = -[(\alpha+1)/\alpha]\delta(\mathbf{x})^{-\alpha}$, and for $\alpha = 0$, let $V(\mathbf{x}) = \ln\, \delta(\mathbf{x})$. Then, $|\nabla V|^2 = (\alpha+1)^2\delta^{-2(\alpha+1)}$, and when $\Delta\delta \le 0$,

$$-\Delta V = (\alpha+1)^2\delta^{-(\alpha+2)} + (\alpha+1)\delta^{-(\alpha+1)}(-\Delta\delta) \ge (\alpha+1)^2\delta^{-(\alpha+2)},$$

whence

**Corollary 6.2.6** *Let $\Omega \subset \mathbb{R}^n$, $n \ge 2$, be such that $\Delta\delta \le 0$ in the distributional sense. Then, if $\alpha > -1$,*

$$(\alpha+1)^2\int_\Omega \frac{|u|^2}{\delta(\mathbf{x})^{\alpha+2}} d\mathbf{x} \le 4\int_\Omega \frac{|\nabla u|^2}{\delta(\mathbf{x})^\alpha} d\mathbf{x}, \qquad u \in C_0^\infty(\Omega).$$

This yields

**Corollary 6.2.7** *Let $\Omega \subset \mathbb{R}^n$, $n \ge 2$, be such that $\Delta\delta \le 0$ in the distributional sense. Then*

$$\int_\Omega |\Delta u|^2 d\mathbf{x} \ge \frac{9}{16}\int_\Omega \frac{|u|^2}{\delta^4} d\mathbf{x}, \qquad u \in C_0^\infty(\Omega), \tag{6.2.6}$$

*Proof* We first claim that

$$\int_\Omega |\Delta u|^2 d\mathbf{x} = \sum_{j=1}^n \int_\Omega |\nabla(\partial_j u)|^2 d\mathbf{x}, \quad u \in C_0^\infty(\Omega). \tag{6.2.7}$$

where $\partial_j := \partial/\partial x_j$. For, with the notation $u_j = \partial_j u$, $u_{jk} = \partial_j\partial_k u$ and $u_{jkl} = \partial_j\partial_k\partial_l u$, we have

$$\begin{aligned}
\int_\Omega |\Delta u|^2 d\mathbf{x} &= \sum_{j,k=1}^n \int_\Omega u_{jj}\overline{u}_{kk} d\mathbf{x} \\
&= \sum_{j=k}\int_\Omega |u_{jj}|^2 d\mathbf{x} - \sum_{j\neq k}\int_\Omega u_j\overline{u}_{jkk} d\mathbf{x} \\
&= \sum_{j=k}\int_\Omega |u_{jj}|^2 d\mathbf{x} + \sum_{j\neq k}\int_\Omega u_{jk}\overline{u}_{jk} d\mathbf{x}
\end{aligned}$$

and hence (6.2.7). On applying Corollary 6.2.6 twice to (6.2.7), the first time with $\alpha = 0$ and then with $\alpha = 2$, the corollary follows. □

*Remark 6.2.8* If $\Omega$ is convex, then we proved in Theorem 2.3.2 that the hypothesis of Corollary 6.2.7 is satisfied. Also in Corollary 3.7.11, we showed that $\Omega$ being weakly mean convex with null cut locus is sufficient for the hypothesis to be satisfied. The weak mean convexity of $\Omega$ was shown in Proposition 2.5.4 to be equivalent to $\Delta\delta \leq 0$ in $G(\Omega)$.

We can also use Corollaries 6.2.6 and 6.2.7 to obtain inequalities which are analogous to the HSM inequalities of Chap. 4.

**Proposition 6.2.9** *Let $\Omega \subsetneq \mathbb{R}^n$, $n \geq 3$, be convex with inradius $\delta_0 := \sup_\Omega \delta < \infty$. Then, for all $u \in C_0^\infty(\Omega)$,*

$$\int_\Omega \left( |\Delta u|^2 - \frac{9}{16}\frac{|u|^2}{\delta^4} \right) d\mathbf{x} \geq \frac{\lambda_0^2}{4\delta_0^2} \int_\Omega \frac{|u|^2}{\delta^2} d\mathbf{x} + \frac{\lambda_0^4}{\delta_0^4} \int_\Omega |u|^2 d\mathbf{x}, \tag{6.2.8}$$

*where $\lambda_0$ is the first zero in $(0, \infty)$ of $J_0(x) - 2xJ_1(x)$.*

*Proof* From (6.2.7) and the Avkhadiev/Wirths result reproduced in Theorem 3.6.12,

$$\begin{aligned}
\int_\Omega |\Delta u|^2 d\mathbf{x} &\geq \sum_{j=1}^n \left\{ \frac{1}{4} \int_\Omega \frac{|\partial_j u|^2}{\delta^2} d\mathbf{x} + \frac{\lambda_0^2}{\delta_0^2} \int_\Omega |\partial_j u|^2 d\mathbf{x} \right\} \\
&= \frac{1}{4} \int_\Omega \frac{|\nabla u|^2}{\delta^2} d\mathbf{x} + \frac{\lambda_0^2}{\delta_0^2} \int_\Omega |\nabla u|^2 d\mathbf{x} \\
&\geq \frac{9}{16} \int_\Omega \frac{|u|^2}{\delta^4} d\mathbf{x} + \frac{\lambda_0^2}{\delta_0^2} \left\{ \frac{1}{4} \int_\Omega \frac{|u|^2}{\delta^2} d\mathbf{x} + \frac{\lambda_0^2}{\delta_0^2} \int_\Omega |u|^2 d\mathbf{x} \right\},
\end{aligned}$$

whence (6.2.8). □

The assumption that $\Omega$ be convex can be dropped in favour of weak mean convexity if we use Corollary 3.7.16 instead of the Akhadiev/Wirths theorem.

**Proposition 6.2.10** *Let $\Omega$ be a weakly mean convex domain in $\mathbb{R}^n$, $n \geq 2$, with null cut locus. Then*

$$\int_\Omega \left( |\Delta u|^2 - \frac{9}{16}\frac{|u|^2}{\delta^4} \right) d\mathbf{x} \geq \frac{\lambda(n, \Omega)^2}{4} \int_\Omega \frac{|u|^2}{\delta^2} d\mathbf{x} + \lambda(n, \Omega)^4 \int_\Omega |u|^2 d\mathbf{x}, \tag{6.2.9}$$

*where $\lambda(n, \Omega) = (1/2) \inf_G \delta^{-1}(-\Delta\delta) \geq 2(n-1) \inf_{\partial\Omega} |H|^2$.*

From Corollary 4.3.2, if $\Omega \subsetneq \mathbb{R}^n$, $n \geq 3$, is convex, there exist a constant $K$ such that

$$\int_\Omega \left( |\nabla u|^2 - \frac{1}{4}\frac{|u|^2}{\delta^2} \right) d\mathbf{x} \geq K \left( \int_\Omega |u|^{\frac{2n}{n-2}} d\mathbf{x} \right)^{\frac{n-2}{n}} \tag{6.2.10}$$

for all $u \in C_0^\infty(\Omega)$. From this follows

**Proposition 6.2.11** *Let $\Omega \subsetneqq \mathbb{R}^n, n \geq 5$, be convex. Then*

$$\int_\Omega \left( |\Delta u|^2 - \frac{9}{16} \frac{|u|^2}{\delta^4} \right) d\mathbf{x} \geq K \left( \int_\Omega |u|^{2n/(n-4)} d\mathbf{x} \right)^{(n-4)/n}. \tag{6.2.11}$$

*for all $u \in C_0^\infty(\Omega)$.*

*Proof* The proposition follows from the use of (6.2.7), (6.2.10) and the Sobolev inequality

$$\left( \int_\Omega |u|^{2n/(n-4)} d\mathbf{x} \right)^{(n-4)/n} \leq C_n \sum_{j=1}^n \left( \int_\Omega \left| \frac{\partial u}{\partial x_j} \right|^{2n/(n-2)} d\mathbf{x} \right)^{(n-2)/n}$$

obtained from (1.3.6) with $p = 2n/(n-2)$. □

A similar inequality to (6.2.11) can be obtained for bounded weakly mean convex domains from Theorem 4.4.4.

## 6.3 The Rellich Inequality in $L^p(\mathbb{R}^n)$, $n \geq 2$

In [45], Davies and Hinz obtain a Rellich-type inequality in $L^p(\Omega)$, $1 \leq p < \infty$, when $\Omega$ is a bounded region in a complete Riemann manifold. A consequence of their result is the inequality

$$\int_{\mathbb{R}^n} \frac{|u(\mathbf{x})|^p}{|\mathbf{x}|^\beta} d\mathbf{x} \leq c(d,m,p,\beta)^p \int_{\mathbb{R}^n} \frac{|\Delta^m u(\mathbf{x})|^p}{|\mathbf{x}|^{\beta-2mp}} d\mathbf{x}, \quad u \in C_0^\infty(\mathbb{R}^n \setminus \{0\}), \tag{6.3.1}$$

for $2(1+(m-1)p) < \beta < n$, with an explicit constant $c(d,m,p,\beta)$ which is shown to be sharp. A special case is the Rellich inequality in $L^p(\mathbb{R}^n \setminus \{0\})$ for $n > 2p$. An earlier proof of this Rellich inequality was in fact established by Okazawa in [125]; see also [92]. However the proof we shall give is that in [45], because it is more in line with our overall approach. Okazawa's main concern is with an analysis of the operator $-\Delta + 1/|\mathbf{x}|^2$ in $L^p(\mathbb{R}^n)$, and determining when it is $m$-accretive and $m$-sectorial in Kato's sense.

The first lemma is a basic tool in the approach, and is motivated by Theorem 1.2.8.

**Lemma 6.3.1** *If $V \geq 0$, $\Delta V < 0$, and there is a constant $c \geq 0$ such that for all $u \in C_0^\infty(\Omega)$*

$$c \int_\Omega |\Delta V||u|^p d\mathbf{x} \leq p(p-1) \int_{\{\mathbf{x}\in\Omega, u(\mathbf{x})\neq 0\}} V|u|^{p-2}|\nabla u|^2 d\mathbf{x}, \tag{6.3.2}$$

*then*

$$(1+c)^p \int_\Omega |\Delta V||u|^p d\mathbf{x} \leq p^p \int_\Omega \frac{V^p}{|\Delta V|^{p-1}} |\Delta u|^p d\mathbf{x}, \qquad u \in C_0^\infty(\Omega). \tag{6.3.3}$$

*If* $p = 1$ *then the lemma holds for* $c = 0$.

*Proof* The first step is to show that we may assume that the functions $u$ are real-valued. To prove this, we use the following identity from [40]: for all $z \in \mathbb{C}$,

$$|z|^p = A \int_{-\pi}^{\pi} |\mathrm{Re}[z] \cos\theta + \mathrm{Im}[z] \sin\theta|^p d\theta, \;\; A := \left( \int_{-\pi}^{\pi} |\cos\theta|^p d\theta \right)^{-1}. \tag{6.3.4}$$

It is proved by putting $z = r(\cos\gamma + i \sin\gamma)$, $r = |z|$, and simplifying. Suppose that (6.3.3) has been proved for real-valued functions. Then, on setting $u = u_1 + iu_2$, we have by (6.3.4), and changing the order of integration,

$$\begin{aligned} &(1+c)^p \int_\Omega |\Delta V||u|^p d\mathbf{x} \\ &= \quad (1+c)^p A \int_{-\pi}^{\pi} \int_\Omega |\Delta V||u_1 \cos\theta + u_2 \sin\theta|^p d\mathbf{x} d\theta \\ &\leq \quad p^p A \int_{-\pi}^{\pi} \int_\Omega \frac{V^p}{|\Delta V|^{p-1}} |\Delta[u_1 \cos\theta + u_2 \sin\theta]|^p d\mathbf{x} d\theta, \\ &= \quad p^p \int_\Omega \frac{V^p}{|\Delta V|^{p-1}} |\Delta u|^p d\mathbf{x}, \qquad u \in C_0^\infty(\Omega), \end{aligned}$$

on using (6.3.4) again. The claim is therefore verified, and we assume hereafter that $u$ is real-valued.

Let $\varepsilon > 0$ and set $u_\varepsilon := (|u|^2 + \varepsilon^2)^{p/2} - \varepsilon^p$. Then $0 \leq u_\varepsilon \in C_0^\infty$ and

$$\begin{aligned} \int_\Omega |\Delta V| u_\varepsilon d\mathbf{x} &= -\int_\Omega (\Delta V) u_\varepsilon d\mathbf{x} = -\int_\Omega V \Delta u_\varepsilon d\mathbf{x} \\ &= -\int_\Omega V \left\{ p(p-2)u^2(u^2+\varepsilon^2)^{(p-4)/2} + p(u^2+\varepsilon^2)^{(p-2)/2} \right\} |\nabla u|^2 d\mathbf{x} \\ &\quad - p \int_\Omega Vu(u^2+\varepsilon^2)^{(p-2)/2} \Delta u d\mathbf{x}. \end{aligned}$$

Hence

$$\begin{aligned} &\int_\Omega \left[ |\Delta V| u_\varepsilon + V \left\{ p(p-2)u^2(u^2+\varepsilon^2)^{(p-4)/2} + p(u^2+\varepsilon^2)^{(p-2)/2} \right\} \right] |\nabla u|^2 d\mathbf{x} \\ &\qquad \leq p \int_\Omega V|u|(u^2+\varepsilon^2)^{(p-2)/2} |\Delta u| d\mathbf{x}. \end{aligned}$$

We now let $\varepsilon \to 0$. The integrand on the right is bounded by $V(\max|u|^2+1)^{(p-1)/2}\max|\Delta u|$, which is integrable since $u \in C_0^\infty(\Omega)$, and so the integral tends to $\int_\Omega V|u|^{p-1}|\Delta u|d\mathbf{x}$, by dominated convergence. The integrand on the left is non-negative and tend to $|\Delta V||u|^p + p(p-1)V|u|^{p-2}|\nabla u|^2$ pointwise, only for $u(\mathbf{x}) \neq 0$ when $p < 2$, otherwise for all $\mathbf{x}$. It follows by Fatou's lemma that

$$\int_\Omega |\Delta V||u|^p d\mathbf{x} + \int_{\{\mathbf{x}\in\Omega;u(\mathbf{x})\neq 0\}} \left\{p(p-1)V|u|^{p-2}|\nabla u|^2\right\} d\mathbf{x} \\ \leq p\int_\Omega V|u|^{p-1}|\Delta u|d\mathbf{x},$$

and on substituting (6.3.2), followed by Hölder's inequality,

$$(1+c)\int_\Omega |\Delta V||u|^p d\mathbf{x} \leq p\int_\Omega V|u|^{p-1}|\Delta u|d\mathbf{x} \\ \leq p\left(\int_\Omega |\Delta V||u|^p d\mathbf{x}\right)^{\frac{p-1}{p}}\left(\int_\Omega \frac{V^p}{|\Delta V|^{p-1}}|\Delta u|^p d\mathbf{x}\right)^{\frac{1}{p}}.$$

The lemma follows from this. □

The next lemma meets the requirement (6.3.2).

**Lemma 6.3.2** *Let $p \in (1,\infty)$. If $0 < V \in C(\Omega)$, $\Delta V < 0$ and $\Delta V^\delta \leq 0$ for some $\delta > 1$, then*

$$(\delta-1)\int_\Omega |\Delta V||u|^p d\mathbf{x} \leq p^2 \int_{\{\mathbf{x}\in\Omega, u(\mathbf{x})\neq 0\}} V|u|^{p-2}|\nabla u|^2 d\mathbf{x} < \infty$$

*for all $u \in C_0^\infty(\Omega)$.*

*Proof* We shall use

$$0 \geq \Delta(V^\delta) = \delta V^{\delta-2}\{(\delta-1)|\nabla V|^2 + V\Delta V\} \tag{6.3.5}$$

and hence

$$(\delta-1)|\nabla V|^2 \leq V|\Delta V|.$$

Under the conditions imposed on $V$, this needs to be justified by regularisation. However, for the application we have in mind, it is sufficient to assume that $V \in C^2(\Omega)$. From Theorem 1.2.8

$$(\delta-1)\int_\Omega |\Delta V||u|^2 d\mathbf{x} \leq 4(\delta-1)\int_\Omega \frac{|\nabla V|^2}{|\Delta V|}|\nabla u|^2 d\mathbf{x}$$

$$\leq 4 \int_\Omega V|\nabla u|^2 d\mathbf{x} = 4 \int_{\{\mathbf{x}\in\Omega; u(\mathbf{x})\neq 0\}} V|\nabla u|^2 d\mathbf{x}, \tag{6.3.6}$$

the last equality following since $\{\mathbf{x} \in \Omega; u(\mathbf{x}) = 0, |\nabla u(\mathbf{x})| \neq 0\}$ is of measure zero. The lemma is therefore proved in the case $p = 2$.

For $p \neq 2$, put $v_\varepsilon = (u^2 + \varepsilon^2)^{p/4} - \varepsilon^{p/2}$, and let $\varepsilon \to 0$. Since $0 \leq v_\varepsilon \leq |u|$, the left-hand side of (6.3.6), with $u$ replaced by $v_\varepsilon$, tends to $(\delta - 1) \int_\Omega |\Delta V||u|^p d\mathbf{x}$ by dominated convergence. Also,

$$\begin{aligned} |\nabla v_\varepsilon(\mathbf{x})|^2 V(\mathbf{x}) &= \{\frac{p}{2} u(\mathbf{x})(u^2(\mathbf{x}) + \varepsilon^2)^{(p-4)/4} \nabla u(\mathbf{x})\}^2 V(\mathbf{x}) \\ &\to \frac{p^2}{4} |u(\mathbf{x})|^{p-2} |\nabla u(\mathbf{x})|^2 V(\mathbf{x}) \end{aligned}$$

if $u(\mathbf{x}) \neq 0$. It follows as in the proof of Lemma 6.3.1, through the use of Fatou's lemma, that the right-hand side of (6.3.6) tends to

$$\int_{\{\mathbf{x}\in\Omega; u(\mathbf{x})\neq 0\}} V|u|^{p-2}|\nabla u|^2 d\mathbf{x}$$

and this completes the proof. □

By Lemma 6.3.2, we may put $c = [(p-1)/p](\delta - 1)$ in Lemma 6.3.1 to obtain Theorem 4 in [45], which is an extension of the case $p = 2$ in [29], Theorem 5; thus

**Theorem 6.3.3** *If* $0 < V \in C(\Omega)$ *with* $\Delta V < 0$ *and* $\Delta(V^\delta) \leq 0$ *for some* $\delta > 1$, *then*

$$\int_\Omega |\Delta V||u|^p d\mathbf{x} \leq \frac{p^{2p}}{[(p-1)\delta + 1]^p} \int_\Omega \frac{V^p}{|\Delta V|^{p-1}} |\Delta u|^p d\mathbf{x},$$

*for all* $u \in C_0^\infty(\Omega)$.

The choice $V(\mathbf{x}) = |\mathbf{x}|^{-(\alpha-2)}$, $\delta = (n-2)/(\alpha-2)$ yields

**Corollary 6.3.4** *Let* $2 < \alpha < n$. *Then, for all* $u \in C_0^\infty(\mathbb{R}^n \setminus \{0\})$,

$$\int_{\mathbb{R}^n} \frac{|u(\mathbf{x})|^p}{|\mathbf{x}|^\alpha} d\mathbf{x} \leq c(n,p,\alpha)^p \int_{\mathbb{R}^n} \frac{|\Delta u(\mathbf{x})|^p}{|\mathbf{x}|^{\alpha-2p}} d\mathbf{x}, \tag{6.3.7}$$

*where*

$$c(n,p,\alpha) = \frac{p^2}{(n-\alpha)[(p-1)n + \alpha - 2p]}. \tag{6.3.8}$$

The special case of Rellich's inequality is given prominence and extended to functions in the Sobolev space $W^{2,p}(\mathbb{R}^n) = W_0^{1,p}(\mathbb{R}^n)$; this means, in particular, that the inequality holds for all functions in $C_0^\infty(\mathbb{R}^n)$.

**Corollary 6.3.5** *Let $1 < p < \infty$ and $n > 2p$. Then, for all $u \in W^{1,p}(\mathbb{R}^n)$*

$$\int_{\mathbb{R}^n} \frac{|u(\mathbf{x})|^p}{|\mathbf{x}|^{2p}} d\mathbf{x} \le c(n,p)^p \int_{\mathbb{R}^n} |\Delta u(\mathbf{x})|^p d\mathbf{x}, \tag{6.3.9}$$

*where*

$$c(n,p) = \frac{p^2}{n(p-1)(n-2p)} \tag{6.3.10}$$

*is sharp.*

*Proof* We first note that $C_0^\infty(\mathbb{R}^n \setminus \{0\})$ is dense in $W^{2,p}(\mathbb{R}^n)$. To see this let $\phi \in C_0^\infty[0,1)$ be such that $\phi(r) = 1$, for $0 \le r \le 1/2$, and $0$ for $r > 1/2$, and set $\phi_\varepsilon(r) := \phi(r/\varepsilon)$, where $\varepsilon > 0$. Then, if $u \in C_0^\infty(\mathbb{R}^n)$, we have that $[1-\phi_\varepsilon]u \in C_0^\infty(\mathbb{R}^n \setminus \{0\})$ and

$$\|\phi_\varepsilon u\|_{W^{2,2}(\mathbb{R}^n)} \to 0$$

as $\varepsilon \to 0$ if $n > 2p$; this establishes the assertion.

It remains to show that $c(n,p)$ is sharp. Consider $u(\mathbf{x}) = |\mathbf{x}|^{-\sigma}$ near the origin with $0 < \sigma < \frac{n-2p}{p}$. Then $u \in W^{2,p}(B_1)$ for $B_1 := B_1(0)$, the unit ball in $\mathbb{R}^n$. A calculation gives

$$|\Delta u(\mathbf{x})| = \sigma(n-\sigma-2)|\mathbf{x}|^{-2}u(\mathbf{x}).$$

In fact, for $\sigma = \frac{n-2p}{p}$

$$|\Delta u(\mathbf{x})| = c(n,p)^{-1}|\mathbf{x}|^{-2}|u(\mathbf{x})|, \qquad \mathbf{x} \neq 0.$$

Let $\sigma_k := \frac{n-2p-\frac{1}{k}}{p}$ and $u_k(\mathbf{x}) := |\mathbf{x}|^{-\sigma_k}\chi_{B_1} \in W^{2,p}(B_1)$. Then,

$$\begin{aligned} c(n,p)^{-1} &\le \inf_{u\in W^{2,p}(\mathbb{R}^n)} \frac{\|\Delta u\|_p}{\||\mathbf{x}|^{-2}u\|_p} \le \inf_{k\in\mathbb{N}} \frac{\|\Delta u_k\|_p}{\||\mathbf{x}|^{-2}u_k\|_p} \\ &\le \lim_{k\to\infty} \sigma_k(n-\sigma_k-2) = c(n,p)^{-1}, \end{aligned}$$

implying that $c(n,p)$ is sharp. □

## 6.4 The Rellich Inequality with Magnetic Potentials

### 6.4.1 A General Theorem

We proved in Sect. 1.2.5 that there is no valid Hardy inequality in $L^n(\mathbb{R}^n)$. In the case $n = 2$, it is a consequence of the fact that, for any $\gamma > 0$, $\cos(\gamma \ln |\mathbf{x}|)$ and $\sin(\gamma \ln |\mathbf{x}|)$ are linearly independent, oscillatory solutions of

$$-\Delta\varphi = \frac{\gamma^2}{|\mathbf{x}|^2}\varphi, \qquad \mathbf{x} \in \mathbb{R}^2.$$

The function $\psi(\mathbf{x}) := \chi_A(\mathbf{x}) \sin(\gamma \ln |\mathbf{x}|)$, where $\chi_A$ is the characteristic function of the annulus $A := \{e^{\pi/\gamma} < |\mathbf{x}| < e^{2\pi/\gamma}\}$, lies in the closure of $C_0^\infty(\mathbb{R}^2 \setminus \{0\})$ in the $W^{1,2}(\mathbb{R}^2)$ norm, and so if Hardy's inequality is valid, it must be satisfied by $\psi$. However,

$$\begin{aligned} &\int_{\mathbb{R}^2} [|\nabla\psi(\mathbf{x})|^2 - \gamma^2 \tfrac{|\psi(\mathbf{x})|^2}{|\mathbf{x}|^2}] d\mathbf{x} \\ &\quad = \int_{\mathbb{R}^2} [-\Delta \sin(\gamma \ln |\mathbf{x}|) - \tfrac{\gamma^2}{|\mathbf{x}|^2} \sin(\gamma \ln |\mathbf{x}|)]\overline{\psi}(\mathbf{x}) d\mathbf{x} = 0, \end{aligned} \tag{6.4.1}$$

which is a contradiction since $\gamma$ can be arbitrarily small.

For any magnetic potential $\mathbf{A} : \mathbb{R}^n \to \mathbb{R}^n$ in $L^2_{loc}(\mathbb{R}^n; \mathbb{R}^n)$, the diamagnetic inequality (5.3.2) applied to the Hardy inequality yields

$$C \int_{\mathbb{R}^n} \frac{|u(\mathbf{x})|^2}{|\mathbf{x}|^2} d\mathbf{x} \le \int_{\mathbb{R}^n} |(\nabla + i\mathbf{A})u(\mathbf{x})|^2 d\mathbf{x}, \qquad u \in C_0^\infty(\mathbb{R}^n), \tag{6.4.2}$$

with $C = (n-2)^2/4$, and therefore no non-trivial information for $n = 2$ is gathered. However, we saw in Sect. 5.5 that if $\mathbf{A}$ is an Aharonov-Bohm magnetic potential with non-integer flux, there is a valid inequality (6.4.2) with $C > 0$ when $n = 2$. To summarize Theorem 5.5.1, the magnetic field $\mathbf{B} = \mathrm{curl}\mathbf{A} = 0$ in $\mathbb{R}^2 \setminus \{0\}$, and $\mathbf{A}$ is gauge equivalent to

$$\mathbf{A}(r, \theta) := \frac{\Psi}{r}(-\sin\theta, \cos\theta), \qquad \Psi := \Psi(\mathbf{A}), \tag{6.4.3}$$

where $\mathbf{x} = (r\cos\theta, r\sin\theta) \in \mathbb{R}^2 \setminus \{0\}$, and $\Psi(\mathbf{A})$ is the flux of $\mathbf{A}$. One then has the Laptev-Weidl inequality

$$\int_{\mathbb{R}^2} \frac{|u(\mathbf{x})|^2}{|\mathbf{x}|^2} d\mathbf{x} \le \mathrm{dist}(\Psi, \mathbb{Z})^2 \int_{\mathbb{R}^2} |(\nabla + i\mathbf{A})u(\mathbf{x})|^2 d\mathbf{x}, \tag{6.4.4}$$

with sharp constant. If $\Psi \in \mathbb{Z}$, the magnetic Laplacian $-\Delta_{\mathbf{A}}$ is unitarily equivalent to $-\Delta$, as operators in $L^2(\mathbb{R}^2)$, and there is no valid inequality.

Motivated by the work of Laptev and Weidl [99], Rellich-type inequalities for magnetic Laplacians $-\Delta_{\mathbf{A}}$ with magnetic potentials having similar characteristics to those of the Aharonov-Bohm potential were obtained in [54]. The main theorem established makes it possible to analyse the Rellich inequality in the cases $n = 2$ and $n = 4$ when (6.1.1) is trivial.

The theorem and proof that follow uncover the basic elements of Rellich's approach in [129], Theorem 1, p. 91, and is based on [54]. Polar coordinates in $\mathbb{R}^n$ will be denoted by $(r, \omega)$ with $r := |\mathbf{x}|$, $\omega = \mathbf{x}/|\mathbf{x}|$ for $\mathbf{x} \in \mathbb{R}^n$. We shall denote the $L^2(\mathbb{R}^n)$ norm by $\|\cdot\|$.

**Theorem 6.4.1** *Let $\Lambda_\omega$ be a non-negative, self-adjoint operator with domain $\mathcal{D}(\Lambda_\omega) \subseteq L^2(\mathbb{S}^{n-1}; d\omega)$, whose spectrum is discrete, consisting of isolated eigenvalues $\lambda_m$, $m \in \mathcal{I}$, for some countable index set $\mathcal{I}$. Let*

$$L_r := -\frac{\partial^2}{\partial r^2} - \frac{n-1}{r}\frac{\partial}{\partial r} \tag{6.4.5}$$

*and define the operator $D := L_r + \frac{1}{r^2}\Lambda_\omega$ on its domain in $L^2(\mathbb{R}^n)$ given by*

$$\mathcal{D}_0 := \{f \in C_0^\infty(\mathbb{R}^n \setminus \{0\}) : f(r,\cdot) \in \mathcal{D}(\Lambda_\omega) \; for \; r > 0, Df \in L^2(\mathbb{R}^n)\}.$$

*Then, for all $f \in \mathcal{D}_0$ such that $|\cdot|^{-\alpha/2}Df \in L^2(\mathbb{R}^n)$, we have that*

$$\int_{\mathbb{R}^n} \frac{|Df(\mathbf{x})|^2}{|\mathbf{x}|^\alpha} d\mathbf{x} \geq C(n,\alpha) \int_{\mathbb{R}^n} \frac{|f(\mathbf{x})|^2}{|\mathbf{x}|^{\alpha+4}} d\mathbf{x}, \tag{6.4.6}$$

*where*

$$C(n,\alpha) = \inf_{m\in\mathcal{I}} \left\{ \lambda_m + \left(\frac{n+\alpha}{2}\right)\left(\frac{n-\alpha-4}{2}\right)\right\}^2. \tag{6.4.7}$$

*Remark 6.4.2* If $D = -\Delta$, $\Lambda_\omega$ is the Laplace-Beltrami operator on $\mathbb{S}^{n-1}$. In that case $\lambda_m = m(m+n-2)$, $m \in \mathcal{I} = \{0, 1, \ldots\}$, and (6.4.6) reduces to

$$\int_{\mathbb{R}^n} \frac{|\Delta f|^2}{|\mathbf{x}|^\alpha} d\mathbf{x} \geq C(n,\alpha) \int_{\mathbb{R}^n} \frac{|f(\mathbf{x})|^2}{|\mathbf{x}|^{\alpha+4}} d\mathbf{x},$$

for $f \in C_0^\infty(\mathbb{R}^n \setminus \{0\})$ and $\alpha \in \mathbb{R}$. Hence, with $\alpha = 0$,

$$\int_{\mathbb{R}^n} |\Delta f|^2 d\mathbf{x} \geq C(n,0) \int_{\mathbb{R}^n} \frac{|f(\mathbf{x})|^2}{|\mathbf{x}|^4} d\mathbf{x},$$

for $f \in C_0^\infty(\mathbb{R}^n \setminus \{0\})$, where

$$C(2,0) = 0, \quad C(3,0) = \frac{9}{16}$$

and

$$C(n,0) = \left(\frac{n(n-4)}{4}\right)^2;$$

Rellich's inequality is therefore recovered. Also Lemma 2 of Allegretto [6] is recovered:

$$C(n,\alpha) \geq \frac{(n+\alpha)^2(n-\alpha-4)^2}{16}$$

for $\alpha \geq 0$ and $n \geq 1 + \sqrt{(3+\alpha)(1+\alpha)}$.

*Proof of Theorem 6.4.1* Since the spectrum of $\Lambda_\omega$ is assumed to be discrete, its normalized eigenvectors $u_m$, $m \in \mathcal{I}$ (with the eigenvalues $\{\lambda_m\}$ repeated according to multiplicity) form an orthonormal basis of $L^2(\mathbb{S}^{n-1}; d\omega)$. For $f \in \mathcal{D}_0$, set

$$F_m(r) := \int_{\mathbb{S}^{n-1}} f(r,\omega)\overline{u_m(\omega)}d\omega. \tag{6.4.8}$$

Then $F_m \in C_0^\infty(\mathbb{R}_+)$ and, on using Parseval's identity, we obtain

$$\begin{aligned}\sum_{m\in\mathcal{I}} \|L_r F_m\|^2_{L^2(\mathbb{R}_+, r^{n-1}dr)} &= \int_0^\infty \sum_{m\in\mathcal{I}} |L_r F_m(r)|^2 r^{n-1} dr \\ &= \|L_r f\|^2 < \infty,\end{aligned} \tag{6.4.9}$$

with

$$L_r F_m(r) = \int_{\mathbb{S}^{n-1}} L_r f(r,\omega)\overline{u_m(\omega)}d\omega.$$

Also, $L_r f$, $Df \in L^2(\mathbb{R}^n)$ imply that $|\cdot|^{-2}\Lambda_\omega f \in L^2(\mathbb{R}^n)$ and

$$\begin{aligned}\sum_{m\in\mathcal{I}} \lambda_m^2 \int_0^\infty |F_m(r)|^2 r^{n-5} dr &= \int_0^\infty \sum_{m\in\mathcal{I}} \lambda_m^2 |F_m(r)|^2 r^{n-5} dr \\ &= \||\cdot|^{-2}\Lambda_\omega f\|^2 < \infty.\end{aligned} \tag{6.4.10}$$

In fact, if $|\cdot|^{-\alpha/2}Df \in L^2(\mathbb{R}^n)$, then $|\cdot|^{-2-\alpha/2}\Lambda_\omega f \in L^2(\mathbb{R}^n)$ and

$$\sum_{m\in\mathcal{I}}\int_0^\infty |F_m(r)|^2 r^{n-\alpha-5}dr = \||\cdot|^{-2-\alpha/2}f\|^2 < \infty, \tag{6.4.11}$$

$$\sum_{m\in\mathcal{I}}\int_0^\infty |L_rF_m(r)|^2 r^{n-\alpha-1}dr = \||\cdot|^{-\alpha/2}L_rf\|^2 < \infty, \tag{6.4.12}$$

$$\sum_{m\in\mathcal{I}}\lambda_m^2\int_0^\infty |F_m(r)|^2 r^{n-\alpha-5}dr = \||\cdot|^{-2-\alpha/2}\Lambda_\omega f\|^2 < \infty. \tag{6.4.13}$$

To prove the theorem, we start with

$$\begin{aligned}\int_{\mathbb{R}^n}|Df|^2\frac{d\mathbf{x}}{|\mathbf{x}|^\alpha} &= \int_{\mathbb{R}^n}|L_rf|^2\frac{d\mathbf{x}}{|\mathbf{x}|^\alpha} + 2\mathrm{Re}\left[\int_{\mathbb{R}^n}L_rf\,\overline{\Lambda_\omega f}\frac{d\mathbf{x}}{|\mathbf{x}|^{\alpha+2}}\right]\\ &\quad+\int_{\mathbb{R}^n}|\Lambda_\omega f|^2\frac{d\mathbf{x}}{|\mathbf{x}|^{\alpha+4}}.\end{aligned} \tag{6.4.14}$$

The choices $p = 2,\ \varepsilon = t/2+1$ in Theorem 1.2.1 lead to the Hardy-type inequality

$$\int_0^\infty |F'(r)|^2 r^{t+2}dr \geq \left(\frac{t+1}{2}\right)^2\int_0^\infty |F(r)|^2 r^t dr,\quad t\in\mathbb{R}, \tag{6.4.15}$$

for all $F \in C_0^1(0,\infty)$. On integrating by parts, we obtain

$$\begin{aligned}&\textstyle\int_0^\infty |L_rF_m(r)|^2 r^{n-\alpha-1}dr\\ &= \textstyle\int_0^\infty \big(|F_m''(r)|^2 + \frac{2(n-1)}{r}\mathrm{Re}[F_m''\overline{F_m'}]\\ &\quad+\textstyle\frac{(n-1)^2}{r^2}|F_m'(r)|^2\big)r^{n-\alpha-1}dr\\ &\geq \textstyle\left(\frac{n-\alpha-2}{2}\right)^2\int_0^\infty |F_m'(r)|^2 r^{n-\alpha-3}dr\\ &\quad\textstyle-(n-1)(n-\alpha-2)\int_0^\infty |F_m'(r)|^2 r^{n-\alpha-3}dr\\ &\quad\textstyle+(n-1)^2\int_0^\infty |F_m'(r)|^2 r^{n-\alpha-3}dr\\ &= \textstyle\left[(n-1)(\alpha+1)+\left(\frac{n-\alpha-2}{2}\right)^2\right]\int_0^\infty |F_m'(r)|^2 r^{n-\alpha-3}dr\\ &\geq \textstyle\left(\frac{n+\alpha}{2}\right)^2\left(\frac{n-\alpha-4}{2}\right)^2\int_0^\infty |F_m(r)|^2 r^{n-\alpha-5}dr.\end{aligned}$$

Thus, from (6.4.11) and (6.4.12),

$$\int_{\mathbb{R}^n}\frac{|L_rf|^2}{|\mathbf{x}|^\alpha}d\mathbf{x} \geq \left(\frac{n+\alpha}{2}\right)^2\left(\frac{n-\alpha-4}{2}\right)^2\int_{\mathbb{R}^n}\frac{|f|^2}{|\mathbf{x}|^{\alpha+4}}d\mathbf{x}. \tag{6.4.16}$$

Since

$$\int_{\mathbb{S}^{n-1}} L_r f(r,\omega)\overline{u_m(\omega)}d\omega = L_r F_m(r)$$

and

$$\int_{\mathbb{S}^{n-1}} \Lambda_\omega f(r,\omega)\overline{u_m(\omega)}d\omega = \lambda_m F_m(r),$$

it follows from Parseval's identity that in (6.4.14),

$$\int_{\mathbb{R}^n} \frac{L_r f\overline{\Lambda_\omega f}}{|\mathbf{x}|^{\alpha+2}}d\mathbf{x} = \int_0^\infty \sum_{m\in\mathcal{I}} \lambda_m \overline{F_m(r)} L_r F_m(r) r^{n-\alpha-3}dr. \tag{6.4.17}$$

Integration by parts yields

$$\begin{aligned}
&2\mathrm{Re}\left[\int_0^\infty \overline{F_m(r)} L_r F_m(r) r^{n-\alpha-3}dr\right]\\
&= 2\mathrm{Re}\left[\int_0^\infty F_m'\left(\overline{F_m'}r^{n-\alpha-3} + (n-\alpha-3)\overline{F_m}r^{n-\alpha-4}\right)dr\right]\\
&\quad +(n-1)(n-\alpha-4)\int_0^\infty |F_m|^2 r^{n-\alpha-5}dr\\
&= 2\int_0^\infty |F_m'|^2 r^{n-\alpha-3}dr\\
&\quad +\{-(n-\alpha-3)+(n-1)\}(n-\alpha-4)\int_0^\infty |F_m|^2 r^{n-\alpha-5}dr\\
&\geq \left\{2\left(\tfrac{n-\alpha-4}{2}\right)^2 + (n-\alpha-4)(\alpha+2)\right\}\int_0^\infty |F_m|^2 r^{n-\alpha-5}dr\\
&\quad \text{(by (6.4.15))},\\
&= \tfrac{1}{2}(n-\alpha-4)(n+\alpha)\int_0^\infty |F_m|^2 r^{n-\alpha-5}dr.
\end{aligned}$$

This gives in (6.4.17)

$$\begin{aligned}
2\mathrm{Re}\left[\int_{\mathbb{R}^n} \frac{L_r f\overline{\Lambda_\omega f}}{|\mathbf{x}|^{\alpha+2}}d\mathbf{x}\right] \geq& \frac{1}{2}(n-\alpha-4)(n+\alpha)\\
&\times \sum_{m\in\mathcal{I}} \lambda_m \int_0^\infty |F_m(r)|^2 r^{n-\alpha-5}dr.
\end{aligned} \tag{6.4.18}$$

Finally in (6.4.14), by (6.4.13),

$$\int_{\mathbb{R}^n} \frac{|\Lambda_\omega f|^2}{|\mathbf{x}|^{\alpha+4}}d\mathbf{x} = \sum_{m\in\mathcal{I}} \lambda_m^2 \int_0^\infty |F_m(r)|^2 r^{n-\alpha-5}dr. \tag{6.4.19}$$

The theorem follows on substituting (6.4.16), (6.4.18) and (6.4.19) in (6.4.14) and noting (6.4.11). □

## 6.4.2 An Inequality for $D = -\Delta_{\mathbf{A}}$

We now apply Theorem 6.4.1 to the magnetic Laplacian associated with a magnetic potential $\mathbf{A}$ which is of Aharonov-Bohm type when $n = 2$, and has analogous characteristics for other values of $n$. In order to handle the case $n = 4$, it will be necessary to discuss the case $n = 3$, which we sketch, referring the reader to [54] for further details. For values of $n > 4$ and higher order Rellich inequalities, see [142]. The anomalous $n = 2$ and $n = 4$ results for the Rellich inequality (6.1.1) will be consequences of the main theorem.

### The Case $n = 2$

The magnetic potential $\mathbf{A}$ is now assumed to satisfy (6.4.3), with non-integer flux $\Psi$. With $\mathbf{e}_r := (\cos\theta, \sin\theta)$ and $\mathbf{e}_\theta := (-\sin\theta, \cos\theta)$, we have

$$\nabla_{\mathbf{A}} := \nabla + i\mathbf{A} = \mathbf{e}_r \frac{\partial}{\partial r} + \mathbf{e}_\theta \frac{1}{r}\left(\frac{\partial}{\partial \theta} + i\Psi\right) \tag{6.4.20}$$

and

$$-\Delta_{\mathbf{A}} = -\frac{\partial^2}{\partial r^2} - \frac{1}{r}\frac{\partial}{\partial r} + \frac{1}{r^2}\left(i\frac{\partial}{\partial \theta} - \Psi\right)^2. \tag{6.4.21}$$

Thus, in the notation of Theorem 6.4.1, $\Lambda_\omega = \Lambda_\theta$ is the non-negative self-adjoint operator in $L^2(0, 2\pi)$ defined by $\Lambda_\theta = K_\theta^2$, where

$$K_\theta u(\theta) = i\frac{du}{d\theta} - \Psi u(\theta) \tag{6.4.22}$$

with domain

$$\{u : u \in AC[0, 2\pi], u(0) = u(2\pi)\},$$

where $AC[0, 2\pi]$ denotes the set of functions which are absolutely continuous on $[0, 2\pi]$. Clearly $K_\theta$ has eigenvalues $m - \Psi$, $m \in \mathbb{Z}$, and the corresponding normalized eigenfunctions are

$$u_m(\theta) = \frac{1}{\sqrt{2\pi}} e^{-im\theta}, \tag{6.4.23}$$

which constitute an orthonormal basis of $L^2(\mathbb{S}^1)$.

For $m \in \mathbb{Z}$, $U : f \mapsto e^{-im\theta} f$ is unitary on $L^2(\mathbb{R}^2)$ and

$$U^{-1}\nabla_{\mathbf{A}} U = \nabla_{\tilde{\mathbf{A}}},$$

where $\tilde{\mathbf{A}} = \frac{(\Psi - m)}{r}\mathbf{e}_\theta$. The magnetic potentials $\mathbf{A}$, $\tilde{\mathbf{A}}$ are gauge equivalent and their fluxes differ by $m$. Therefore we may assume that $\Psi \in [0, 1)$.

Since $\mathcal{D}_0 = C_0^\infty(\mathbb{R}^2 \setminus \{0\})$, we have from Theorem 6.4.1

**Corollary 6.4.3** *For all* $f \in C_0^\infty(\mathbb{R}^2 \setminus \{0\})$,

$$\int_{\mathbb{R}^2} |\Delta_{\mathbf{A}} f(\mathbf{x})|^2 \frac{d\mathbf{x}}{|\mathbf{x}|^\alpha} \geq C(2, \alpha) \int_{\mathbb{R}^2} |f(\mathbf{x})|^2 \frac{d\mathbf{x}}{|\mathbf{x}|^{\alpha+4}}, \tag{6.4.24}$$

*where*

$$C(2, \alpha) = \inf_{m \in \mathbb{Z}} \left\{ (m + \Psi)^2 - \frac{(\alpha + 2)^2}{4} \right\}^2. \tag{6.4.25}$$

*If* $\Psi \notin \mathbb{Z}$ ($\Psi \in (0, 1)$ *without loss of generality), we have*

$$\begin{aligned} C(2, 0) &= \min\{(\Psi^2 - 1)^2, \Psi^2(\Psi - 2)^2\} \\ &= \begin{cases} (\Psi^2 - 1)^2 & \text{if } \Psi \in [\frac{1}{2}, 1), \\ \Psi^2(\Psi - 2)^2 & \text{if } \Psi \in [0, \frac{1}{2}). \end{cases} \end{aligned} \tag{6.4.26}$$

*Remark 6.4.4* If $\Psi \in \mathbb{Z}$, then $C(2, 0) = 0$. However, if $F_1 = F_{-1} = 0$ in (6.4.8), i.e.,

$$\int_0^{2\pi} f(r, \theta) \cos\theta d\theta = \int_0^{2\pi} f(r, \theta) \sin\theta d\theta = 0,$$

then the infimum in (6.4.25) is over $m \in \mathbb{Z} \setminus \{-1, 1\}$ and this gives $C(2, 0) = 1$. Hence, Rellich's result in ([129], p. 91) for $n = 2$, as noted in Remark 6.2.5, is recovered.

### The Case $n = 3$

In spherical polar coordinates, we define the orthonormal vectors

$$\begin{aligned} \mathbf{e}_0 &:= \tfrac{\mathbf{x}}{|\mathbf{x}|} = (\cos\theta_1, \sin\theta_1 \cos\theta_2, \sin\theta_1 \sin\theta_2), \\ \mathbf{e}_1 &:= (-\sin\theta_1, \cos\theta_1 \cos\theta_2, \cos\theta_1 \sin\theta_2), \\ \mathbf{e}_2 &:= (0, -\sin\theta_2, \cos\theta_2), \end{aligned}$$

where $r = |\mathbf{x}| \in (0, \infty)$, $\theta_1 \in (0, \pi)$, and $\theta_2 \in (0, 2\pi)$. We now take

$$\mathbf{A} := \frac{1}{r \sin\theta_1} \psi(\theta_2)\mathbf{e}_2, \qquad \psi \in L^\infty(0, 2\pi), \qquad \psi(0) = \psi(2\pi),$$

on $\mathbb{R}^3 \setminus \mathcal{L}_3$, where $\mathcal{L}_3 = \{(r, \theta_1, \theta_2) : r\sin\theta_1 = 0\}$. It is in Poincaré gauge and $\text{curl}\mathbf{A} = 0$ in $\mathbb{R}^3 \setminus \mathcal{L}_3$.

Then

$$\nabla_{\mathbf{A}} = \nabla + i\mathbf{A} = \mathbf{e}_0 \frac{\partial}{\partial r} + \mathbf{e}_1 \frac{1}{r}\frac{\partial}{\partial\theta_1} + \mathbf{e}_2 \frac{1}{r\sin\theta_1}\left(\frac{\partial}{\partial\theta_2} + i\psi(\theta_2)\right) \tag{6.4.27}$$

and

$$-\Delta_{\mathbf{A}} = -\frac{\partial^2}{\partial r^2} - \frac{2}{r}\frac{\partial}{\partial r} + \frac{1}{r^2}\Lambda(\theta_1, \theta_2), \tag{6.4.28}$$

where

$$\Lambda(\theta_1, \theta_2) = -\frac{\partial^2}{\partial\theta_1^2} - \cot\theta_1 \frac{\partial}{\partial\theta_1} + \frac{1}{\sin^2\theta_1} K_{\theta_2}^2 \tag{6.4.29}$$

and

$$K_{\theta_2} = i\frac{\partial}{\partial\theta_2} - \psi(\theta_2). \tag{6.4.30}$$

The self-adjoint operator $K_{\theta_2}$ in $L^2(\mathbb{S}^1)$ has eigenvalues $k - \Psi$, $k \in \mathbb{Z}$, with $\Psi := \frac{1}{2\pi}\int_0^{2\pi} \psi(\theta_2)d\theta_2$, and corresponding eigenvectors

$$u_k(\theta_2) = \frac{1}{\sqrt{2\pi}} \exp\left[-i\left(\theta_2(k - \Psi) + \int_0^{\theta_2} \psi(\eta)d\eta\right)\right]$$

which form an orthonormal basis of $L^2(\mathbb{S}^1)$. Identifying $L^2(\mathbb{S}^2)$ with $\bigoplus_{k\in\mathbb{Z}} \left(L^2(0, \pi; \sin\theta_1 d\theta_1) \bigotimes \{u_k\}\right)$, we shall take the operator $\Lambda_\omega$ of Theorem 6.4.1 to be

$$\Lambda_\omega = \bigoplus_{k\in\mathbb{Z}} \left(\Lambda_k(\theta_1) \bigotimes I_k\right), \tag{6.4.31}$$

where $I_k$ denotes the identity on $\{u_k\}$ and $\Lambda_k(\theta_1)$ is a self-adjoint realisation of the operator $\Lambda_k^0(\theta_1)$ defined on $C_0^\infty(0, \pi)$ by

$$\Lambda_k^0(\theta_1)u = \left(-\frac{d^2}{d\theta_1^2} - \cot\theta_1 \frac{d}{d\theta_1} + \frac{(k - \Psi)^2}{\sin^2\theta_1}\right)u. \tag{6.4.32}$$

Before we are able to apply Theorem 6.4.1 we must first make a suitable choice of the operators $\Lambda_k(\theta_1)$ for all $k \in \mathbb{Z}$ and determine their eigenvalues. The information required is contained in the next two lemmas. They require knowledge of the following topics:

- the Hermann Weyl **limit point/ limit circle** characterisation of a singular formally self-adjoint second-order differential expression $L$, say, defined on an interval $(a, b)$;
- the **essential self-adjointness** of an operator realisation of $L$ in $L^2(a, b)$;
- the **Friedrichs extensions** defined by quadratic forms associated with $L$.

A brief description of these notions follows as an aid to the understanding of the lemmas, but see [48] or [83] for further details.

The differential expression $L$ is in the limit-point case at the end point $a$ if, for any $\lambda \in \mathbb{C}\setminus\mathbb{R}$, there exists a unique (up to constant multiples) solution $u$ of $(L-\lambda)u = 0$ which is square integrable in a neighbourhood of $a$, i.e., $u \in L^2(a, X)$ for $X \in (a, b)$. Otherwise, all solutions are in $L^2(a, X)$ for all $\lambda \in \mathbb{C}$, and this constitutes the limit-circle case. The end-point $b$ is characterised similarly. The operator $T$ defined by $L$ on $C_0^\infty(a, b)$ is said to be essentially self-adjoint if its closure is self-adjoint, and this is so if and only if the end points $a, b$ are both in the limit-point case. If $T$ is not essentially self-adjoint, other ways have to be found to generate self-adjoint operators from $T$. A favourite candidate is the Friedrichs extension determined by the quadratic form associated with $T$; see Sect. 1.5. The quadratic form has to be semi-bounded for the Friedrichs extension to be defined, a requirement which is met in the application made here, in fact $T \geq 0$. In the case of $T$ not being essentially self-adjoint, the domains of self-adjoint extensions are determined by boundary conditions at the end points of the interval $(a, b)$.

**Lemma 6.4.5** *For $\mu \in [0, \infty)$, the associated Legendre equation*

$$\frac{d^2u}{d\theta^2} + \cot\theta \frac{du}{d\theta} + \left(\lambda - \frac{\mu^2}{\sin^2\theta}\right) u = 0, \quad \lambda \in \mathbb{C}, \tag{6.4.33}$$

*is in the limit-circle case at 0 and $\pi$ if $\mu \in [0, 1)$ and in the limit-point case at 0 and $\pi$ otherwise.*

*Let*

$$\mathcal{D}_\mu := \left\{u : u, \sin\theta \frac{du}{d\theta} \in AC_{loc}(0, \pi), u, L_\mu u \in L^2((0, \pi); \sin\theta d\theta)\right\}, \tag{6.4.34}$$

*where*

$$L_\mu := -\frac{d^2}{d\theta^2} - \cot\theta \frac{d}{d\theta} + \frac{\mu^2}{\sin^2\theta},$$

*and denote the restriction of $L_\mu$ to $C_0^\infty(0, \pi)$ by $\Lambda_\mu^0$. Then $\Lambda_\mu^0$ is non-negative. It is essentially self-adjoint if and only if $\mu \in [1, \infty)$ and for $\mu \in [0, 1)$ its Friedrichs* *extension* $\Lambda_\mu$ *is the realisation of* $L_\mu$ *on the following domains:*

- *if* $\mu \in (0,1)$,

$$\mathcal{D}(\Lambda_\mu) = \left\{u : u \in \mathcal{D}_\mu, \sin^\mu \theta \, u(\theta) \to 0 \text{ as } \theta \to 0 \text{ and } \pi\right\}; \tag{6.4.35}$$

- *if* $\mu = 0$,

$$\mathcal{D}(\Lambda_\mu) = \left\{u : u \in \mathcal{D}_\mu, u(\theta)/|\ln(\cot \frac{\theta}{2})| \to 0 \text{ as } \theta \to 0 \text{ and } \pi\right\}; \tag{6.4.36}$$

- *if* $\mu \in [1,\infty)$, $\Lambda_\mu$ *is the closure of* $\Lambda_\mu^0$ *and*

$$\mathcal{D}(\Lambda_\mu) = \mathcal{D}_\mu. \tag{6.4.37}$$

*For* $\mu \in [0,1)$, $\Lambda_\mu \geq \mu(\mu+1)$.

*Proof* On substituting $x = \cos\theta$, (6.4.33) becomes

$$\tau_\mu u := -\frac{d}{dx}\{(1-x^2)\frac{du}{dx}\} + \frac{\mu^2}{1-x^2}u = \lambda u, \quad x \in (-1,1) \tag{6.4.38}$$

and $L^2((0,\pi); \sin\theta d\theta)$ becomes $L^2(-1,1)$. Denote the restriction of $\tau_\mu$ to $C_0^\infty(-1,1)$ by $T_\mu^0$. Clearly $T_\mu^0 \geq 0$.

Define the functions

$$f(x) = (1-x^2)^{\mu/2}$$
$$g(x) = f(x)h(x), \quad h(x) := \left|\int_0^x (1-t^2)^{-1-\mu} dt\right|.$$

It is shown in [54] that $f$, $g$ are respectively **principal** and **non-principal** solutions of the equation $\tau_\mu u = \mu(\mu+1)u$. If $\mu \geq 1$, $g$ is proved to be neither in $L^2(-1,0)$ nor $L^2(0,1)$; hence both end-points $\mp 1$ are in the limit-point case and consequently $T_\mu^0$ is essentially self-adjoint; denote the closure of $T_\mu^0$ by $T_\mu$. For $\mu \in [0,1)$, $f$ and $g$ are both in $L^2(-1,1)$ and therefore the two end-points are in the limit-circle case. To characterise the Friedrichs extension $T_\mu$ of $T_\mu^0$, Rosenberger's Theorem 3 in [130] is applied. Results of Kalf in [82] are used to prove the final result $\Lambda_\mu \geq \mu(\mu+1)$ □

**Lemma 6.4.6** *The operator* $T_\mu$ *in the proof of Lemma 6.4.5 has a discrete spectrum consisting of eigenvalues*

$$\nu_j(\mu) = \big(j-\mu\big)\big(j+1-\mu\big), \quad j \in \mathbb{Z}', \tag{6.4.39}$$

*where* $\mathbb{Z}' = \{j \in \mathbb{Z} : (j-\mu)(j+1-\mu) \geq 0\}$.

*Proof* For any $\lambda \in \mathbb{C} \setminus \mathbb{R}$, there exist solutions $\psi_1, \psi_2$ of $T_\mu u = \lambda u$ which satisfy $\psi_1 \in L^2(-1,0)$, $\psi_2 \in L^2(0,1)$, and which are unique, up to constant multiples. These are so-called **Titchmarsh-Weyl** solutions. The spectrum of $T_\mu$ is discrete, the eigenvalues being the zeros of the Wronskian $\psi_2\psi_1' - \psi_1\psi_2'$. The lemma is proved with the help of asymptotic formulae for the Titchmarsh-Weyl solutions and their first derivatives obtained from [26]. We refer to [54], Lemma 2 for details. □

**Corollary 6.4.7** *For all $f \in C_0^\infty(\mathbb{R}^3 \setminus \mathcal{L}_3)$, we have*

$$\int_{\mathbb{R}^3} |\Delta_{\mathbf{A}} f(\mathbf{x})|^2 \frac{d\mathbf{x}}{|\mathbf{x}|^\alpha} \geq C(3,\alpha) \int_{\mathbb{R}^3} |f(\mathbf{x})|^2 \frac{d\mathbf{x}}{|\mathbf{x}|^{\alpha+4}}, \tag{6.4.40}$$

*where*

$$C(3,\alpha) := \inf_{m\in\mathbb{Z}'} \left\{ (m-\Psi)(m-\Psi+1) - \frac{(3+\alpha)(1+\alpha)}{4} \right\}^2, \tag{6.4.41}$$

*and* $\mathbb{Z}' = \{m \in \mathbb{Z} : (m-\Psi)(m-\Psi+1) \geq 0\}$.

*Proof* We may clearly suppose that $|\cdot|^{-\alpha/2}\Delta_{\mathbf{A}} f \in L^2(\mathbb{R}^3)$. Then $\Lambda_\omega f(r,\omega) = \Lambda(\theta_1,\theta_2) f(r,\omega) \in L^2(\mathbb{R}^3)$ and $\Lambda(\theta_2,\theta_2) f(r,\omega) \in L^2(\mathbb{S}^2)$ for all $r \in (0,\infty)$. If $\Lambda_k(\theta_1)$ denotes provisionally the formal operator in (6.4.32) and

$$F_k(r,\theta_1) := \int_0^{2\pi} f(r,\theta_1,\theta_2)\overline{u_k(\theta_2)}\, d\theta_2,$$

where the $u_k$ are the functions in (6.4.23), then we have

$$\int_0^{2\pi} |\Lambda(\theta_1,\theta_2) f(r,\theta_1,\theta_2)|^2 d\theta_2 = \sum_{k\in\mathbb{Z}} |\Lambda_k(\theta_1) F_k(r,\theta_1)|^2.$$

Hence, for any $k \in \mathbb{Z}$ and $r \in (0,\infty)$, $\Lambda_k(\theta_1) F_k(r,\theta_1) \in L^2((0,\pi); \sin\theta_1 d\theta_1)$. Since $F_k(r,\cdot) \in L^2((0,\pi); \sin\theta_1 d\theta_1)$ and the boundary conditions given in Lemma 6.4.5 are satisfied, it follows that $F_k(r,\cdot)$ lies in the domain of the operator $\Lambda_k(\theta_1)$ and $f(r,\cdot) \in \mathcal{D}(\Lambda_\omega)$.

From Lemmas 6.4.5 and 6.4.6 the eigenvalues of the operator $\Lambda_k$ in (6.4.31) are

$$\nu_j(k) = (j - |k-\Psi|)(j+1-|k-\Psi|), \tag{6.4.42}$$

for $j \in \mathbb{Z}$ such that $\nu_j(k) \geq 0$. Denote the corresponding normalised eigenvectors by $P_{j,k}(\theta_1;\Psi)$. Then

$$Y_{j,k;\Psi}(\theta_1,\theta_2) := P_{j,k}(\theta_1;\Psi) u_k(\theta_2), \quad j,k \in \mathbb{Z},$$

are the eigenvectors of $\Lambda_\omega$ corresponding to the eigenvalues $\nu_j(k)$, and form an orthonormal basis of $L^2(\mathbb{S}^2)$. The corollary follows from Theorem 6.4.1. □

*Remark 6.4.8* When $\alpha = 0$ and $\Psi = 0$, (6.4.40) holds on $C_0^\infty(\mathbb{R}^3 \setminus \{0\})$ and $C(3,0) = \frac{9}{16}$. Corollary 6.4.7 therefore gives the Rellich inequality in this case, and recovers the constant obtained by Rellich [129], Theorem 1, p. 91. Note also that $C(3,0) = 0$ when $\Psi = 1/2$.

### The Case $n = 4$

In this case we define the orthonormal vectors

$$\begin{aligned}
\mathbf{e}_0 &= \tfrac{\mathbf{x}}{|\mathbf{x}|} = (\cos\theta_1, \sin\theta_1\cos\theta_2, \sin\theta_1\sin\theta_2\cos\theta_3, \sin\theta_1\sin\theta_2\sin\theta_3)\\
\mathbf{e}_1 &= (-\sin\theta_1, \cos\theta_1\cos\theta_2, \cos\theta_1\sin\theta_2\cos\theta_3, \cos\theta_1\sin\theta_2\sin\theta_3)\\
\mathbf{e}_2 &= (0, -\sin\theta_2, \cos\theta_2\cos\theta_3, \cos\theta_2\sin\theta_3)\\
\mathbf{e}_3 &= (0, 0, -\sin\theta_3, \cos\theta_3),
\end{aligned}$$

where $\theta_1, \theta_2 \in (0,\pi)$, $\theta_3 \in (0, 2\pi)$. In this case,

$$\nabla = \mathbf{e}_0\frac{\partial}{\partial r} + \mathbf{e}_1\left(\frac{1}{r}\frac{\partial}{\partial\theta_1}\right) + \mathbf{e}_2\left(\frac{1}{r\sin\theta_1}\frac{\partial}{\partial\theta_2}\right) + \mathbf{e}_3\left(\frac{1}{r\sin\theta_1\sin\theta_2}\frac{\partial}{\partial\theta_3}\right).$$

We now take

$$\mathbf{A} := \frac{1}{r\sin\theta_1\sin\theta_2}\psi(\theta_3)\mathbf{e}_3, \qquad \Psi \in L^\infty(0,2\pi), \qquad \psi(0) = \psi(2\pi), \tag{6.4.43}$$

in $\mathbb{R}^4 \setminus \mathcal{L}_4$, where $\mathcal{L}_4 := \{\mathbf{x} = (r,\theta_1,\theta_2,\theta_3) : r\sin\theta_1\sin\theta_2 = 0\}$. We then have

$$-\Delta_{\mathbf{A}} = L_r + \frac{1}{r^2}\Lambda$$

for $L_r$ defined in (6.4.5), where

$$\Lambda = -\frac{\partial^2}{\partial\theta_1^2} - 2\cot\theta_1\frac{\partial}{\partial\theta_1} + \frac{1}{\sin^2\theta_1}\left\{-\frac{\partial^2}{\partial\theta_2^2} - \cot\theta_2\frac{\partial}{\partial\theta_2} + \frac{K_{\theta_3}^2}{\sin^2\theta_2}\right\} \tag{6.4.44}$$

with

$$K_{\theta_3} := i\frac{\partial}{\partial\theta_3} - \psi(\theta_3).$$

On repeating the procedure described in the case of $n = 2$ and using the same notation, we take $\Lambda_\omega$ to be

$$\Lambda_\omega = \bigoplus_{j,k\in\mathbb{Z}} \left(\Lambda_{j,k}(\theta_1) \bigotimes I_{j,k}\right), \tag{6.4.45}$$

where $\Lambda_{j,k}(\theta_1)$ is the self-adjoint operator generated by

$$\Lambda_{j,k}(\theta_1) = -\frac{\partial^2}{\partial\theta_1^2} - 2\cot\theta_1 \frac{\partial}{\partial\theta_1} + \frac{\nu_j(k)}{\sin^2\theta_1} \tag{6.4.46}$$

in $L^2((0,\pi); \sin^2\theta_1 d\theta_1)$. The operator $\Lambda_{j,k}(\theta_1)$ is again chosen to be the Friedrichs extension of the operator defined on $C_0^\infty(0,\pi)$. To apply Theorem 6.4.1 we need

**Lemma 6.4.9** *The Friedrichs extension of the operator* $\Lambda_{j,k}(\theta_1)\big|_{C_0^\infty(0,\pi)}$ *in* $L^2((0,\pi); \sin^2\theta_1 d\theta_1)$ *has eigenvalues*

$$\gamma_\ell(j,k) = (\ell - [\nu_j(k)+1/4]^{\frac{1}{2}})(\ell - [\nu_j(k)+1/4]^{\frac{1}{2}}+1) - \frac{3}{4}, \qquad \ell \in \mathbb{Z}'' \tag{6.4.47}$$

*where* $\nu_j(k)$ *is given by (6.4.42) and* $\mathbb{Z}'' := \{\ell \in \mathbb{Z} : \gamma_\ell(j,k) \geq 0\}$.

*Proof* On substituting $x = \cos\theta_1$, the equation $\Lambda_{j,k}(\theta_1)u = \lambda u$ becomes

$$(1-x^2)\frac{d^2u}{dx^2} - 3x\frac{du}{dx} + \left(\lambda - \frac{\nu_j(k)}{1-x^2}\right)u = 0.$$

Further, set $w = (1-x^2)^{\frac{1}{4}}u$ to obtain

$$(1-x^2)\frac{d^2w}{dx^2} - 2x\frac{dw}{dx} + \left(\lambda + \frac{3}{4} - \frac{\nu_j(k)+\frac{1}{4}}{1-x^2}\right)w = 0 \tag{6.4.48}$$

with $L^2(-1,1)$ for the underlying Hilbert space. The problem is therefore reduced to that for (6.4.38) with $\lambda + \frac{3}{4}$ instead of $\lambda$ and $\nu_j(k) + \frac{1}{4}$ for $\mu^2$, and the lemma follows from Lemma 6.4.6. □

From (6.4.42), $\nu_j(k) = (j - |k-\Psi|)(j+1-|k-\Psi|)$, which implies that

$$\nu_j(k) + \frac{1}{4} = \left(j - |k-\Psi| + \frac{1}{2}\right)^2.$$

Thus, from (6.4.47), $\gamma_\ell(j,k) = (\ell - j + |k - \Psi|)^2 - 1$ if $j - |k - \Psi| + 1/2 \geq 0$ and $\gamma_\ell(j,k) = (\ell + j - |k - \Psi| + 1)^2 - 1$ otherwise. These can be enumerated as

$$\lambda_m = (m - \Psi)^2 - 1, \qquad m \in \mathbb{Z}',$$

where $\mathbb{Z}' := \{m : (m - \Psi)^2 \geq 1\}$.

It follows by an argument similar to that in the proof of Corollary 6.4.7 that for any $f \in C_0^\infty(\mathbb{R}^4 \setminus \{0\})$ with $\Delta_{\mathbf{A}} f \in L^2(\mathbb{R}^4)$, we have $f(r,\cdot) \in \mathcal{D}(\Lambda_\omega)$. Hence, from Theorem 6.4.1,

**Corollary 6.4.10** *Let $f \in C_0^\infty(\mathbb{R}^4 \setminus \mathcal{L}_4)$. Then*

$$\int_{\mathbb{R}^4} |\Delta_{\mathbf{A}} f(\mathbf{x})|^2 \frac{d\mathbf{x}}{|\mathbf{x}|^\alpha} \geq C(4,\alpha) \int_{R^4} |f(\mathbf{x})|^2 \frac{d\mathbf{x}}{|\mathbf{x}|^{\alpha+4}} \tag{6.4.49}$$

*where*

$$C(4,\alpha) := \inf_{m \in \mathbb{Z}'} \left\{ \left[ (m - \Psi)^2 - 1 - \frac{\alpha(\alpha+4)}{4} \right]^2 \right\},$$

*and $\mathbb{Z}' := \{m \in \mathbb{Z} : (m - \Psi)^2 \geq 1\}$. In particular, when $\alpha = 0$ and $\Psi \in (0,1)$,*

$$C(4,0) = \min\{[(1-\Psi)^2 - 1]^2, [(-2-\Psi)^2 - 1]^2\} > 0.$$

When $\Psi = 0$, (6.4.49) is satisfied on $C_0^\infty(\mathbb{R}^4 \setminus \{0\})$. The inequality is trivial if $C(4,0) = 0$, but if $F_1 = F_{-1} = 0$ (see (6.4.8)), then the infimum is attained for $m = \pm 2$, giving $C(4,0) = 9$, which is an analogue for $n = 4$ of the result for $n = 2$ in Remark 6.4.4.

## 6.5 Eigenvalues of a Biharmonic Operator with an Aharonov-Bohm Magnetic Field

We now apply results from the previous section to give bounds for the number of eigenvalues of biharmonic operators given formally by $\Delta_{\mathbf{A}}^2 - V$, with the Aharonov-Bohm type magnetic potential $\mathbf{A}$ considered there. In particular, upper bounds of Cwikel-Lieb-Rosenblum type will be obtained; cf. [55].

### *6.5.1 Some Inequalities*

The following inequalities play a pivotal roll in the subsequent analysis.

**Theorem 6.5.1** *For $D$ and $\mathcal{D}_0$ defined in Theorem 6.4.1*

$$\begin{aligned}&\|Df\|^2 + \max_m\{\lambda_m(2-\lambda_m)\}\int_{\mathbb{R}^n}\frac{|f(\mathbf{x})|^2}{|\mathbf{x}|^4}d\mathbf{x}\\&\geq \sup_{r\in(0,\infty)}\{r^{n-2}\int_{S^{n-1}}|\tfrac{\partial f}{\partial r}|^2 d\omega + 2\min_m\{\lambda_m\}r^{n-4}\int_{S^{n-1}}|f|^2 d\omega\}\end{aligned} \tag{6.5.1}$$

*for $f \in \mathcal{D}_0$.*

*Proof* For $L_r$ given by (6.4.5) and $F_m(r)$ by (6.4.8), we have, on using Parseval's identity, that for all $f \in \mathcal{D}_0$,

$$\begin{aligned}&\int_{\mathbb{R}^n}|Df|^2 d\mathbf{x}\\&= \int_{\mathbb{R}^n}|L_r f|^2 d\mathbf{x} + 2\mathrm{Re}[\int_{\mathbb{R}^n}L_r f\overline{\Lambda_\omega f}\frac{d\mathbf{x}}{|\mathbf{x}|^2}]\\&\quad+\int_{\mathbb{R}^n}|\Lambda_\omega f|^2\frac{d\mathbf{x}}{|\mathbf{x}|^4}\\&=\sum_m\{\int_0^\infty |L_rF_m|^2 r^{n-1}dr + 2\mathrm{Re}[\lambda_m\int_0^\infty \overline{F_m}L_rF_m r^{n-3}dr]\\&\quad+\lambda_m^2\int_0^\infty |F_m(r)|^2 r^{n-5}dr\}\\&=:\sum_m\{I_1 + 2\lambda_m I_2 + \lambda_m^2 I_3\}.\end{aligned} \tag{6.5.2}$$

It follows that

$$\begin{aligned}I_{1,m} &= \int_0^\infty\left[|F_m''|^2 + 2\tfrac{n-1}{r}\mathrm{Re}\{F_m''\overline{F_m'}\} + \tfrac{(n-1)^2}{r^2}|F_m'|^2\right]r^{n-1}dr\\&=\int_0^\infty\left[|F_m''|^2 + \tfrac{n-1}{r^2}|F_m'|^2\right]r^{n-1}dr,\\I_{2,m} &= \int_0^\infty\left[|F_m'|^2 r^{-2} + (n-4)|F_m|^2 r^{-4}\right]r^{n-1}dr,\end{aligned}$$

and

$$I_{3,m} = \int_0^\infty \frac{|F_m|^2}{r^4}r^{n-1}dr.$$

Thus,

$$\|Df\|^2 = \sum_m\Big\{\int_0^\infty\big(|F_m''|^2 + \tfrac{n-1+2\lambda_m}{r^2}|F_m'|^2 + \tfrac{2(n-4)\lambda_m+\lambda_m^2}{r^4}|F_m|^2\big)r^{n-1}dr\Big\}. \tag{6.5.3}$$

Since $F_m \in C_0^\infty(0, \infty)$,

$$2\text{Re}\int_0^r t^{n-4}\overline{F_m(t)}F_m'(t)dt = r^{n-4}|F_m(r)|^2 - (n-4)\int_0^r t^{n-5}|F_m(t)|^2 dt$$

and

$$2\text{Re}\int_0^r t^{n-2}\overline{F_m'(t)}F_m''(t)dt = r^{n-2}|F_m'(r)|^2 - (n-2)\int_0^r t^{n-3}|F_m'(t)|^2 dt,$$

which imply that

$$r^{n-4}|F_m(r)|^2 \le \int_0^r |F_m'(t)|^2 t^{n-3} dt + (n-3)\int_0^r t^{n-5}|F_m(t)|^2 dt$$

and

$$r^{n-2}|F_m'(r)|^2 \le \int_0^r |F_m''(t)|^2 t^{n-1} dt + (n-1)\int_0^r t^{n-3}|F_m'(t)|^2 dt.$$

By substituting these inequalities into (6.5.3) and using Parseval's identity, we may conclude that, for $0 < r < \infty$,

$$\begin{aligned}\|Df\|^2 &\ge \sum_m \{r^{n-2}|F_m'(r)|^2 + 2\lambda_m r^{n-4}|F_m(r)|^2 \\ &\quad + \int_0^\infty \tfrac{\lambda_m(\lambda_m-2)}{r^4}|F_m(r)|^2 r^{n-1}dr\} \\ &\ge r^{n-2}\int_{S^{n-1}} |\tfrac{\partial f}{\partial r}|^2 d\omega + 2\min_m\{\lambda_m\} r^{n-4}\int_{S^{n-1}} |f|^2 d\omega \\ &\quad - \max_m\{\lambda_m(2-\lambda_m)\}\int_{\mathbb{R}^n} \tfrac{|f(\mathbf{x})|^2}{|\mathbf{x}|^4}d\mathbf{x},\end{aligned}$$

whence (6.5.1). □

**Corollary 6.5.2** *For $C(n,\alpha)$ defined in (6.4.7) and all $f \in \mathcal{D}_0$,*

$$\begin{aligned}&\left\| r^{n-2}\|\tfrac{\partial f}{\partial r}\|_{L^2(S^{n-1})} + 2\min_m\{\lambda_m\} r^{n-4}\|f\|_{L^2(S^{n-1})}\right\|_{L^\infty(0,\infty)} \\ &\qquad \le \|Df\|^2 + \max_m\{\lambda_m(2-\lambda_m)\}\||\mathbf{x}|^{-2}f\|^2 \\ &\qquad \le \left(1 + \tfrac{\max_m\{\lambda_m(2-\lambda_m)\}}{C(n,0)}\right)\|Df\|^2,\end{aligned} \tag{6.5.4}$$

*if (for the last inequality),*

$$C(n,0) := \inf_{m\in\mathcal{I}}\{\lambda_m + \frac{n(n-4)}{4}\}^2 \ne 0 \tag{6.5.5}$$

*and* $\max_m\{\lambda_m(2-\lambda_m)\} \ge 0$.

*Proof* The proof follows from (6.4.6) and Theorem 6.5.1. □

Note that $\max\{\lambda_m(2-\lambda_m)\} \le 1$, with equality attained only if some $\lambda_m = 1$. In particular, when $n = 4$ and $\min_m \lambda_m > 0$, then

$$\| \|f\|_{L^2(S^3)} \|_{L^\infty(0,\infty)} \le C\|Df\|^2$$

for a positive constant $C$. Hence, for radial $f \in \mathcal{D}_0$, it follows that $f \in L^\infty(0,\infty)$

We shall assume that $n = 2, 3,$or $4$, in order to make use of results already established. From Sect. 6.4, we see that for $n = 2, 4$, $C(n, 0) > 0$ and $\min\{\lambda_m\} > 0$ if $\Psi \in (0, 1)$. For $n = 3$, $\min\{\lambda_m\} > 0$ if $\Psi \in (0, 1)$ and $C(3, 0) > 0$ if $\Psi \in [0, \frac{1}{2}) \cup (\frac{1}{2}, 1)$. Therefore, by Corollary 6.5.2, we have

**Corollary 6.5.3** *If $\Psi \in (0, 1)$ when $n = 2, 4$, and $\Psi \in (0, \frac{1}{2}) \cup (\frac{1}{2}, 1)$ when $n = 3$, it follows that for all $f \in \mathcal{D}_0$,*

$$\|r^{n-2}\|\partial f/\partial r\|^2_{L^2(\mathbb{S}^{n-1})}\|_{L^\infty(0,\infty)}, \quad \|r^{n-4}\|f\|_{L^2(\mathbb{S}^{n-1})}\|_{L^\infty(0,\infty)} \le C\|\Delta_{\mathbf{A}} f\|^2 \tag{6.5.6}$$

*for some positive constant $C$.*

### 6.5.2 Forms and Operators

We shall assume hereafter that $n = 2, 3$, or $4$. For larger values of $n$ and higher order operators, see [142]. Define

$$\mathcal{D}_0' := C_0^\infty(\mathbb{R}^n \setminus \mathcal{L}_n);$$

note that

$$\mathcal{D}_0' \subseteq \mathcal{D}_0$$

and consequently, Theorem 6.5.1 and Corollary 6.5.2 apply for $f \in \mathcal{D}_0'$.

Let $S_{\mathbf{A}}^2$ denote the Friedrichs extension of the restriction of $\Delta_{\mathbf{A}}^2$ to $\mathcal{D}_0'$. The form domain $\mathcal{Q}(S_{\mathbf{A}}^2) = \mathcal{H}(S_{\mathbf{A}})$ of $S_{\mathbf{A}}^2$, is the completion of $\mathcal{D}_0'$ with respect to $[\|\Delta_{\mathbf{A}} f\|^2 + \|f\|^2]^{\frac{1}{2}}$. Therefore $\mathcal{H}(S_{\mathbf{A}})$ is the Hilbert space defined by the inner product

$$\begin{aligned}(\varphi, \psi)_{S_{\mathbf{A}}} &= ((S_{\mathbf{A}} + i)\varphi, (S_{\mathbf{A}} + i)\psi)_{L^2(\mathbb{R}^n)} \\ &= (S_{\mathbf{A}}\varphi, S_{\mathbf{A}}\psi)_{L^2(\mathbb{R}^n)} + (\varphi, \psi)_{L^2(\mathbb{R}^n)}, \qquad \varphi, \psi \in \mathcal{D}(S_{\mathbf{A}}),\end{aligned}$$

which induces the graph norm associated with $S_{\mathbf{A}} : \mathcal{D}(S_{\mathbf{A}}) \to L^2(\mathbb{R}^n)$, where $\mathcal{D}(S_{\mathbf{A}})$ denotes the domain of $S_{\mathbf{A}}$.

**Lemma 6.5.4** *Assume the hypothesis of Corollary 6.5.3. Let $B_+$ be the operator of multiplication by the function $b_+$, where*

$$0 \le b_+ \in L^1(\mathbb{R}_+; L^\infty(\mathbb{S}^{n-1}); r^3 dr) \equiv L^1(\mathbb{R}_+; r^3 dr) \otimes L^\infty(\mathbb{S}^{n-1}).$$

*Then, $B_+^{\frac{1}{2}} : \mathcal{H}(S_{\mathbf{A}}) \to L^2(\mathbb{R}^n)$ is bounded and $B_+^{\frac{1}{2}}(S_{\mathbf{A}} + i)^{-1}$ is compact on $L^2(\mathbb{R}^n)$.*

*Proof* For $\varphi \in \mathcal{D}_0' = C_0^\infty(\mathbb{R}^n \setminus \mathcal{L}_n)$,

$$\begin{aligned} |(B_+\varphi, \varphi)| &= \int_{\mathbb{S}^{n-1}} \int_0^\infty b_+(r,\omega)|\varphi(r,\omega)|^2 r^{n-1} dr d\omega \\ &\le \int_0^\infty \|b_+\|_{L^\infty(\mathbb{S}^{n-1})} r^3 dr \sup_{0<r<\infty} \left( r^{n-4} \int_{\mathbb{S}^{n-1}} |\varphi|^2 d\omega \right) \\ &\le C \|b_+\|_{L^1(\mathbb{R}_+; L^\infty(\mathbb{S}^{n-1}); r^3 dr)} \|S_{\mathbf{A}}\varphi\|^2 \end{aligned} \tag{6.5.7}$$

by Corollary 6.5.2. Thus, $\mathcal{D}(S_{\mathbf{A}})$ lies in the form domain of $B_+$ and $B_+^{\frac{1}{2}} : \mathcal{H}(S_{\mathbf{A}}) \to L^2(\mathbb{R}^n)$ is bounded.

Let $\varphi_\ell \rightharpoonup 0$ in $L^2(\mathbb{R}^n)$ and set $\psi_\ell = (S_{\mathbf{A}} + i)^{-1}\varphi_\ell$. Then, $\psi_\ell \in \mathcal{D}(S_{\mathbf{A}})$ and $\psi_\ell \rightharpoonup 0$ in $\mathcal{H}(S_{\mathbf{A}})$. Given $\varepsilon > 0$, choose $\tilde{b}_+$ such that

$$\tilde{b}_+ \in C_0^\infty(\mathbb{R}_+; L^\infty(\mathbb{S}^{n-1})), \quad \operatorname{supp} \tilde{b}_+ \subset \Omega_\varepsilon = B(0; k_\varepsilon) \setminus B(0; 1/k_\varepsilon),$$
$$\|\tilde{b}_+\|_{L^\infty(\mathbb{R}^n)} < k_\varepsilon, \quad \text{and} \quad \left\| \|b_+ - \tilde{b}_+\|_{L^\infty(\mathbb{S}^{n-1})} \right\|_{L^1(\mathbb{R}_+; r^3 dr)} < \varepsilon$$

for some $k_\varepsilon > 1$.

Furthermore,

$$\begin{aligned} &\|B_+^{\frac{1}{2}}(S_{\mathbf{A}} + i)^{-1}\varphi_\ell\|^2 = \|B_+^{\frac{1}{2}}\psi_\ell\|^2 = (B_+\psi_\ell, \psi_\ell) \\ &= \int_{\mathbb{R}^n} \tilde{b}_+ |\psi_\ell|^2 d\mathbf{x} + \int_{\mathbb{R}^n} (b_+ - \tilde{b}_+)|\psi_\ell|^2 d\mathbf{x} \\ &\le k_\varepsilon \int_{\Omega_\varepsilon} |\psi_\ell|^2 d\mathbf{x} \\ &\quad + \| \|b_+ - \tilde{b}_+\|_{L^\infty(\mathbb{S}^{n-1})} \|_{L^1(\mathbb{R}_+; r^3 dr)} \sup_{0<r<\infty} \left\{ r^{n-4} \int_{\mathbb{S}^{n-1}} |\psi_\ell|^2 d\omega \right\} \\ &\le k_\varepsilon \int_{\Omega_\varepsilon} |\psi_\ell|^2 d\mathbf{x} + \varepsilon C \|S_{\mathbf{A}}\psi_\ell\|^2 \end{aligned} \tag{6.5.8}$$

by Corollary 6.5.3.

For $u \in \mathcal{D}_0' = C_0^\infty(\mathbb{R}^n \setminus \mathcal{L}_n)$

$$\begin{aligned} \|\nabla_{\mathbf{A}} u\|^2 &= (-\Delta_{\mathbf{A}} u, u) \le \|\Delta_{\mathbf{A}} u\| \|u\| \\ &\le \frac{1}{2} \left( \|\Delta_{\mathbf{A}} u\|^2 + \|u\|^2 \right) \\ &= \frac{1}{2} \|(S_{\mathbf{A}} + i)u\|^2. \end{aligned}$$

Hence,

$$\|\nabla_{\mathbf{A}}\psi_l\| \le \frac{1}{\sqrt{2}}\|\varphi_l\|,$$

and by the diamagnetic inequality

$$\|\nabla|\psi_l|\| \le \|\nabla_{\mathbf{A}}\psi_l\| \le \frac{1}{\sqrt{2}}\|\varphi_l\|.$$

It follows that the sequence $\{|\psi_\ell|\}$ must be bounded in $H^1(\mathbb{R}^n)$. Since $H^1(\Omega_\varepsilon)$ is compactly embedded in $L^2(\Omega_\varepsilon)$, it follows that $\psi_\ell \to 0$ in $L^2(\Omega_\varepsilon)$. The result now follows from (6.5.8) and the fact that $\varepsilon$ can be chosen arbitrarily small. □

*Remark 6.5.5* The compactness of $B_+^{\frac{1}{2}}(S_{\mathbf{A}}+i)^{-1} : L^2(\mathbb{R}^n) \to L^2(\mathbb{R}^n)$ established in Lemma 6.5.4 implies that $B_+^{\frac{1}{2}}$ is $S_{\mathbf{A}}$-compact, and consequently, by [48] (Corollary III.7.7), $B_+^{\frac{1}{2}}$ has $S_{\mathbf{A}}$-bound zero. This implies that the form $(B_+u, u)$ is relatively bounded with respect to the form $(S_{\mathbf{A}}u, S_{\mathbf{A}}u)$ with relative bound zero. Therefore, $\Delta_{\mathbf{A}}^2 \pm B_+$ is defined in the form sense, and has form domain $\mathcal{D}(S_{\mathbf{A}})$ by Kato's Second Representation Theorem; see [48, 83].

**Lemma 6.5.6** *Let $n = 4$ and suppose that $\Psi \in (0, 1)$. For*

$$0 \le V \in L^1(\mathbb{R}_+; L^\infty(\mathbb{S}^3), r^3dr),$$

*let $B_-$ be a nonnegative self-adjoint operator with form domain $\mathcal{D}(S_{\mathbf{A}})$ which satisfies the following condition: given $\varepsilon > 0$, there is $k(\varepsilon)$ such that for all $\varphi \in \mathcal{D}(S_{\mathbf{A}})$,*

$$\begin{aligned}(B_-\varphi, \varphi) \le \varepsilon \int_0^\infty \int_{S^3} r\left|\tfrac{\partial}{\partial r}\varphi(r,\omega)\right|^2 d\omega dr \\ +k(\varepsilon)\int_0^\infty \int_{S^3} V(r,\omega)|\varphi(r,\omega)|^2 d\omega dr.\end{aligned} \tag{6.5.9}$$

*Then $B_-^{\frac{1}{2}}(S_{\mathbf{A}}+i)^{-1}$ is compact on $L^2(\mathbb{R}^4)$.*

*Proof* As in the proof of Lemma 6.5.4, given $\delta > 0$, we may choose $\tilde{V}$ such that for some $k_\delta > 1$,

$$\tilde{V} \in C_0^\infty(\mathbb{R}_+; L^\infty(\mathbb{S}^3)), \quad \text{supp}\,\tilde{V} \subset \Omega_\delta = B(0;k_\delta) \setminus B(0; 1/k_\delta),$$

$$\|\tilde{V}\|_{L^\infty(\mathbb{R}^4)} < k_\delta, \text{ and } \left\|\|V-\tilde{V}\|_{L^\infty(\mathbb{S}^3)}\right\|_{L^1((0,\infty);r^3dr)} < \delta.$$

Let $\varphi_\ell \rightharpoonup 0$ in $L^2(\mathbb{R}^4)$ with $\|\varphi_\ell\| \le 1$, and set $\psi_\ell = (S_{\mathbf{A}} + i)^{-1}\varphi_\ell$. Then, $\psi_\ell \rightharpoonup 0$ in $\mathcal{H}(S_{\mathbf{A}})$ and, on using (6.5.9),

$$\begin{aligned}\|B_-^{\frac{1}{2}}(S_{\mathbf{A}} + i)^{-1}\varphi_\ell\| &\le \varepsilon \textstyle\int_0^\infty \int_{S^3} r\left|\frac{\partial}{\partial r}\psi_\ell(r,\omega)\right|^2 d\omega dr \\ &\quad + k(\varepsilon)\big\{k_\delta \textstyle\int_{\Omega_\delta} |\psi_\ell(\mathbf{x})|^2 d\mathbf{x} \\ &\quad + \delta C \sup_{0<r<\infty} \textstyle\int_{\mathbb{S}^3} |\psi_\ell(r,\omega)|^2 d\omega\big\} \\ &\le \varepsilon \textstyle\int_0^\infty \int_{S^3} r\left|\frac{\partial}{\partial r}\psi_\ell(r,\omega)\right|^2 d\omega dr \\ &\quad + k(\varepsilon)\big\{k_\delta \textstyle\int_{\Omega_\delta} |\psi_\ell(\mathbf{x})|^2 d\mathbf{x} + \delta C\|S_{\mathbf{A}}\psi_\ell\|^2\big\},\end{aligned}$$

by (6.5.6). Now, note that for the case $n = 4$ and $I_{1,m}$ defined in the proof of Theorem 6.5.1,

$$3\int_0^\infty \int_{S^3} r\left|\frac{\partial}{\partial r}\psi_\ell(r,\omega)\right|^2 d\omega dr \le \sum_m I_{1,m} \le \|S_{\mathbf{A}}\psi_\ell\|^2,$$

by (6.5.2) and since $\min\{\lambda_m\} > 0$. Consequently,

$$\|B_-^{\frac{1}{2}}(S_{\mathbf{A}} + i)^{-1}\varphi_\ell\| \le \tfrac{\varepsilon}{3}\|\varphi_\ell\|^2 + k(\varepsilon)\big\{k_\delta \textstyle\int_{\Omega_\delta} |\psi_\ell(\mathbf{x})|^2 d\mathbf{x} + \delta C\|\varphi_\ell\|^2\big\}.$$

We therefore conclude, as in the proof of Lemma 6.5.4, that

$$\limsup_{\ell\to\infty} \|B_-^{\frac{1}{2}}(S_{\mathbf{A}} + i)^{-1}\varphi_\ell\| \le \varepsilon + Ck(\varepsilon)\delta.$$

Since $\delta$ and $\varepsilon$ are arbitrary, the lemma follows. □

At this point it should be helpful to explore examples of multiplication operators $B_-$ that satisfy the hypothesis of Lemma 6.5.6.

**Lemma 6.5.7** *Let $b(r) \ge 0$ on $(0,\infty)$ and*

$$\int_0^\infty \int_r^\infty b(s)s^2 ds dr < \infty, \qquad \int_0^\infty r\Big(\int_r^\infty b(s)s^2 ds\Big)^2 dr < \infty. \tag{6.5.10}$$

*Then, there is a function $W \in L^1((0,\infty); r^3 dr)$ such that, for any $\varepsilon > 0$,*

$$\int_0^\infty b(r)|\varphi(r)|^2 r^3 dr \le \varepsilon \int_0^\infty r|\varphi'(r)|^2 dr + k(\varepsilon)\int_0^\infty W(r)|\varphi(r)|^2 r^3 dr, \tag{6.5.11}$$

*for all $\varphi \in C_0^\infty(0,\infty)$ and some constant $k(\varepsilon) > 0$. We can take*

$$r^3 W(r) = r\left(\int_r^\infty b(s)s^2 ds\right)^2 + \int_r^\infty b(s)s^2 ds. \tag{6.5.12}$$

*Proof* Let

$$r^{\frac{3}{2}}\sqrt{\omega(r)} = \int_r^\infty b(s)s^2 ds. \tag{6.5.13}$$

According to Opic and Kufner [126], Theorem 5.9, p. 63, the inequality

$$\int_0^\infty b(r)|\varphi(r)|^2 r^3 dr \le c\int_0^\infty \frac{d}{dr}(r|\varphi(r)|^2) r^{\frac{3}{2}}\sqrt{\omega(r)}dr \tag{6.5.14}$$

is satisfied for some $c > 0$ if and only if

$$C := \sup_{0<r<\infty}\left[\int_r^\infty t^2 b(t)dt \cdot \sup_{0<t<r}\left\{[t^{\frac{3}{2}}\sqrt{\omega(t)}]^{-1}\right\}\right] < \infty$$

with $c = C$ the best possible constant for (6.5.14); this is derived from Theorem 1.2.3 on taking Remark 1.2.4 into account. On choosing (6.5.13), it follows that $C \le 1$. From (6.5.14) with $c \le 1$

$$\begin{aligned}\textstyle\int_0^\infty b(r)|\varphi(r)|^2 r^3 dr &\le 2\textstyle\int_0^\infty r|\varphi(r)\varphi'(r)|r^{\frac{3}{2}}\sqrt{\omega(r)}dr\\ &\quad+\textstyle\int_0^\infty |\varphi(r)|^2 r^{\frac{3}{2}}\sqrt{\omega(r)}dr\\ &\le \varepsilon\textstyle\int_0^\infty r|\varphi'(r)|^2 dr + \frac{1}{\varepsilon}\int_0^\infty |\varphi(r)|^2\omega(r)r^4 dr\\ &\quad+\textstyle\int_0^\infty |\varphi(r)|^2 r^{\frac{3}{2}}\sqrt{\omega(r)}dr.\end{aligned}$$

The choice (6.5.12) yields (6.5.11) with $k(\varepsilon) = \varepsilon^{-1} + 1$ and $W \in L^1((0,\infty); r^3 dr)$ in view of (6.5.10). □

**Theorem 6.5.8** *Assume the hypothesis of Lemma 6.5.4, and when $n = 4$, assume the hypothesis of Lemma 6.5.6. Then we have the following:*

(i) *The form $(S_{\mathbf{A}}u, S_{\mathbf{A}}v)$ is closed with core $\mathcal{D}_0'$ and $S_{\mathbf{A}}^2$ is the associated self-adjoint operator.*
(ii) *The symmetric form $\mathfrak{t}_{\mathbf{A}}[u,v] = (S_{\mathbf{A}}u, S_{\mathbf{A}}v) + (B_+u, v)$ is closed and bounded below with core $\mathcal{D}_0'$. Let $T_{\mathbf{A}}^2 = S_{\mathbf{A}}^2 + B_+$ denote the operator associated with $\mathfrak{t}_{\mathbf{A}}$. It has form domain $\mathcal{Q}(\mathfrak{t}_{\mathbf{A}}) = \mathcal{Q}(S_{\mathbf{A}}^2) = \mathcal{D}(S_{\mathbf{A}})$ and $\sigma_{ess}(T_{\mathbf{A}}^2) = \sigma_{ess}(S_{\mathbf{A}}^2) = [0,\infty)$.*
(iii) *For $T_{\mathbf{A}}$ defined as the positive square root of $T_{\mathbf{A}}^2$ and $n = 4$, $B_-^{\frac{1}{2}}(T_{\mathbf{A}} + i)^{-1}$ is compact on $L^2(\mathbb{R}^4)$ and $T_{\mathbf{A}}^2 - B_-$ is defined in the form sense with form domain $\mathcal{D}(S_{\mathbf{A}})$. Moreover,*

$$\sigma_{ess}(S_{\mathbf{A}}^2 + B_+ - B_-) = \sigma_{ess}(S_{\mathbf{A}}^2) = [0,\infty).$$

*Proof* The proof of (i) follows as in [83], Examples VI.2.13 & VI.1.23.

The first part of (ii) follows from Remark 6.5.5. The fact that $\mathcal{Q}(\mathfrak{t}_{\mathbf{A}}) = \mathcal{Q}(S_{\mathbf{A}}^2) = \mathcal{D}(S_{\mathbf{A}})$ follows from Kato's Second Representation Theorem; see [83], p. 331. Since $B_+^{\frac{1}{2}}(S_{\mathbf{A}}+i)^{-1}$ is compact in $L^2(\mathbb{R}^n)$ by Lemma 6.5.4, Theorem IV.4.4 of [48] applies (with $p_2 = 0$) showing that Theorem IV.4.2 (vi) in [48] holds: equivalently, in the language of Sect. 1.5.1, the form $(B_+\cdot,\cdot)$ is compact relative to the form $(S_{\mathbf{A}}\cdot, S_{\mathbf{A}}\cdot)$. This fact implies that $\sigma_{ess}(T_{\mathbf{A}}^2) = \sigma_{ess}(S_{\mathbf{A}}^2)$.

To show (iii), we begin by observing that, for $f \in \mathcal{D}(S_{\mathbf{A}})$,

$$\|S_{\mathbf{A}}f\|^2 \le \|T_{\mathbf{A}}f\|^2 = \|S_{\mathbf{A}}f\|^2 + (B_+f, f),$$

implying that for some $C > 0$,

$$\|(S_{\mathbf{A}}+i)f\|^2 \le \|(T_{\mathbf{A}}+i)f\|^2 = C\|(S_{\mathbf{A}}+i)f\|^2,$$

by (6.5.7). Then with $f = (T_{\mathbf{A}}+i)^{-1}g$, we have that

$$\|(S_{\mathbf{A}}+i)(T_{\mathbf{A}}+i)^{-1}g\| \le \|g\|,$$

so that from Lemma 6.5.6 we have that $B_-^{\frac{1}{2}}(T_{\mathbf{A}}+i)^{-1}$ is compact on $L^2(\mathbb{R}^4)$. The remainder of the proof of part (iii) follows that of part (i) above. □

### 6.5.3 *Estimating the Number of Eigenvalues*

**Theorem 6.5.9** *Let the hypotheses of Lemmas 6.5.4 and 6.5.6 be satisfied. Then*

(i) $L_{\mathbf{A}} := S_{\mathbf{A}}^2 + B_+ - B_-$ *is a self-adjoint operator defined in the form sense;*
(ii) $B_-^{\frac{1}{2}}(T_{\mathbf{A}}+i)^{-1}$ *is compact in* $L^2(\mathbb{R}^4)$, *where* $T_{\mathbf{A}}^2 = S_{\mathbf{A}}^2 + B_+$;
(iii) $\sigma_{ess}(L_{\mathbf{A}}) = [0,\infty)$;
(iv) *if* $\Psi \in (0,1)$ *and* $n = 4$, *there exists a positive constant* $C = C(\Psi)$ *such that the number* $N(L_{\mathbf{A}})$ *of negative eigenvalues of* $L_{\mathbf{A}}$ *satisfies*

$$N(L_{\mathbf{A}}) \le C(\Psi) \Big\| \|V\|_{L^\infty(\mathbb{S}^3)} \Big\|_{L^1((0,\infty);r^3dr)}, \tag{6.5.15}$$

*where* $V$ *is given in Lemma 6.5.6 and the constant* $C(\Psi)$ *is dependent upon the distance of* $\Psi$ *from the boundary values* 0 *and* 1.

*Proof* Parts (i)–(iii) are covered in Theorem 6.5.8 and are included here for completeness.

For part (iv), we see from (6.5.3) that for $n = 2, 3, 4$,

$$\|\Delta_{\mathbf{A}}f\|^2 = \sum_m \int_0^\infty \overline{F_m} D_m F_m r^{n-1} dr,$$

where $F_m$ is given by (6.4.8) and

$$\begin{aligned} D_m = & \tfrac{1}{r^{n-1}}\tfrac{d^2}{dr^2}\left(r^{n-1}\tfrac{d^2}{dr^2}\right) - \tfrac{(n-1)+2\lambda_m}{r^{n-1}}\tfrac{d}{dr}\left(r^{n-3}\tfrac{d}{dr}\right) \\ & + \tfrac{2(n-4)\lambda_m+\lambda_m^2}{r^4}. \end{aligned} \tag{6.5.16}$$

Define

$$W(r) := \|V(r,\cdot)\|_{L^\infty(\mathbb{S}^3)}.$$

Thus, when $n = 4$, since

$$B_- \le \frac{\varepsilon}{r^3}\frac{d}{dr}(r\frac{d}{dr}) + k(\varepsilon)W(r)$$

from (6.5.9), we have

$$\begin{aligned} \Delta_{\mathbf{A}}^2 + B_+ - B_- &\ge \Delta_{\mathbf{A}}^2 - B_- \\ &\ge \bigoplus_{m\in\mathbb{Z}''}\left\{\left[D_m + \tfrac{\varepsilon}{r^3}\tfrac{d}{dr}(r\tfrac{d}{dr}) - k(\varepsilon)W(r)\right]\otimes \mathbb{I}_m\right\} \end{aligned} \tag{6.5.17}$$

where

$$\mathbb{Z}'' := \{m \in \mathbb{Z} : (m-\Psi)^2 \ge 1\},$$

$\mathbb{I}_m$ is the identity on the orthonormal basis $\{u_m\}_{m\in\mathbb{Z}''}$ of $L^2(\mathbb{S}^3)$, and $\lambda_m = (m-\Psi)^2 - 1$ as shown in Sect. 6.4.2. In (6.5.17)

$$D_m + \frac{\varepsilon}{r^3}\frac{d}{dr}(r\frac{d}{dr}) = \frac{1}{r^3}\frac{d^2}{dr^2}(r^3\frac{d^2}{dr^2}) - \frac{3+2\lambda_m-\varepsilon}{r^3}\frac{d}{dr}(r\frac{d}{dr}) + \frac{\lambda_m^2}{r^4}.$$

We also have that

$$\Delta^2 + \frac{c}{r^4} = \bigoplus_{|m|\ge 1}\{[D_m^0 + \frac{c}{r^4}]\otimes \mathbb{I}_m\}$$

in which

$$D_m^0 + \frac{c}{r^4} = \frac{1}{r^3}\frac{d^2}{dr^2}(r^3\frac{d^2}{dr^2}) - \frac{3+2\lambda_m^0}{r^3}\frac{d}{dr}(r\frac{d}{dr}) + \frac{(\lambda_m^0)^2+c}{r^4},$$

with $\lambda_m^0 = m^2 - 1$. If $m \in \mathbb{Z}''$, then either $m \ge 1$, in which case

$$\lambda_m \ge \lambda_m^0 + \Psi^2, \qquad \lambda_m^2 \ge (\lambda_m^0)^2 + \Psi^4,$$

or $m \leq -2$ and thus

$$\lambda_m \geq (m+1)^2 - 1 + (1-\Psi)^2 = \lambda^0_{m+1} + (1-\Psi)^2,$$
$$\lambda^2_m \geq (\lambda^0_{m+1})^2 + (1-\Psi)^4.$$

As a consequence, for $m \geq 1$

$$D_m + \frac{\varepsilon}{r^3}\frac{d}{dr}\Big(r\frac{d}{dr}\Big) \geq D^0_m + \frac{c}{r^4}$$

if $\varepsilon < 2\Psi^2$ and $c < \Psi^4$. For $m \leq -2$

$$D_m + \frac{\varepsilon}{r^3}\frac{d}{dr}\Big(r\frac{d}{dr}\Big) \geq D^0_{m+1} + \frac{c}{r^4}$$

if $\varepsilon < 2(1-\Psi)^2$ and $c < (1-\Psi)^4$. Hence, if $\varepsilon < 2\min\{\Psi^2, (1-\Psi)^2\}$ and $c < \min\{\Psi^4, (1-\Psi)^4\}$, then

$$\begin{aligned} &N\Big(\bigoplus_{m\geq 1}\big[D_m + \tfrac{\varepsilon}{r^3}\tfrac{d}{dr}\big(r\tfrac{d}{dr}\big) - k(\varepsilon)W(r)\big] \otimes \mathbb{I}_m\Big) \\ &\quad \leq N\Big(\bigoplus_{m\geq 1}\big[D^0_m + \tfrac{c}{r^4} - k(\varepsilon)W(r)\big] \otimes \mathbb{I}_m\Big) \end{aligned}$$

and

$$\begin{aligned} &N\Big(\bigoplus_{m\leq -2}\big[D_m + \tfrac{\varepsilon}{r^3}\tfrac{d}{dr}\big(r\tfrac{d}{dr}\big) - k(\varepsilon)W(r)\big] \otimes \mathbb{I}_m\Big) \\ &\quad \leq N\Big(\bigoplus_{m\leq -1}\big[D^0_m + \tfrac{c}{r^4} - k(\varepsilon)W(r)\big] \otimes \mathbb{I}_m\Big). \end{aligned}$$

Now, Theorem 1.2 of Laptev and Netrusov [97] and the last two inequalities imply (6.5.15); cf. the proof of Theorem 5.6.4. □

**Theorem 6.5.10** *Let $\Psi$ satisfy Corollary 6.5.3, $V(\mathbf{x}) \geq 0$, and*

$$V \in L^1(\mathbb{R}_+; L^\infty(\mathbb{S}^{n-1}), r^3 dr).$$

*Then, the operator $S^2_{\mathbf{A}} - V$ is defined in the form sense and has essential spectrum $[0, \infty)$. Moreover, for $\lambda_m$ given in Sect. 6.4.2,*

$$N(S^2_{\mathbf{A}} - V) \leq {\sum}' \frac{4}{|4\lambda_m + n(n-4)|\sqrt{n^2 + 8\lambda_m}} \int_0^\infty r^3 \|V(r,\cdot)\|_{L^\infty(\mathbb{S}^{n-1})} dr,$$

*where ${\sum}'$ indicates that all summands less than 1 are omitted.*

*Proof* The fact that $S^2_{\mathbf{A}} - V$ is defined in the form sense and has essential spectrum $[0, \infty)$ follows as in Lemma 6.5.4 and Theorem 6.5.8.

For all $f \in \mathcal{D}_0' = C_0^\infty(\mathbb{R}^n \setminus \mathcal{L}_n)$ and

$$F_m(r) := \int_{\mathbb{S}^{n-1}} f(r,\omega)\overline{u_m(\omega)}d\omega,$$

we have from (6.5.3) with $n = 2, 3, 4$,

$$\begin{aligned}\|\Delta_{\mathbf{A}} f\|^2 &= \sum_m \{ \int_0^\infty \big(|F_m''|^2 + \tfrac{n-1+2\lambda_m}{r^2}|F_m'|^2 + \tfrac{2(n-4)\lambda_m+\lambda_m^2}{r^4}|F_m|^2\big) r^{n-1}dr\} \\ &\geq \sum_m \{ \int_0^\infty \big(\tfrac{\frac{1}{4}(n-2)^2+n-1+2\lambda_m}{r^2}|F_m'|^2 + \tfrac{2(n-4)\lambda_m+\lambda_m^2}{r^4}|F_m|^2\big) r^{n-1}dr\}\end{aligned}$$

by Hardy's inequality. On making the substitutions

$$c(n,\lambda_m) := n^2 + 8\lambda_m \quad \text{and} \quad \varphi_m(r) := \frac{\sqrt{c(n,\lambda_m)}}{2} r^{(n-3)/2} F_m(r),$$

we have that

$$\|\Delta_{\mathbf{A}} f\|^2 \geq \sum_m \int_0^\infty \big[|\varphi_m'|^2 + \tfrac{(n-3)(n-5)+16\lambda_m(\lambda_m+2(n-4))c(n,\lambda_m)^{-1}}{4r^2}|\varphi_m|^2\big]dr.$$

Therefore, for $f \in \mathcal{D}_0'$ and

$$K(n,\lambda_m) := (n-3)(n-5) + 16\lambda_m(\lambda_m + 2(n-4))(n^2+8\lambda_m)^{-1},$$

it follows that

$$((\Delta_{\mathbf{A}}^2 - V)f, f) \geq \sum_m \int_0^\infty \big[|\varphi_m'|^2 + \tfrac{K(n,\lambda_m)}{4r^2}|\varphi_m|^2 - \tfrac{4r^2}{n^2+8\lambda_m}W(r)|\varphi_m|^2\big]dr, \tag{6.5.18}$$

with $W(r) := \|V(r,\cdot)\|_{L^\infty(\mathbb{S}^{n-1})}$. Bargmann's estimate from [24] (see the proof of Theorem 5.6.3) for the number of negative eigenvalues applies to the Sturm-Liouville operator associated with the integral on the right-hand side of (6.5.18), i.e.,

$$\tau(n,m) := -\frac{d^2}{dr^2} + \frac{K(n,\lambda_m)}{4r^2} - \frac{4r^2}{n^2+8\lambda_m}W(r), \quad n = 2,3,4,$$

if

$$K(n,\lambda_m) > -1. \tag{6.5.19}$$

In that case,

$$N(\tau(n,m)) < \frac{4}{(n^2+8\lambda_m)\sqrt{K(n,\lambda_m)+1}} \int_0^\infty r^3 W(r)dr.$$

We first note that

$$K(n,\lambda_m)+1=[4\lambda_m+n(n-4)]^2/(n^2+8\lambda_m)\geq 0$$

since $\min\{\lambda_m\}>0$. In fact, it is easy to show that the strict inequality (6.5.19) holds with this hypothesis on substituting the values of $\lambda_m$, namely

$$\begin{aligned}
\lambda_m &= (m-\Psi)^2,\ m\in\mathbb{Z}, && \text{for } n=2;\\
\lambda_m &= (m-\Psi)(m-\Psi+1),\ m\in\mathbb{Z}', && \text{for } n=3;\\
\lambda_m &= (m-\Psi)^2-1,\ m\in\mathbb{Z}'', && \text{for } n=4.
\end{aligned} \tag{6.5.20}$$

In view of (6.5.18), the proof is complete. □

We now are able to use these results to give explicit criteria for the absence of negative eigenvalues.

**Corollary 6.5.11** *Assume the hypothesis of Theorem 6.5.10. Then $S_{\mathbf{A}}^2-V$ has no negative eigenvalues if for $n=2$,*

$$\int_0^\infty r^3\|V(r,\cdot)\|_{L^\infty(\mathbb{S}^{n-1})}dr<\begin{cases}2\Psi(2-\Psi)\sqrt{3-4\Psi+2\Psi^2} & \text{for } \Psi\in(0,\frac{1}{2}],\\ 2(1-\Psi^2)\sqrt{1+2\Psi^2} & \text{for } \Psi\in(\frac{1}{2},1];\end{cases} \tag{6.5.21}$$

*for $n=3$,*

$$\int_0^\infty r^3\|V(r,\cdot)\|_{L^\infty(\mathbb{S}^{n-1})}dr<\begin{cases}|\Psi(1+\Psi)-\frac{3}{4}|\sqrt{9+8\Psi(1+\Psi)} & \text{for } \Psi\in[0,\frac{1}{2}],\\ |\Psi^2-3\Psi+\frac{5}{4}|\sqrt{25-24\Psi+8\Psi^2} & \text{for } \Psi\in(\frac{1}{2},1);\end{cases} \tag{6.5.22}$$

*for $n=4$,*

$$\int_0^\infty r^3\|V(r,\cdot)\|_{L^\infty(\mathbb{S}^{n-1})}dr<\begin{cases}2^{\frac{3}{2}}\Psi(2+\Psi)\sqrt{2+2\Psi+\Psi^2} & \text{for } \Psi\in(0,\frac{1}{2}],\\ 2^{\frac{3}{2}}((2-\Psi)^2-1)\sqrt{1+(2-\Psi)^2} & \text{for } \Psi\in(\frac{1}{2},1).\end{cases} \tag{6.5.23}$$

*Proof* Define

$$B(\lambda_m,n):=\frac{1}{4}|4\lambda_m+n(n-4)|\sqrt{n^2+8\lambda_m}.$$

Then by Theorem 6.5.10 there will be no eigenvalues if

$$\int_0^\infty r^3\|V(r,\cdot)\|_{L^\infty(\mathbb{S}^{n-1})}dr<\min_m\{B(\lambda_m,n)\}$$

for $m\in\mathbb{Z}$ further restricted according to (6.5.20).

The functions $B(x,n)$, $n = 2,3,4$, are minimized on $[0,\infty)$ for some $x \in (0,2)$ and accordingly, in order to minimize $B(\lambda, n)$ we may restrict our attention to those $\lambda_m$ given in (6.5.20) that lie in the interval $(0,2)$. Noting that $\lambda_m = \lambda_m(\Psi)$, the estimate (6.5.21) follows from the fact that

$$\min_{m\in\mathbb{Z}} B(\lambda_m, 2) = \min_{\Psi\in(0,1)} \{B(\lambda_0, 2), B(\lambda_{-1}, 2)\};$$

(6.5.22) follows from the fact that

$$\min_{m\in\mathbb{Z}} B(\lambda_m, 3) = \min_{\Psi\in[0,1)} \{B(\lambda_{-1}, 3), B(\lambda_1, 3)\};$$

and (6.5.23) follows from the fact that

$$\min_{m\in\mathbb{Z}} B(\lambda_m, 4) = \min_{\Psi\in(0,1)} \{B(\lambda_1, 4), B(\lambda_{-2}, 4)\}.$$

□

# References


1. Adams, R.A.: Sobolev Spaces. Academic Press, New York (1975)
2. Adimurthi, Tintarev, K.: Hardy inequalities for weighted Dirac operator. Ann. Mat. Pura Appl. **189**, 241–251 (2010)
3. Aharonov, Y., Bohm, D.: Significance of electromagnetic potentials in quantum theory. Phys. Rev. **115**, 485–491 (1959)
4. Aharonov, Y., Bohm, D.: Further considerations on electromagnetic potentials in quantum theory. Phys. Rev. **123**, 1511–1524 (1961)
5. Ahlfors, L.V.: Complex Analysis. An Introduction to the Theory of Analytic Functions of One Complex Variable. International Series in Pure and Applied Mathematics, 3rd edn. McGraw-Hill Book Co., New York (1978)
6. Allegretto, W.: Nonoscillation theory of elliptic equations of order $2n$. Pac. J. Math. **64**(1), 1–16 (1976)
7. Ancona, A.: On strong barriers and an inequality of Hardy for domains in $\mathbb{R}^2$. J. Lond. Math. Soc. **34**, 274–290 (1986)
8. Armitage, D.H., Kuran, Ü.: The convexity of a domain and the superharmonicity of the signed distance function. Proc. Am. Math. Soc. **93**(4), 598–600 (1985)
9. Aubin, T.: Problemes isoperimetriques et espaces de Sobolev. J. Differ. Geom. **11**, 573–598 (1976)
10. Avkhadiev, F.G.: Hardy type inequalities in higher dimensions with explicit estimate of constants. Lobachevskii J. Math. **21**, 3–31 (2006). http://ljm.ksu.ru
11. Avkhadiev, F.G., Laptev, A.: Hardy inequalities for non-convex domains. In: Around the Research of Vladimir Maz'ya. I, International Mathematical Series (N. Y.), vol. 11, pp. 1–12. Springer, New York (2010)
12. Avkhadiev, F.G., Wirths, K.: Unified Poincaré and Hardy inequalities with sharp constants for convex domains. Z. Angew. Math. Mech. **87**(8–9), 632–642 (2007)
13. Avron, J., Herbst, I., Simon, B.: Schrödinger operators with magnetic fields. I. General Interactions. Duke Math. J. **45**(4), 847–883 (1978)
14. Balinsky, A.A.: Hardy type inequalities for Aharonov-Bohm magnetic potentials with multiple singularities, Math. Res. Lett. **10**, 169–176 (2003)
15. Balinsky, A., Evans, W.D.: On the zero modes of Pauli operators. J. Funct. Anal. **179**, 120–135 (2001)
16. Balinsky, A., Evans, W.D.: Some recent results on Hardy-type inequalities. Appl. Math. Inf. Sci. **4**(2), 191–208 (2010)
17. Balinsky, A., Evans, W.D., Hundertmark, D., Lewis, R.T.: On inequalities of Hardy-Sobolev type. Banach J. Math. Anal. **2**(2), 94–106 (2008)




A.A. Balinsky et al., *The Analysis and Geometry of Hardy's Inequality*, Universitext, DOI 10.1007/978-3-319-22870-9

18. Balinsky, A., Evans, W.D., Lewis, R.T.: On the number of negative eigenvalues of Schrodinger operators with an Aharonov-Bohm magnetic field. Proc. R. Soc. Lond. **457**, 2481–2489 (2001)
19. Balinsky, A., Evans, W.D., Lewis, R.T.: Sobolev, Hardy and CLR inequalities associated with Pauli operators in $\mathbb{R}^3$. J. Phys. A **34**(5), L19–L23 (2001)
20. Balinsky, A., Evans, W.D., Lewis, R.T.: Hardy's inequality and curvature. J. Funct. Anal. **262**, 648–666 (2012) [Available online 13 October 2011]
21. Balinsky, A., Evans, W.D., Umeda, T.: The Dirac-Hardy and Dirac-Sobolev inequalities in $L^1$. Publ. Res. Inst. Math. Sci. **47**(3), 791–801 (2011)
22. Balinsky, A., Laptev, A., Sobolev, A.: Generalized Hardy inequality for the magnetic Dirichlet forms. J. Stat. Phys. **116**(114), 507–521 (2004)
23. Barbatis, G., Filippas, S., Tertikas, A.: A unified approach to improved $L^p$ Hardy inequalities with best constants. Trans. Am. Math. Soc. **356**(6), 2169–2196 (2004)
24. Bargmann, V.: On the number of bound states in a central field of force. Proc. Nat. Acad. Sci. USA **38**, 961–966 (1952)
25. Batelaan, H., Tonomura, A.: The Aharonov–Bohm effects: variations on a subtle theme. Phys. Today **62**(9), 38–43 (2009)
26. Bateman, H.: Higher Transcendental Functions, vol. I. McGraw-Hill, New York (1953)
27. Benguria, R.D., Van Den Bosch, H.: A criterion for the existence of zero modes for the Pauli operator with fastly decaying fields. J. Math. Phys. **56**, 052104 (2015)
28. Benguria, R.D., Frank, R.L., Loss, M.: The sharp constant in the Hardy-Sobolev-Maz'ya inequality in the three dimensional upper half-space. Math. Res. Lett. **15**(4), 613–622 (2008)
29. Bennett, D.M.: An extension of Rellich's inequality. Proc. Am. Math. Soc. **106**, 987–993 (1989)
30. Brezis, H., Marcus, M.: Hardy's inequalities revisited. Dedicated to Ennio De Giorgi. Ann. Scuola Norm. Sup. Pisa Cl. Sci. **25**(4), 217–237 (1997)
31. Brezis, H., Vázquez, J.V.: Blow-up solutions of some nonlinear elliptic problems. Revista Matemática de la Universidad Complutense de Madrid **10**(2), 443–469 (1997)
32. Brown, R.C., Edmunds, D.D., Rákosník, J.: Remarks on inequalities of Poincaré type. Czech. Math. J. **45**, 351–377 (1995)
33. Bunt, L.H.N.: Bijdrage tot de Theorie der convexe Puntverzamelingen. Thesis, University of Groningen, Amsterdam (1934)
34. Caldiroli, P., Musina, R.: Rellich inequality with weights (English summary). Calc. Var. PDE. **45**(1–2), 147–164 (2012)
35. Cannarsa, P., Sinestrari, C.: Semiconcave functions, Hamilton Jacobi equations and optimal control. In: Progress in Nonlinear Differential equations and their Applications, vol. 58, Birkhäuser, Boston (2004)
36. Chambers, R.G.: Shift of an electron interference pattern by enclosed magnetic flux. Phys. Rev. Lett. **5**(3), (1960)
37. Chisholm, R.S., Everitt, W.N.: On bounded integral operators in the space of square integrable functions. Proc. R. Soc. Edinb. A **69**, 199–204 (1971)
38. Conlon, J.P.: A new proof of the Cwikel-Lieb-Rosenbljum bound. Rocky Mt. J.Math. 117–122 (1985)
39. Cwikel, M.: Weak type estimates for singular values and the number of bound states of Schrödinger operators. Ann. Math. **106**, 93–100 (1977)
40. Davies, E.B.: One-Parameter Semigroups. Academic, London (1980)
41. Davies, E.B.: Some norm bounds and quadratic form inequalities for Schrödinger operators (II). J. Oper. Theory **12**, 177–196 (1984)
42. Davies, E.B.: Heat Kernels and Spectral Theory. Cambridge Tracts in Mathematics, vol. 92. Cambridge University Press, Cambridge (1989)
43. Davies, E.B.: Spectral Theory and Differential Operators. Cambridge Studies in Advanced Mathematics, vol. 42. Cambridge University Press, Cambridge (1995)
44. Davies, E.B.: The Hardy constant. Q. J. Math. Oxf. (2) **46**, 417–431 (1995)

45. Davies, E.B., Hinz, A.M.: Explicit constants for Rellich inequalities in $L_p(\Omega)$. Math. Z. **227**(3), 511–523 (1998)
46. Dolbeault, J., Esteban, M.J., Séré, E.: On the eigenvalues of operators with gaps. Application to Dirac operators. J. Funct. Anal. **174**(1), 208–226 (2000)
47. Dolbeault, J., Esteban, M.J., Loss, M., Vega, L.: An analytic proof of Hardy-like inequalities related to the Dirac operator. J. Funct. Anal. **216**, 1–21 (2004)
48. Edmunds, D.E., Evans, W.D.: Spectral Theory and Differential Operators. Oxford University Press, Oxford (1987) [OX2 GDP]
49. Edmunds, D.E., Evans, W.D.: Hardy Operators, Function Spaces, and Embeddings. Springer Monographs in Mathematics. Springer, Berlin/Heidelberg/New York (2004)
50. Ehrenberg, W., Siday, R.: The refractive index in electron optics and the principles of dynamics. Proc. Phys. Soc. B **62**, 821 (1949)
51. Esteban, M.J., Loss, M.: Self-adjointness of Dirac operators via Hardy-Dirac inequalities. J. Math. Phys. **48**, 112107 (2007)
52. Evans, L.C., Gariepy, R.F.: Measure Theory and Fine Properties of Functions. Studies in Advanced Mathematics. CRC, Boca Raton/London/New York/Washington, DC (1992)
53. Evans, W.D., Harris, D.J.: Sobolev embeddings for generalized ridged domains, Proc. Lond. Math. Soc. **54**, 141–175 (1987)
54. Evans, W.D., Lewis, R.T.: On the Rellich inequality with magnetic potentials. Math. Z. **251**, 267–284 (2005)
55. Evans, W.D., Lewis, R.T.: Counting eigenvalues of biharmonic operators with magnetic fields. Math. Nachrichten **278**(12–13), 1524–1537 (2005)
56. Evans, W.D., Lewis, R.T.: Hardy and Rellich inequalities with remainders. J. Math. Inequal. **1**(4), 473–490 (2007)
57. Falconer, K.: Fractal Geometry: Mathematical Foundations and Applications. Wiley, New York (2007)
58. Federer, H.: Curvature measures. Trans. Am. Math. Soc. **93**(3), 418–491 (1959)
59. Federer, H., Fleming, W.: Normal and integral currents. Ann. Math. **72**, 458–520 (1960)
60. Filippas, S., Maz'ya, V., Tertikas, A.: On a question of Brezis and Marcus. Calc. Var. **25**(4), 491–501 (2006)
61. Filippas, S., Maz'ya, V., Tertikas, A.: Critical Hardy-Sobolev inequalities. J. Math. Pures Appl. **87**, 37–56 (2007)
62. Frank, R.L., Loss, M.: Hardy-Sobolev-Maz'ya inequalities for arbitrary domains. J. Math. Pures Appl. **97**, 39–54 (2012)
63. Frank, R.L., Lieb, E., Seiringer, R.: Hardy-Lieb-Thirring inequalities for fractional Schrödinger operators. J. Am. Math. Soc., **21**, 925–950 (2008)
64. Frank, R.L., Lieb, E.H., Seiringer, R.: Equivalence of Sobolev inequalities and Lieb-Thirring inequalities. In: Exner, P. (ed.) Proceedings of the XVIth International Congress on Mathematical Physics, Prague, 2009, pp. 523–535, World Scientific (2010)
65. Fremlin, D.H.: Skeletons and central sets. Proc. Lond. Math. Soc. **74**, 701–720 (1997)
66. Frohlich, J., Lieb, E., Loss, M.: Stability of Coulomb systems with magnetic fields I. The one-electron atom. Commun. Math. Phys. **104**(2), 251–270 (1986)
67. Gagliardo, E.: Proprietaàdi alcune classi di funzionidi piu variabili. Ricerche Mat. **7**, 102–137 (1958)
68. Gilbarg, D., Trudinger, N.S.: Elliptic Partial Differential Equations of Second Order. Springer Classics in Mathematics. Springer, Berlin/Heidelberg/New York (2001). Reprint of the 1998 Edition
69. Gkikas, K.T.: Hardy-Sobolev inequalities in unbounded domains and heat kernel estimates. J. Funct. Anal. **264**(3), 837–893 (2013)
70. Hadwiger, H.: Vorlesungen über Inhalt, Oberfläche und Isoperimetrie. Springer, Berlin/Göttingen/Heidelberg (1957)
71. Hajłasz, P.: Pointwise Hardy inequalities. Proc. Am. Math. Soc. **127**(2), 417–423 (1999)
72. Hardy, G.H.: Note on a theorem of Hilbert. Math. Zeit. **6**, 314–317 (1920)
73. Hardy, G.H.: An inequality between integrals. Messenger Math. **54**, 150–156 (1925)

74. Hardy, G.H.: A Mathematician's Apology. Cambridge University Press, Cambridge (1992). With a foreword by C.P. Snow. Reprint of the 1967 edition
75. Hardy, G.H., Littlewood, J.E., Pólya, G.: Inequalities, 2nd edn. Cambridge University Press, Cambridge (1952)
76. Helffer, B., Mohamed, A.: Caractérisation du spectre essentiel de l'opérateur de Schrödinger avec un champ magnétique. Ann. Inst. Fourier **38**(2), 95–112 (1988)
77. Herbst, I.: Spectral theory of the operator $(p^2+m^2)^{1/2}-ze^2/r$. Commun. Math. Phys. **53**(3), 285–294 (1977)
78. Hoffmann-Ostenhof, M., Hoffmann-Ostenhof, T., Laptev, A.: A geometrical version of Hardy's inequality. J. Funct. Anal. **189**, 539–548 (2002)
79. Hörmander, L.: The Analysis of Linear Partial Differential Operators I. Springer, Berlin/Heidelberg (1983)
80. Hörmander, L.: Notions of Convexity. Birkhäuser, Boston/Basel/Berlin (1994)
81. Itoh, J.-I., Tanaka, M.: The Lipschitz continuity of the distance function to the cut locus. Trans. Am. Math. Soc. **353**(1), 21–40 (2001)
82. Kalf, H.: A characterization of the Friedrichs extension of Sturm-Liouville operators. J. Lond. Math. Soc. **17**, 511–521 (1978)
83. Kato, T.: Perturbation Theory for Linear Operators, 2nd edn. Springer, Berlin/Heidelberg/New York (1976)
84. Kato, T.: Schrödinger operators with singular potentials. Isr. J. Math. **13**(1–2), 135–148 (1972)
85. Kinnunen, J., Martio, O.: Hardy's inequality for Sobolev functions. Math. Res. Lett. **4**, 489–500 (1997)
86. Kinnunen, J., Korte, R.: Characterizations of Hardy's Inequality. Around the research of Vladimir Mazy'a I, pp. 239–254. Springer, New York (2010)
87. Koskela, P., Zhong, X.: Hardy's inequality and the boundary size. Proc. Am. Math. Soc. **131**(4), 1151–1158 (2002)
88. Krantz, S.: Complex Analysis: The Geometric Viewpoint. Carus Mathematical Monographs, vol. 23. Mathematical Association of America, Washington, DC (1990)
89. Kregar, A.: Aharonov-Bohm Effect, University of Ljubljana, Department of Physics, March 2011
90. Kreyszig, E.: Differential Geometry. Dover, New York (1991)
91. Kufner, A., Maligranda, L., Persson, L.-E.: The Hardy Inequality – About Its History and Some Related Results. Vydavatelský servis, Pilsen (2007)
92. Kovalenko, V.F., Perelmuter, M.A., Semenov, Ya.A.: Schrödinger operators with $L_w^{1/2}(R^l)$ potentials. J. Math. Phys. **22**(5), 1033–1044 (1981)
93. Kuratowski, K.: Topology, vol. II. Academic, New York (1968)
94. Landau, E.: A note on a theorem concerning series of positive terms. J. Lond. Math. Soc. **1**, 38–39 (1926)
95. Landau, L.D., Lifshitz, E.M.: Quantum Mechanics (Nonrelativistic Theory). Pergamon, Oxford (1977)
96. Laptev, A.: Spectral inequalities for partial differential equations and their applications. AMS/IP Stud. Adv. Math. **51**, 629–643 (2012)
97. Laptev, A., Netrusov, Yu.: On the negative eigenvalues of a class of Schrödinger operators. Differential Operators and Spectral Theory. Am. Math. Soc. Transl. 2 **189**, 173–186 (1999)
98. Laptev, A., Sobolev, A.V.: Hardy inequalities for simply connected planar domains. Am. Math. Soc. Transl. Ser. 2 **225**, 133–140 (2008)
99. Laptev, A., Weidl, T.: Hardy inequalities for magnetic Dirichlet forms. Oper. Theory: Adv. Appl. **108**, 299–305 (1999)
100. Lehrbäck, J.: Pointwise Hardy inequalities and uniformly fat sets. Proc Am. Math. Soc. **136**(6), 2193–2200 (2008)
101. Lehrbäck, J.: Weighted Hardy inequalities and the size of the boundary. Manuscripts Math. **127**, 249–273 (2008)

102. Lehrbäck, J., Tuominen, H.: A note on the dimensions of Assouad and Aikawa. J. Math. Soc. Jpn. **65.2**, 343–356 (2013)
103. Levin, D., Solomyak, M.: The Rozenbljum-Lieb-Cwikel inequality for Markov generators. J. Anal. Math. **71**, 173–193 (1997)
104. Lewis, J.L.: Uniformly fat sets. Trans. Am. Math. Soc. **308**, 177–196 (1988)
105. Lewis, R.T.: Singular elliptic operators of second order with purely discrete spectra. Trans. Am. Math. Soc. **271**, 653–666 (1982)
106. Lewis, R.T.: Spectral properties of some degenerate elliptic differential operators. In: Operator Theory: Advances and Applications, vol. 219, pp. 139–156. Springer, Basel (2012)
107. Lewis, R.T., Li, J., Li, Y.: A geometric characterization of a sharp Hardy inequality. J. Funct. Anal. **262**(7), 3159–3185 (2012)
108. Li, Y., Nirenberg, L.: The distance to the boundary, Finsler geometry, and the singular set of viscosity solutions of some Hamilton-Jacobi equations. Commun. Pure Appl. Math. **18**(1), 85–146 (2005)
109. Li, P., Yau, S.-T.: On the Schrödinger equation and the eigenvalue problem. Commun. Math. Phys. **88**, 309–318 (1983)
110. Lieb, E.H.: Bounds on the number of eigenvalues of Laplace and Schrödinger operators. Bull. Am. Math. Soc. **82**, 751–753 (1976)
111. Lieb, E.H., Loss, M.: Analysis. Graduate Studies in Mathematics, vol. 14, 2nd edn. American Mathematical Society, Providence (2001)
112. Lieb, E.H., Seiringer, R.: The Stability of Matter in Quantum Mechanics. Cambridge University Press, New York (2010)
113. Luukainen, J.: Assouad dimension: Antifractal metrization, porous sets and homogeneous measures. J. Korean Math. Soc. **1**, 23–76 (1998)
114. MacLaurin, C.: A second letter to Martin Folkes, Esq.; concerning the roots of equations, with the demonstration of other rules in algebra. Philos. Trans. **36**, 59–96 (1729)
115. Mantegazza, C., Mennucci, A.C.: Hamilton-Jacobi equations and distance functions on Riemannian manifolds. Appl. Math. Optim. **47**, 1–25 (2003)
116. Marcus, M., Mizel, V., Pinchover, Y.: On the best constant for Hardy's inequality in $\mathbb{R}^n$. Trans. Am. Math. Soc. **350**(8), 3237–3255 (1998)
117. Matskewich, T., Sobolevskii, P.: The best possible constant in generalized Hardy's inequality for convex domain in $\mathbb{R}^n$. Nonlinear Anal. Theory Methods Appl. **28**(9), 1601–1610 (1997)
118. Maz'ya, V.G.: Classes of domains and embedding theorems for function spaces. Dokl. Akad. Nauk. SSSR **133**, 527–530 (1960). English transl.: Sov. Math. Dokl. **1**, 882–885
119. Maz'ya, V.G.: Sobolev Spaces. Springer, Berlin (1985)
120. Motzkin, T.S.: Sur quelques propriétés charactéristiques des ensembles convexes. Atti Real. Accad. Naz. Lincei Rend. Cl. Sci. Fis. Mat. Natur. Serie VI **21**, 562–567 (1935)
121. Muckenhoupt, B.: Hardy's inequality with weights. Stud. Math. **44**, 31–38 (1972)
122. Nenciu, G., Nenciu, I.: On confining potentials and essential self-adjointness for Schrödinger operators on bounded domains in $\mathbb{R}^n$. Ann. Henri Poincaré **10**, 377–394 (2009)
123. Newton, I.: Sive de compositione et resolutione arithmetica liber. Arithmetica universalis (1707)
124. Nirenberg, L.: On elliptic partial differential equations. Ann. Sc. Norm. Pisa **13**, 1–48 (1959)
125. Okazawa, N.: $L^p$-theory of Schrödinger operators with strongly singular potentials. Jpn J. Math. **22**(2), 200–239 (1996)
126. Opic, B., Kufner, A.: Hardy-type Inequalities. Pitman Research Notes in Mathematics, Series, vol. 219. Longman Science & Technology, Harlow (1990)
127. Psaradakis, G.: $L^1$ Hardy inequalities with weights. J. Geom. Anal. **23**(4), 1703–1728 (2013)
128. Rellich, F.: Halbeschränkte Differentialoperatoren höherer Ordnung. In: Gerretsen, J.C.H., de Groot, J. (eds.) Proceedings of the International Congress of Mathematicians 1954, vol. III, pp.243–250. Noordhoff, Groningen (1956)
129. Rellich, F., Berkowitz, J.: Perturbation Theory of Eigenvalue Problems. Gordon and Breach, New York/London/Paris (1969)

130. Rosenberger, R.: A new characterization of the Friedrichs extension of semi-bounded Sturm-Liouville operators. J. Lond. Math. Soc. (2) **31**, 501–510 (1985)
131. Rosenbljum, G.V.: The distribution of the discrete spectrum for singular differential operators. Soviet Math. Dokl. **13**, 245–249 (1972)
132. Schmidt, K.M.: A short proof for Bargmann-type inequalities. R. Soc. Lond. Proc. Ser. A Math. Phys. Eng. Sci. **458**(2027), 2829–2832 (2002)
133. Schmincke, U.-W.: Essential selfadjointness of a Schrödinger operator with strongly singular potential. Math. Z. **124**, 47–50 (1972)
134. Seiringer, R.: Inequalities for Schrödinger operators and applications to the stability of matter problem. Lectures given in Tucson, Arizona, 16–20 March 2009
135. Shen, Zh.: Eigenvalue asymptotics and exponential decay of eigenfunctions of Schrödinger operators with magnetic fields, Trans. Am. Math. Soc. **348**, 4465–4488 (1996)
136. Simon, B.: Essential self-adjointness of Schrödinger operators with singular potentials. Rat. Anal. Mech. **52**(1), 44–48 (1973)
137. Sobolev, S.L.: On a theorem of functional analysis. Mat. Sb. Am. Math. Soc. Transl. II Ser. **34**, 39–68 (1938); **46**, 471–497 (1963)
138. Solomyak, M.Z.: A remark on the Hardy inequalities. Integr. Equ. Oper. Theory **19**, 120–124 (1994)
139. Stein, E.M.: Singular Integrals and Differentiability properties of Functions. Princeton University Press, Princeton (1970)
140. Talenti, G.: Best constant in Sobolev inequality. Ann. Mat. Pura Appl. **110**, 353–372 (1976)
141. Thaller, B.: The Dirac Equation. Springer, Berlin (1992)
142. Thomas, J.C.: Some Problems Associated with Sum and Integral Inequalities. Ph.D. thesis, Cardiff University, Wales (2007)
143. Thorpe, J.: Elementary Topics in Differential Geometry. Undergraduate Texts in Mathematics, Springer, New York (1994)
144. Tidblom, J.: A geometrical version of Hardy's inequality for $W_0^{1,p}(\Omega)$. Proc. Am. Math. Soc. **132**(8), 2265–2271 (2004)
145. Tidblom, J.: Improved $L^p$ Hardy Inequalities. Ph.D. thesis, Stockholm University (2005)
146. Ward, A.: On Essential Self-Adjointness, Confining Potential and the $L_p$ Hardy inequality. Ph.D. thesis, Massey University, Albany (2014)
147. Weder, R.A.: Spectral properties of one-body relativistic spin-zero Hamiltonians. Ann. Inst. H. Poincaré Sect. A (NS) **20**, 211–220 (1974)
148. Weder, R.A.: Spectral analysis of pseudodifferential operators. J. Funct. Anal. **20**(4), 319–337 (1975)
149. Wen, G.-C.: Conformal Mappings and Boundary-Value Problems. Translations of Mathematical Monographs, vol. 166. American Mathematical Society, Providence (1992)
150. Whittaker, E.T., Watson, G.N.: A Course of Modern Analysis, 4th edn. The University Press, Cambridge (1940)
151. Yafaev, D.: Sharp constants in the Hardy-Rellich inequalities. J. Funct. Anal. **168**, 121–144 (1999)

# Author Index





A.A. Balinsky et al., *The Analysis and Geometry of Hardy's Inequality*, Universitext, DOI 10.1007/978-3-319-22870-9

# Subject Index





A.A. Balinsky et al., *The Analysis and Geometry of Hardy's Inequality*,
Universitext, DOI 10.1007/978-3-319-22870-9

# Notation





A.A. Balinsky et al., *The Analysis and Geometry of Hardy's Inequality*, Universitext, DOI 10.1007/978-3-319-22870-9

Printed by Printforce, the Netherlands